# Information Systems Engineering and Management

Volume 47

The book series "Information Systems Engineering and Management" (ISEM) publishes innovative and original works in the various areas of planning, development, implementation, and management of information systems and technologies by enterprises, citizens, and society for the improvement of the socio-economic environment.

The series is multidisciplinary, focusing on technological, organizational, and social domains of information systems engineering and management. Manuscripts published in this book series focus on relevant problems and research in the planning, analysis, design, implementation, exploration, and management of all types of information systems and technologies. The series contains monographs, lecture notes, edited volumes, pedagogical and technical books as well as proceedings volumes.

Some topics/keywords to be considered in the ISEM book series are, but not limited to: Information Systems Planning; Information Systems Development; Exploration of Information Systems; Management of Information Systems; Blockchain Technology; Cloud Computing; Artificial Intelligence (AI) and Machine Learning; Big Data Analytics; Multimedia Systems; Computer Networks, Mobility and Pervasive Systems; IT Security, Ethics and Privacy; Cybersecurity; Digital Platforms and Services; Requirements Engineering; Software Engineering; Process and Knowledge Engineering; Security and Privacy Engineering, Autonomous Robotics; Human-Computer Interaction; Marketing and Information; Tourism and Information; Finance and Value; Decisions and Risk; Innovation and Projects; Strategy and People.

Indexed by Google Scholar. All books published in the series are submitted for consideration in the Web of Science.

For book or proceedings proposals please contact Alvaro Rocha (amrrocha@gmail.com).

Ngoc Dang Khoa Nguyen · Imran Ali ·
Sanjoy Kumar Paul · Álvaro Rocha
Editors

# Digital Transformation and Sustainability: Methods and Applications

Springer

*Editors*
Ngoc Dang Khoa Nguyen
School of Business and Law
Central Queensland University
Melbourne, VIC, Australia

Imran Ali
School of Business and Law
Central Queensland University
Melbourne, VIC, Australia

Sanjoy Kumar Paul
UTS Business School
University of Technology Sydney
Ultimo, NSW, Australia

Álvaro Rocha
ISEG Lisbon School of Economics
and Management
Universidade de Lisboa
Lisbon, Portugal

ISSN 3004-958X ISSN 3004-9598 (electronic)
Information Systems Engineering and Management
ISBN 978-3-031-91277-1 ISBN 978-3-031-91278-8 (eBook)
https://doi.org/10.1007/978-3-031-91278-8



This Springer imprint is published by the registered company Springer Nature Switzerland AG
The registered company address is: Gewerbestrasse 11, 6330 Cham, Switzerland

# Preface

Our book addresses how digital innovations—Artificial Intelligence (AI), blockchain, robotics, and Internet of Things (IoT)—are leveraged to advance, yet sometimes complicated, sustainable development, particularly within today's Volatile, Uncertain, Complex, and Ambiguous (VUCA) global context, reflecting both opportunities and underlying challenges.

This book examines the evolving role of digital transformation in advancing sustainability across sectors. It provides interdisciplinary insights for researchers, practitioners, and policymakers engaged at the intersection of technology and sustainable development. As a rigorous academic resource, it supports advanced studies in information systems, business, management, and supply chains. It spans methodologically the research process—from question formulation to data analysis—drawing on case studies, qualitative methods, quantitative modeling, and systematic reviews. Eighteen chapters explore how digital technologies reshape industries, governance, and societal outcomes.

Chapter "Unveiling the Dark Side of Digital Transformation Towards Sustainability" investigates ethical dilemmas arising from digital transformation within the financial sector through thematic analysis of data gathered via focus group discussions, highlighting complexities and proposing strategies for ethical and responsible digital adoption.

Chapter "Deciphering Workplace Tech Adoption: Investigating Influential Factors in Shaping Employee Behaviour Within the Healthcare Sector" assesses quantitatively psychological and organizational factors influencing healthcare professionals' acceptance of workplace technologies, employing structured questionnaires and structural equation modeling to propose practical strategies for enhanced adoption and improved healthcare outcomes.

Chapter "Quantifying the Social Impact of Carbon Projects and Generative AI Tool for United Nations Social Development Goals Claims Verification" introduces an advanced analytical framework combining United Nations' Sustainable Development Goals (SDGs) indicators, statistical modeling, and AI-based verification to rigorously quantify and validate the social impacts of carbon credit projects, ensuring accountability, transparency and credibility.

Chapter "Looking Beyond the COVID-19 Pandemic: Navigating Strategic Shift for Achieving the New Normal in the Global Infrastructure Sector" systematically reviews the literature on digital and sustainability-focused strategies implemented by infrastructure managers post-COVID-19, emphasizing proactive, long-term transformations to foster resilience within construction supply chains amid future disruptions.

Chapter "Fostering Supply Chain Resilience in VUCA World with Blockchain Technology: A Bibliometric and Systematic Review" presents a bibliometric and systematic review examining blockchain applications in enhancing supply chain resilience within VUCA business environments, identifying frameworks, critical challenges, and practical strategies to optimize blockchain integration.

Chapter "Blockchain and Cryptocurrencies: Revolutionizing Finance in the Digital Age" employs mixed methods to analyze blockchain technology's transformative influence on financial markets, exploring transaction efficiency, regulatory considerations, and investor-specific challenges associated with cryptocurrencies.

Chapter "Digital Transformation for Supplier Selection: A Pragmatic Constructivist Case Study at a Restaurant Chain in Vietnam" conducts a pragmatic constructivist case study of a Vietnamese restaurant chain to analyze digital transformation strategies for supplier selection, demonstrating how technological interventions can enhance transparency, operational efficiency, fair competition, and inter-departmental coordination.

Chapter "Transforming Agri-Food Supply Chains with Blockchain: Insights from Australian Case Studies" analyzes Australian case studies to examine blockchain integration within agri-food supply chains, detailing its effectiveness in fostering transparency, traceability, and trust while addressing challenges related to scalability, data privacy, and supplier compliance.

Chapter "Strategic Approach to Facilitate Integration of IoT in the Rural Healthcare Management of Bangladesh" employs a mixed-method approach, integrating structured interviews, Analytical Hierarchy Process (AHP), SWOT analysis, and Delphi validation, to develop robust IoT-based telemedicine models aimed at bridging healthcare access gaps in rural Bangladesh.

Chapter "From X to Z: Engaging Different Generations of Customers in Refurbishing Business Models in Electronics" applies the qualitative analysis of semi-structured interviews to examine generational (X, Y, Z) differences in customer engagement with refurbished electronics, identifying cognitive, affective, and behavioral strategies crucial for effective circular economy business models.

Chapter "Navigating the AI Landscape for Sustainable Business Growth" uses mixed-method analysis, combining qualitative insights and quantitative sampling, to assess how artificial intelligence influences organizational productivity, decision-making and employee engagement, providing strategic guidance for sustainable and technology-driven organizational growth.

Chapter "Exploring the Impact of Consumer Behaviour as a Mediator on the Payment Preferences Among Distinct Occupational Segments in West Bengal, India" quantitatively investigates behavioral mediators influencing digital payment adoption across various occupational groups in West Bengal, India, emphasizing consumer behavior as a key factor shaping transaction preferences and technology acceptance.

Chapter "Healing with Technology: The Convergence of AI and Robotics in Healthcare" synthesizes current advancements in AI and robotics integration within healthcare, critically examining their transformative roles in diagnostics, surgical procedures, remote monitoring, and overall system efficiency, alongside ethical considerations and operational challenges.

Chapter "Exploring the Relationship Between Financial Inclusion and Broader Sustainable Development Within Low-Income Sectors" explores quantitatively the correlation between digital financial inclusion initiatives and sustainable development outcomes in low-income communities through Partial Least Squares Structural Equation Modeling (PLS-SEM), emphasizing financial inclusion's socio-economic impact.

Chapter "A Study on Stress Management Techniques and Workforce Sustainability Among IndianOil Refinery Employees in Assam" employs sample survey methodology to evaluate stress management techniques among employees at Indian oil refineries, highlighting demographic considerations crucial for employee well-being, sustainability, and organizational performance.

Chapter "Artificial Intelligence and Firm Performance: The Moderating Role of Circular Economy Intelligence in e-Commerce Firms" utilizes PLS-SEM analysis alongside a systematic literature review to explore the moderating role of circular economy intelligence on the relationship between AI adoption and firm performance within e-commerce enterprises.

Chapter "Technological Innovation in Public Service Delivery: Performance and Uptake of OPGRS in the Landscape of Digital Governance" conducts a systematic evaluation of online public grievance redressal systems adoption and performance, analyzing its effectiveness in enhancing transparency, efficiency, and responsiveness within digital governance frameworks.

Chapter "Industry 4.0 and Sustainable Development Goals: Current Challenges and Solutions" synthesizes existing literature to explore both challenges and strategic solutions in aligning Industry 4.0 technologies with the United Nations' SDGs, highlighting pathways toward integrated sustainable technological innovation.

In a nutshell, this book is an indispensable resource for understanding how digital technologies can be designed, deployed, and governed to support sustainable futures. By blending rigorous analysis with real-world case studies, it contributes to a deeper, more nuanced discourse on digital transformation in an era where sustainability is no longer optional but imperative.

Melbourne, Australia — Ngoc Dang Khoa Nguyen
Melbourne, Australia — Imran Ali
Ultimo, Australia — Sanjoy Kumar Paul
Lisbon, Portugal — Álvaro Rocha

across various occupational groups in West Bengal, India, emphasizing consumer behavior as a key factor shaping transaction preferences and technology acceptance.

Chapter "Healing with Technology: The Convergence of AI and Robotics in Healthcare

# Contents

# Unveiling the Dark Side of Digital Transformation Towards Sustainability

**Abdur Rahim, Md Harun Or Rosid, and Md Motiar Rahman**

**Abstract** Digital transformation has emerged as a key force for improving sustainability in the financial sector. In the contemporary debate on sustainability and digital transformation, there are prevailing narratives of progress, innovation, and positive changes. However, the study of this chapter aims to explore the dark side of digital transformation towards sustainability. The study carried out focus group discussions utilizing purposive and convenient sampling techniques to gather qualitative data from participants. The research employed thematic analysis to extract important insights into the challenges and complexities of digital transformation for sustainability in the financial industry. The findings identified four main themes and sixteen sub-themes. The findings identified several challenges, difficulties, and ethical dilemmas regarding digital transformation towards sustainability. The findings also suggest some measures to minimize these complexities and maximize the benefits of digital transformation towards sustainability in the financial sector. These findings offer a deeper understanding of sustainable digital transformation by offering the financial industry useful guidance to balance growth with responsible environmental and social practices.

**Keywords** Digital transformation · Sustainability · Thematic analysis · Financial institution · Qualitative research

---

A. Rahim (✉) · M. H. O. Rosid
Department of Accounting and Information Systems, University of Barishal, Barishal, Bangladesh
e-mail: abdurrahimbu@gmail.com

A. Rahim · M. M. Rahman
Department of Business Administration, Z. H. Sikder University of Science and Technology, Shariatpur, Bangladesh



N. D. K. Nguyen et al. (eds.), *Digital Transformation and Sustainability: Methods and Applications*, Information Systems Engineering and Management 47,
https://doi.org/10.1007/978-3-031-91278-8_1

## 1 Introduction

A new era of technological advancement has arrived with the rise of Industry 4.0, which has drastically changed industries all over the world, including the financial industry [1]. Besides, with the growing focus on sustainability, the financial sector is going through a significant transition [2]. Now, sustainability in the financial industry has emerged as a critical focus, intertwining economic viability with environmental and social responsibility [3].

Digital transformation is having a big influence on the financial industry, which is why disruptive technologies like blockchain, cloud computing, and machine learning are being integrated to improve company processes and satisfy client needs for customization and flexibility [4]. A significant majority, 83%, of organizations in the financial services industry have adopted cloud-based components for their IT infrastructure and services [5]. Banks are reevaluating conventional business models, emphasizing trust, security, and creative services to meet customer demands [6]. The shift includes many elements, such as blockchain for safe transactions, AI-powered customer service, and mobile banking applications. While technological advancements are crucial, the main challenges lie in organizational culture changes and talent management [7]. Innovative businesses, financial products, and financial instruments are emerging in the financial industry, including fintech innovations and digital banking. Despite the technological shifts, human factors, trust, and security remain key determinants in the evolving financial landscape.

The idea of sustainability is becoming more and more significant as societies become aware of the limited supply of resources and the negative effects of pollution and excessive consumption [8]. It is the process of sustaining natural processes, biodiversity, and production while fulfilling current needs without sacrificing the capacity of future generations to satisfy their own [9]. Sustainability emphasizes the prudent use, conservation, and reuse of resources and includes environmental, economic, and social aspects [10]. Instead of viewing social, environmental, and economic goals as trade-offs, organizations are increasingly using sustainable methods to maximize them all at once. Moving towards a sustainable future necessitates an interdisciplinary approach to contemporary difficulties given the situation of the world economy, social unrest, and environmental stress. To comprehend sustainability, one must take into account the economic and social consequences of sustainable practices as well as the scientific foundation for sustainable resource use [8].

Digital transformation is becoming more widely acknowledged as a major force behind sustainability in several fields. Through sustainable manufacturing, waste management, pollution prevention, and urban sustainability, it can support environmental sustainability [11]. But given its effects on the economy, society, and environment, it is imperative to make sure that digital transformation is sustainable over the long run [12]. Numerous studies identify the advantages of digital transformation toward sustainability [13–15]. However, according to Boston Consulting Group's study, only 30% of digital transformation initiatives achieved or surpassed their intended value goals and produced sustainable organizational changes [16].

The digital industry is responsible for 1.5–3.2 percent of greenhouse gas emissions worldwide [17]. Moreover, digital technologies such as network infrastructure and data centers are responsible for 2% of the world's power consumption [18]. Digital transformation initiatives in financial institutions often fall short of achieving their intended sustainability outcomes [19]. Although there are many advantages to digital transformation, there are also unforeseen drawbacks that may have a detrimental effect on sustainability [20]. There is a lack of research that explores the dark side of digital transformation towards sustainability. Hence, this study aims to investigate the dark side of digital transformation towards sustainability in the context of the financial sector of Bangladesh. Because of its fast expansion, environmental vulnerability, and changing regulatory framework, the financial sector in Bangladesh offers a unique setting for researching digital transformation and sustainability.

This study is essential for giving a solid understanding of the complexities and unforeseen difficulties related to digital transformation in the interest of sustainability. As financial institutions attempt to combine digital technology with sustainability goals, it provides insightful information about the challenges they confront. To minimize these difficulties and maximize the advantages of digital transformation for sustainability, this study also offers practical suggestions. The research offers financial institutions a thorough framework they may use to improve their sustainability performance. The results are especially pertinent to practitioners, regulators, and policymakers because they provide strategic direction for creating strong regulations and control frameworks that support organizational sustainability goals.

The article is structured as follows: to establish a foundation for the investigation, the next section begins with a review of the body of current literature. It then goes on to describe the research methods. The study's findings are then presented and discussed. The study then discusses the results' implications, emphasizing their significance and relevance. The study's strengths and limitations are then discussed in the paper, along with suggestions for further research. Finally, the conclusion of the study is presented.

## 2 Literature Review

### *2.1 Digital Transformation*

Digital transformation is the act of incorporating digital technology into every aspect of enterprises and industries that fundamentally change organizations' operations and the value they provide to customers [21]. It involves creating or modifying business methods and processes to leverage digital technologies effectively [22]. It encompasses reinventing a company's vision, strategy, structure, processes, and culture [23]. This transformation enables organizations to scale their business and innovate at an unprecedented pace [24]. Digital transformation is pivotal for advancing SDGs, and promoting environmental sustainability, social progress, and economic growth [25]. It provides special chances to lessen the effects on the environment and foster innovation, contributing to a climate-neutral economy [25]. Research indicates that digital transformation significantly improves sustainability performance [26]. However, many organizations mistakenly believe that merely implementing automation will suffice for efficiency, leading to costly errors [27]. It can lead to reduced situation awareness and increased error likelihood, particularly in novel situations, necessitating comprehensive design and training strategies [28]. A comprehensive understanding of business goals and risks is essential for successful digital transformation [27]. While digital transformation presents numerous opportunities for growth and sustainability, it also raises concerns [29]. Addressing these concerns is crucial for realizing the full benefits of digitalization [30].

### *2.2 Sustainability*

The concept of sustainability is complex and encompasses elements of the environment, economy, and society and emphasizes the responsible use and development of resources to fulfill current demands without endangering future generations [8]. It aims to optimize social, environmental, and economic goals simultaneously rather than treating them as trade-offs [31]. The concept extends across various fields, including management, design, planning, and environmental services in public institutions, nonprofits, and businesses [32]. Sustainability is not merely an action, but a destination reached through carefully selected pathways [33]. To comprehend sustainability fully, one must understand various interrelated concepts and practices, such as life cycle assessment, industrial ecology, green chemistry, and renewable energy [34]. The pursuit of sustainability involves addressing urgent environmental, economic, and social challenges to ensure the long-term survival and thriving of all species, both locally and globally [35]. Industries, communities, and individuals must adopt sustainable practices that preserve natural resources and promote long-term viability [36]. The unsustainable use of resources necessitates a collective effort to implement strategies that balance economic growth with environmental conservation [37]. Innovations in materials sustainability and eco-design are critical for adapting to changing supply chains and resource availability [38].

## 3 Methodology

### *3.1 Research Design*

This study used qualitative research methods to explore the dark side of digital transformation towards sustainability in the financial sector of Bangladesh. We employed focus group discussions (FGDs) to gather the qualitative dataset. Through focus group discussions (FGDs), we can gather the deep perspectives of relevant stakeholders and enable a thorough comprehension of the experiences, concerns, and expectations associated with sustainability and digital transformation. The study used a semi-structured interview methodology, which combined planned questions and the freedom to dive deeper into other areas of interest and encourage the detailed responses of participants. This approach began with a list of questions and potential questions prepared to guide the interview process.

## 3.2 Sampling and Population

This study was carried out in the context of Bangladesh's scheduled banks. The focus of the inquiry was the employees of these banks. By concentrating on this area, the research most likely sought to identify issues and insights relevant to the financial industry in Bangladesh. The focus group discussion participants were chosen using a combination of convenience and purposive sampling approaches. Purposive sampling was chosen to ensure that participants were selected based on their relevance and expertise in the topic under study [39]. This approach allowed the inclusion of individuals with direct experience in the financial sector and those who are knowledgeable about sustainability and digital transformation efforts. Convenience sampling complemented the purposive approach to account for accessibility and availability constraints. Due to time considerations, some participants were chosen based on their availability to participate in the scheduled discussion sessions.

## 3.3 Data Collection

Data were collected through three focus group discussions with selected participants. Each focus group comprised five participants, allowing for diverse viewpoints while maintaining a manageable group size conducive to open dialogue. The discussions were place virtually, with each session lasting between 70 to 90 min. A five-minute ice-breaking session, which included self-introductions from both sides, was conducted before the start of the conversation to guarantee a comfortable, open, and encouraging atmosphere between the interviewer and the interviewee. Because of the online format's increased accessibility and flexibility, individuals from different places were able to contribute significantly to the discussions. After building rapport, the researcher played the audio recording, described the goal of the discussion, and got the participant's consent.

The focus group discussions were guided by a semi-structured interview protocol that was established after an intensive review of the literature. Even though the original speech was in Bengali, the audiotapes were verbatim transcribed after the data-collecting phase, and the transcripts were kept in English. Ethical approval was obtained to ensure the study met all necessary ethical standards. Every participant was made aware of the goal of the study, the fact that participation was voluntary, and their responses would remain anonymous.

### *3.4 Data Analysis*

Thematic analysis was utilized to analyze the data obtained from the focus group discussions. For thematic analysis, Braun and Clarke's [40] six-step approach was carried out: "becoming familiar with the data, generating initial codes, searching for themes, reviewing themes, defining themes, and writing up findings". The data were then coded and organized into themes using an inductive approach, with a particular focus on recurring concepts, patterns, and unique insights. To find and classify data pertinent to the main concepts being studied, descriptive coding techniques were used. This methodology facilitated the methodical arrangement of responses, enabling the preliminary discovery of themes closely associated with the goals of the study. NVivo version 14 software was utilized for this study.

## 4 Result and Discussion

### *4.1 Profile of the Participants*

Table 1 presents the demographic attributes of participants. All participants are male. The majority (66.7%) are aged 36–40, with others distributed across 25–30 (20%), 31–35 (6.7%), and above 40 (6.7%). Most participants are married (93.3%), with only 6.7% single. All have a master's degree. Slightly more participants are in mid-level management (53.3%) than top-level (46.7%). Nearly half (46.7%) have 11–15 years of experience, with 33.3% having 1–5 years and 20% over 15 years.

### *4.2 Thematic Analysis*

The research findings demonstrate the dark side of digital transformation towards sustainability and the various ways to minimize the challenges and maximize the benefits of digital transformation towards sustainability. Four general thematic clusters emerged through in-depth analysis. Sixteen sub-themes were found within these clusters, offering a thorough understanding of the inherent patterns of these phenomena (Table 2).

**Table 1** Demographic attributes

| Variables | Categories | Frequency | Percentage | Cumulative percentage |
|---|---|---|---|---|
| Gender | Male | 15 | 100 | 100 |
| Age | 25–30 | 3 | 20.0 | 20.0 |
| | 31–35 | 1 | 6.7 | 26.7 |
| | 36–40 | 10 | 66.7 | 93.3 |
| | Above 40 | 1 | 6.7 | 100 |
| Marital status | Single | 1 | 6.7 | 6.7 |
| | Married | 14 | 93.3 | 100 |
| Level of education | Masters | 15 | 100 | 100 |
| Level of management | Mid-Level | 8 | 53.3 | 53.3 |
| | Top Level | 7 | 46.7 | 100 |
| Experience | 1–5 | 5 | 33.3 | 33.3 |
| | 11–15 | 7 | 46.7 | 80.0 |
| | Above 15 | 3 | 20.0 | 100.0 |

*Source* Authors' own work

**Table 2** Themes and sub-themes

| Main theme | Sub-theme |
|---|---|
| The Paradox of progress | Economic growth pressure |
| | Cost of sustainability |
| | Resource overuse |
| | Social displacement |
| | Privacy concerns and cybersecurity risks |
| Ethical Dilemmas | Greenwashing |
| | Divergent stakeholder views |
| | Automation fallout |
| Risk mitigation strategies | Digital risk governance |
| | Proactive regulatory compliance |
| | Cybersecurity and data privacy measures |
| | Stakeholder collaboration |
| Maximizing benefits | Digital innovation for sustainability |
| | Automation and efficiency gains |
| | Talent development and digital literacy |
| | Strategic digital partnerships |

*Source* Authors' own work

### 4.2.1 The Paradox of Progress

The inherent desire for efficiency and growth that comes with digital transformation can make sustainability issues worse. Digital transformation in the financial sector, while driving sustainability, often faces the paradox of creating unintended consequences. Under the main theme, the following five sub-themes were identified.

#### Economic Growth Pressure

Bankers face intense pressure to deliver short-term financial returns to shareholders, often relying on traditional financial products that guarantee profitability but do not necessarily align with sustainability goals. Moreover, the financial landscape in Bangladesh is particularly competitive, so institutions may prioritize market expansion and loan portfolio growth, which often does not align with sustainability goals. This underscores a systemic challenge within the sector, where sustainability is often viewed as a long-term goal, but short-term financial imperatives dominate decision-making. Balancing the need for profitability with a commitment to sustainable digital transformation remains a key challenge for these institutions, as they strive to meet both investor demands and emerging regulatory pressures for sustainable practices [41]. According to participants' responses:

> "We are under pressure to increase revenue streams, which sometimes limits our ability to invest in environmentally sustainable projects" [PP_EGP1].

> "We are leveraging digital transformation to improve operational efficiency and market share but incorporating sustainability features requires additional investment and commitment. It's difficult to justify these investments when we need to maintain high profitability for our shareholders" [PP_EGP2].

#### Cost of Sustainability

From the cost of creating, acquiring, and maintaining new technology to the expense of employee training and business model reform, the high cost of implementing sustainability-driven digital solutions creates a big challenge for organizations [42]. Financial institutions often find themselves grappling with the dilemma of investing in green finance initiatives, such as AI-driven tools for carbon footprint tracking, while balancing immediate profitability concerns. In a market like Bangladesh, where price sensitivity is high, banks face difficulty justifying the significant upfront investments in sustainable platforms, especially when customer demand for these services may not yet be strong enough to guarantee quick returns. Additionally, concerns arise over the uncertain long-term benefits of these investments. Two bank professionals reported:

> "While we understand the value of digital sustainability tools, the cost of developing these platforms in a price-sensitive market like Bangladesh is a challenge. Customers aren't always

> willing to pay premiums for green banking services, so it becomes difficult to balance the investment with potential returns" [PP_CS1].
>
> "Implementing sustainable digital banking solutions is costly. The upfront investment can be daunting, and the return on investment isn't always immediate" [PP_CS2].

#### Resource Overuse

The shift to cloud platforms and digital banking services, while offering operational efficiencies, has led to increased energy consumption. The vast power requirements for data centers often contradict sustainability goals, making it difficult for financial institutions to maintain a green image. The complexity of balancing digital innovation with sustainability objectives becomes apparent as institutions struggle to reduce their environmental footprint. Additionally, the issue of technology waste is pressing. Frequent upgrades in IT systems and the rapid obsolescence of hardware result in substantial amounts of e-waste, which is often inadequately managed due to limited recycling facilities. According to UNCTAD's report, developed nations produce 3.25 kg of digital waste per person, while emerging nations produce less than 1 kg and least developed nations produce only 0.21 kg in 2022 [17]. Banks recognize this challenge, but the lack of infrastructure for proper disposal and recycling of old technologies makes it difficult to align digital transformation with long-term sustainability goals, highlighting the need for systemic solutions. According to three professionals:

> "While we have moved many processes online to cut paper use, the energy consumed by our data centers and digital infrastructure is substantial. It's hard to reconcile our push for green banking with the sheer power demands of these systems" [PP_RO1].
>
> "We update our IT equipment regularly, but disposal remains a challenge. The e-waste generated is a growing concern." [PP_RO2].
>
> "Moving our systems to cloud platforms has reduced some overhead costs, but the energy required for data centers contradicts our sustainability goals" [PP_RO3].

#### Social Displacement

The rapid digital transformation in the financial sector has reduced the need for human involvement in tasks such as loan processing and customer service, leading to workforce downsizing [43]. Hence, the unemployment problem is worsened. Long-serving employees, particularly those who struggle to adapt to new technologies, are often the most affected, highlighting a growing need for reskilling programs to minimize job losses. However, the impact of these programs remains limited, leaving displaced workers with few viable alternatives. Furthermore, the shift to digital banking has exacerbated social inequalities, particularly for older or rural customers who lack the technological literacy or access to benefit from these innovations. Many clients from lower-income backgrounds face barriers in engaging with digital platforms due to limited access to smartphones and reliable internet, deepening the digital divide. Three professionals reported:

> "With the implementation of AI and digital tools, we've significantly reduced the need for manual work. Unfortunately, this has led to job reductions. Many of our senior employees have found it challenging to adapt to the new digital systems" [PP_SD1].

> "The shift to digital banking has been a significant step forward, but we've noticed many of our rural customers and older clients struggling to transition. Digital banking requires access to smartphones and stable internet, which many of our lower-income clients lack" [PP_SD2].

> "Digital solutions are good. But due to these, job reduction will be accelerated. That intensifies our unemployment problems" [PP_SD3].

#### Privacy Concerns and Cybersecurity Risks

As banks increasingly rely on customer data to promote personalized services, including green finance products and sustainability-focused investments, they are gathering vast amounts of personal and financial information. The collection and storage of massive amounts of customer data to support these initiatives have made banks prime targets for hackers. Data breaches, unauthorized access, and cyberattacks like ransomware attacks not only jeopardize customer privacy but also have severe implications for the trustworthiness of sustainability-focused efforts [44]. Clients are becoming more aware and concerned about how their data is being collected, stored, and potentially misused. Several professionals narrated:

> "We collect a lot of customer data to enhance personalized services, but there's always concern about whether this data might be misused" [PP_PC1].

> "While digital transformation allows us to lead in sustainable banking, it also brings significant cybersecurity challenges. With hackers constantly evolving their tactics, we're investing heavily in cyber defenses to ensure that both customer data and our sustainability platforms remain secure from malicious attacks" [PP_PC2].

### 4.2.2 Ethical Dilemmas

Financial organizations face ethical challenges as they pursue sustainability through digital transformation. The main theme represents three sub-themes that show the ethical dilemmas regarding digital transformation towards sustainability within the banking industry. The sub-themes are discussed in the following:

#### Greenwashing

The financial sector is increasingly recognizing the importance of sustainability practices, yet the challenge of greenwashing looms large. Marketing narratives often overshadow the genuine impact of their sustainability initiatives [51]. Internal pressures from regulators and customers further complicate the landscape, as organizations strive to present themselves as environmentally responsible. However, without rigorous measurement of their initiatives, the risk of superficial or misleading sustainability claims is high, undermining long-term trust and credibility. While there is a desire to showcase digital services as sustainable, the high energy costs of maintaining data centers often go unaddressed. The risk of customer distrust is palpable, as consumers become increasingly aware of greenwashing tactics. According to participants:

> "While our bank promotes eco-friendly initiatives, we sometimes feel these claims are more for marketing than actual environmental impact" [ED_G1].

> "While we want to show that we're committed to sustainability, there are challenges in disclosing all the details of our initiatives. For example, we might tout a digital service as eco-friendly without addressing the high energy costs of running data centers" [ED_G2].

> "Sometimes, we express our commitment to sustainability, but the reality may not fully align with these claims" [ED_G3].

### Divergent Stakeholder Views

Organizations face significant challenges in balancing the competing interests of various stakeholders during sustainability-driven digital transformations [45]. Corporate stakeholders, such as shareholders, often prioritize profitability and operational efficiency, urging financial institutions to adopt faster and more cost-effective digital solutions. This focus can conflict with the regulatory frameworks enforced by the government, which typically emphasize sustainability, environmental protection, and social welfare. At the same time, civil society organizations advocate for greater transparency, inclusivity, and ethical practices. These groups highlight the need for banks to prioritize environmental justice and ensure that digital transformation does not exacerbate social inequalities. Financial institutions are therefore caught between the demands of profitability and efficiency from corporate stakeholders and the expectations of social responsibility from NGOs and government regulators. Two participants reported:

> "Our bank often finds itself in the middle of competing interests. While shareholders push for higher profits through digital efficiency, regulators are increasingly pressuring us to comply with environmental and sustainability standards. Balancing these views can be a challenge" [ED_DSV1].

> "We frequently receive feedback from civil society groups, highlighting the need for our digital services to be more inclusive" [ED_DSV2].

### Automation Fallout

The increased adoption of automation in the financial sector has created significant tension between operational efficiency and social responsibility. Automation, through AI-driven processes and digital platforms, has improved service delivery and reduced costs. However, it has simultaneously led to substantial job displacement, particularly among workers in manual or repetitive roles, such as customer service, data entry, and administrative functions. Financial institutions are attempting to mitigate the negative effects by offering retraining and upskilling programs. However, these initiatives are not always sufficient to reskill all displaced workers, leading to persistent concerns about job security and the future of affected employees. One professional noted:

> "As we continue with digital banking, automation is inevitable, but it comes with a social cost. The reduction in workforce is something we can't completely avoid, but we're exploring ways to support affected employees through severance and career transition assistance" [ED_AF1].

### 4.2.3 Risk Mitigation Strategies

Financial institutions require robust risk mitigation strategies as digital transformation advances to protect sustainability objectives. The four sub-themes represent key strategies for managing the inherent risks associated with digital innovations. The sub-themes are discussed below:

#### Digital Risk Governance

Establishing comprehensive frameworks is critical for aligning initiatives with regulatory requirements while simultaneously managing cybersecurity threats and minimizing environmental impacts [46]. Financial institutions recognize the necessity of creating specialized teams dedicated to overseeing digital risk management, ensuring both sustainability and compliance are effectively addressed. Banking executives highlight proactive measures such as the formation of digital risk governance committees aimed at assessing cybersecurity risks linked to sustainability platforms. Regular audits to ensure compliance with environmental and data security regulations are being prioritized, reflecting the increasing complexity of the regulatory landscape in Bangladesh. According to participants:

> "We understand the importance of robust governance, but the challenge lies in balancing risk management with innovation. Setting up dedicated teams to oversee this has been helpful, but it's an ongoing process" [RMS_DRG1].

> "We've recently established a digital risk governance committee. We need to monitor both technological and environmental risks together" [RMS_DRG2].

#### Cybersecurity and Data Privacy

In the financial sector, cybersecurity and data privacy are critical as institutions increasingly adopt digital tools for sustainability efforts. The emphasis on strong encryption protocols highlights the industry's commitment to protecting sensitive financial and environmental data from unauthorized access. Regular security audits are also a cornerstone of these efforts, allowing institutions to identify and mitigate vulnerabilities swiftly, which is essential in an ever-evolving threat landscape. Investment in advanced cybersecurity tools and comprehensive staff training strengthens the organization's overall security posture by ensuring that workers are prepared to identify and address possible threats [46]. Furthermore, maintaining transparency with customers regarding data protection measures fosters trust, which is crucial as financial institutions navigate the complexities of digital transformation. According to three professionals:

> "We prioritize cybersecurity by implementing advanced encryption protocols for all customer and operational data" [RMS_CDP1].

> "Regular security audits are integral to our strategy. By frequently assessing our systems for vulnerabilities, we ensure that any potential risks are identified and addressed promptly, thereby safeguarding both financial and environmental data against cyber threats" [RMS_CDP2].

> "We are transparent about our data protection measures and continuously monitor our systems for any unusual activities to mitigate the risk of data breaches" [RMS_CDP3].

### Stakeholder Collaboration

Engaging stakeholders is a vital part of ensuring the success and resilience of digital sustainability initiatives within the financial sector. Consultation with key stakeholders can significantly reduce risks associated with digital transformation. By holding consultation forums, instituting transparent sustainability reporting mechanisms, and involving local communities, banks can align their initiatives with both regulatory requirements and public expectations. This proactive engagement not only builds trust but also enhances transparency, making stakeholders more invested in the success of these transformations. Additionally, close collaboration with regulators ensures compliance with both financial and environmental standards, further reducing legal risks and enhancing institutional credibility. In the long run, such stakeholder collaboration facilitates smoother implementation, reduces resistance, and strengthens the institution's sustainability credentials, making digital transformations more sustainable and equitable for all involved parties. Two professionals noted:

> "By involving employees and civil society groups early in the development of our digital transformation projects, we can better address potential concerns and incorporate local insights" [RMS_SC1].

> "We organize regular stakeholder consultation forums where we engage with customers and regulators to gather their feedback on our sustainability initiatives" [RMS_SC2].

### Recycling Initiatives

As financial institutions increasingly adopt digital technologies, managing the environmental impact of tech waste becomes a crucial aspect of sustainability efforts. Recycling initiatives are an essential part of ensuring that the shift toward digital transformation does not contribute to long-term environmental harm [47]. E-waste recycling programs can address the disposal of obsolete hardware, such as outdated servers, ATMs, and office equipment. Ensuring that recycling practices are both efficient and aligned with sustainability goals is critical. By collaborating with certified recyclers, financial institutions can minimize their environmental footprint while benefiting from cost reductions associated with equipment reuse. Two participants narrated:

> "As we adopt more digital tools for sustainability, the disposal of old hardware becomes a concern. We've started collaborating with certified recyclers to manage e-waste from outdated servers and computers, ensuring that we minimize the environmental impact of our digital transformation" [RMS_RI1].

> "Our sustainability strategy includes recycling initiatives where we focus on reducing the waste generated from obsolete technology" [RMS_RI2].

### 4.2.4 Maximizing Benefits

Financial institutions are concentrating on optimizing advantages through ongoing developments to benefit from digital transformation for sustainability. The following four sub-themes represent significant strategies to maximize the benefits of digital transformation towards sustainability.

Digital Innovation

Digital innovation plays a pivotal role in advancing sustainability efforts. Digital technologies are transforming the financial landscape and providing unprecedented opportunities to lead sustainable practices. Tools like AI, blockchain, and big data analytics are being increasingly adopted to enhance operational efficiency, transparency, and sustainability outcomes. By using these technologies, financial institutions can optimize resource allocation, reduce carbon emissions, and promote responsible investment strategies. Continuous digital innovation has enormous potential to effectively solve sustainability problems and maximize the impact of financial institutions on environmental and social objectives [48]. However, to maximize these sustainability benefits, an integrated approach must be adopted to align digital innovations with clear sustainability objectives, employee engagement, and regulatory compliance. This alignment promotes an innovation culture that not only responds to the current sustainability challenges but also is resilient and adaptable to future needs. Through digital innovation, financial institutions can not only achieve sustainable objectives but also gain competitive advantage and lead to long-term operational improvement. According to the responses:

> "We're exploring the use of big data analytics to improve our resource allocation, especially in monitoring energy consumption across digital banking services."[MB_DI1].

> "Our focus on digital innovation, such as deploying AI for automated environmental risk assessments, has shown promise."[MB_DI2].

### Automation and Efficiency Gains

In the financial sector, automating sustainability reporting and resource management through digital tools provides substantial benefits. Automation minimizes human error and speeds up decision-making processes. Financial institutions are increasingly adopting digital dashboards and AI-powered tools to track real-time metrics such as energy consumption, carbon emissions, and resource allocation [49]. These automated systems not only contribute to greater operational efficiency but also enhance the accuracy of sustainability reporting, which is critical for regulatory compliance and public transparency. Two bank professionals reported:

> "By automating our sustainability reporting, we've been able to reduce the time spent on manual data collection and analysis. This allows us to focus on more strategic initiatives while ensuring that our reporting is more accurate and timelier."[MB_AEG1].

> "We can implement a real-time dashboard to track our environmental metrics, such as energy consumption and carbon emissions. This automation not only cuts down on human error but also helps us make quicker, data-driven decisions."[MB_AEG2].

### Talent Development and Digital Literacy

Developing employee expertise in both digital tools and sustainability practices is essential to achieving long-term environmental goals while maintaining operational efficiency. By investing in training programs, collaborations with academic institutions, and promoting a culture of ongoing learning, banks are equipping their workforce to adapt to the evolving demands of the industry [50]. These efforts not only enhance innovation but also ensure that employees are aligned with the institution's sustainability objectives. Programs aimed at improving digital literacy provide employees and customers with the skills needed to utilize advanced technologies for sustainable outcomes. According to participants' responses:

> "We've initiated a comprehensive digital training program for employees to ensure they are well-versed in both sustainability initiatives and the digital tools that support them."[MB_TDDT1].

> "By investing in internal workshops focused on the intersection of sustainability and technology, we are creating a culture that encourages innovation in line with our sustainability goals."[MB_TDDT2].

> "Arranging digital literacy programs for customers helps them easily use digital financial services. I think it will be helpful for our customers."[MB_TDDT3].

### Strategic Digital Partnerships

In the financial sector, strategic digital partnerships provide financial institutions with access to cutting-edge digital tools and expertise. By co-developing innovative solutions such as AI-driven sustainability reporting systems or blockchain-enabled transparency tools, financial institutions can more effectively track, measure, and report their environmental impact. Furthermore, partnerships can enable financial institutions to scale green finance products, promote environmental accountability, and drive innovation while sharing the cost burden of implementing new technologies. Two participants noted:

> "Our strategic alliance with fintech companies has allowed us to co-develop blockchain-based tools that enhance transparency in our sustainable investment portfolios."[MB_SDP1].
>
> "Our collaboration with leading tech firms has enabled us to deploy AI-driven advanced tools for improving our sustainability reporting accuracy."[MB_SDP2].

## 5 Implications

The study promotes theoretical viewpoints on the negative aspects of the digital transition in the direction of sustainability. In addition to improving the theoretical discussion on digital transformation and sustainable business practices, it offers a more nuanced understanding of digital transformation towards sustainability. The findings provide actionable insights for financial institutions seeking to improve their sustainability outcomes and navigate the intersection of digital transformation and sustainability.

The findings suggest that financial institutions must carefully balance short-term profitability pressures that often conflict with the long-term investments required for sustainable practices. Organizations should adopt green technology to reduce environmental harm. Digital technology increases energy consumption whereas green technology can be energy efficient. The study suggests that superficial sustainability claims risk eroding trust among stakeholders. For financial institutions, maintaining credibility requires transparent communication and genuine commitments to sustainability. As digital sustainability initiatives increasingly rely on data, institutions face a heightened need for secure, compliant, and transparent data handling practices. Financial institutions should prioritize robust digital risk governance frameworks and enhanced cybersecurity measures to build trust with stakeholders and safeguard sensitive data. The role of stakeholder collaboration and strategic partnerships with technology firms appears instrumental in advancing sustainability initiatives. This finding points to a promising area of research exploring how cross-sector partnerships can enhance sustainability outcomes. Additionally, Organizations should take training programs and workshops for talent development and digital literacy. These implications collectively emphasize that achieving sustainable digital transformation requires a multi-dimensional approach that integrates ethical, financial, environmental, and social considerations.

## 6 Strengths, Limitations, and Suggestions for Further Research

The study provides several key strengths that contribute to the relevance and impact of the debate on digital transformation and sustainability in the financial sector. First, by focusing on Bangladesh's financial industry, the research highlights the emerging market context where sustainability and digital innovation are increasingly ranked as a priority. Moreover, it uncovers the dark side of digital transformation towards sustainability. This qualitative element provides more profound understandings that enhance theoretical frameworks and real-world implementations. The views and experiences of participants are clarified through focus group discussions, providing valuable qualitative data. Participant-interviewer power dynamics, which are supported by the interviewer's professional standing, promote candor and ease the sharing of information. These elements support the study's results' reliability and credibility.

Although this study provides valuable insight into the interaction between digital transformation and sustainability in the financial sector, there are some limitations to be recognized. First, thematic analysis is based primarily on qualitative data, but it can introduce subjectivity and limit the scope of the conclusions. Future research could complement these conclusions with quantitative approaches, such as surveys and empirical data analyses, to provide a broader view of the identified trends and challenges. Second, the study is geographically focused on the Bangladeshi financial sector. Although the conclusions may apply to other emerging markets facing similar sustainability challenges, the relevance of the results to other regions may be limited by contextual factors in Bangladesh. Future research could adopt a comparative approach. In addition, the study does not delve deeply into specific digital technologies such as artificial intelligence, blockchain, or big data analysis, which play a different role in supporting sustainable initiatives. Future research could help to identify how specific digital tools contribute to or complicate sustainability efforts and give a more detailed understanding of the role of technology in sustainability. Finally, only three focus groups with a total of fifteen participants were used in this study, which may have limited the scope and depth of the data acquired. To capture a greater variety of experiences and viewpoints and improve the depth of qualitative data analysis, future research might benefit from bigger sample sizes.

## 7 Conclusion

In conclusion, this study clarifies the complexities and possible drawbacks of digital transformation in the context of sustainability activities in the financial industry. Digital technologies have significant chances to improve sustainability performance, but they also bring with them drawbacks including social displacement, privacy issues, elevated cybersecurity threats, green-washing, and more. The suggested framework gives financial institutions useful advice and strategic insights to help them successfully manage these difficulties, allowing them to take advantage of digital innovation for long-term growth. These results have substantial implications for regulators, policymakers, and business professionals, who may utilize them to create governance frameworks, formulate legislation, and promote a responsible digital adoption culture. In the end, this study advances our knowledge of how to manage digital change to provide long-term results while preserving the financial industry's moral and administrative integrity.

## References


1. Machkour, B., Abriane, A.: Industry 4.0 and its implications for the financial sector. Procedia Comput. Sci. **177**, 496–502 (2020)
2. Svoboda, A.: The role of financial services in the transition to a sustainable economy. J. Strat. Innov. Sustain. **18**(3) (2023)
3. Yu, X.: Integration of ESG in financial institutions: a study on the impact of sustainability reporting. Int. J. Multidiscip. Res. Growth Eval. **4**(5), 468–472 (2023)
4. Mishra, A.B., Ranjan, N., Sharma, K.M., Padhye, P., Chand, K.: Advancement of digital transformation to boost financial services industry. J. Inform. Educ. Res. **3**(2) (2023)
5. Dutta, S.: How Cloud Computing Is a Game Changer for Financial Services in MENA|EY—MENA. https://www.ey.com/en_qa/industries/financial-services/how-cloud-computing-is-a-game-changer-for-financial-services-in-mena (2022). Last accessed 1 Dec 2024
6. Vyas, S., Pareek, A., Ghosal, I.: A comprehensive study: how digital transformation is reshaping banking services. Eur. Econ. Lett. **14**(2), 877–883 (2024). https://doi.org/10.52783/eel.v14i2.1415
7. Marin, R.: The Fusion of Talent Management and Digitalization: A Literature Review (2023)
8. Church, J.M., Tirrell, A., Moomaw, W.R., Ragueneau, O.: Sustainability: from ideas to action in international relations. In: Routledge Handbook of Global Environmental Politics. Routledge, pp. 217–27 (2022)
9. Lanzano, C.: Sustainability. The anthropology of resource extrac sustainabilitytion, pp. 149–66 (2022)
10. Malhotra, A., Melville, N.P., Watson, R.T.: Spurring impactful research on information systems for environmental sustainability. MIS Q. **37**(4), 1265–1274 (2013)
11. Feroz, A.K., Zo, H., Chiravuri, A.: Digital transformation and environmental sustainability: a review and research agenda. Sustainability **13**(3), 1530 (2021)
12. Tupa, J.: Sustainability of digital transformation. In: Proceedings of the 6th European Conference on Industrial Engineering and Operations Management, Lisbon, Portuga (2023)
13. Alojail, M., Khan, S.B.: Impact of digital transformation toward sustainable development. Sustainability **15**(20), 14697 (2023)
14. Ji, Z., Zhou, T., Zhang, Q.: The impact of digital transformation on corporate sustainability: evidence from listed companies in China. Sustainability **15**(3), 2117 (2023)
15. Kunkel, S., Matthess, M.: Digital transformation and environmental sustainability in industry: putting expectations in Asian and African policies into perspective. Environ Sci Policy **112**, 318–329 (2020)
16. BCG.: Companies Can Flip the Odds of Success in Digital Transformations from 30% to 80%. https://www.bcg.com/press/29october2020-companies-can-flip-the-odds-of-success-in-digital-transformations-from-30-to-80 (2020). Last accessed 1 Dec 2024
17. UNCTAD.: Digital Economy Report 2024 Shaping an Environmentally Sustainable and Inclusive Digital Future (2024)

18. ITU.: The Digital Sector's Environmental Dilemma—ITU. https://www.itu.int/hub/2024/11/the-digital-sectors-environmental-dilemma/ (2024). Last accessed 1 Dec 2024
19. Mavlutova, I., Spilbergs, A., Verdenhofs, A., Natrins, A., Arefjevs, I., Volkova, T.: Digital transformation as a driver of the financial sector sustainable development: an impact on financial inclusion and operational efficiency. Sustainability **15**(1), 207 (2022)
20. Rosário, A.T., Dias, J.C.: Sustainability and the digital transition: a literature review. Sustainability **14**(7), 4072 (2022)
21. Gebayew, C., Hardini, I.R., Panjaitan, G.H.A., Kurniawan, N.B.: A systematic literature review on digital transformation. In: 2018 International Conference on Information Technology Systems and Innovation (ICITSI), IEEE, pp. 260–65 (2018)
22. Shalini, S., Devi, T.: Digital Transformation. In: Industry 4.0 Technologies for Education, Auerbach Publications, pp. 67–79 (2022)
23. Gurbaxani, V., Dunkle, D.: Gearing up for successful digital transformation. MIS Q. Executive **18**(3), 6 (2019)
24. Filkins, P.: Private 5G: Empowering Digitalization for Enterprise and Industrial Organizations (2021)
25. Goi, C.L.: Digital transformation: a catalyst for achieving sustainable development goals (SDGs) in Malaysia. In: Perspectives on the Transition Toward Green and Climate Neutral Economies in Asia, IGI Global, pp. 18–29 (2023)
26. Bulović, V., Čović, Z.: The impact of digital transformation on sustainability in fashion retail. In: 2020 IEEE 18th International Symposium on Intelligent Systems and Informatics (SISY), IEEE, pp. 149–54 (2020)
27. Cöster, M., Danielson, M., Ekenberg, L., Gullberg, C., Titlestad, G., Westelius, A., Wettergren, G.: Digital Transformation Understanding Business Goals, Risks, Processes, and Decisions. Open Book Publishers (2023)
28. Endsley, M.R.: Understanding automation failure. J. Cogn. Eng. Decis. Mak. 15553434231222060 (2024)
29. Joel, O.S., Oyewole, A.T., Odunaiya, O.G., Soyombo, O.T.: The impact of digital transformation on business development strategies: trends, challenges, and opportunities analyzed. World J. Adv. Res. Rev. **21**(3), 617–624 (2024)
30. Taneja, S., Dahiya, K.: Digital transformation: catalyzing rural entrepreneurship in the 21st century. In: Utilizing AI and Smart Technology to Improve Sustainability in Entrepreneurship, IGI Global, pp. 88–103 (2024)
31. Kashani, S.J., Hajian, M.: Indicators of sustainability. In: Sustainable Resource Management, Elsevier, pp. 317–34 (2021)
32. Hughes, C.: Sustainability. In: Educating for the Twenty-First Century: Seven Global Challenges, Brill, pp. 63–84 (2018)
33. Van Vuuren, D., Lucas, P., Calvin, K.V., Chung, S.H., Harfoot, M., Hedden, S., Köberle, A.C., Wada, Y., Bouwman, L., He, C.: Pathways Toward Sustainable Development. Cambridge University Press (2019)
34. Constable, D.J.C.: Green and sustainable chemistry–the case for a systems-based, interdisciplinary approach. IScience **24**(12) (2021)
35. Bunch, M.J., Ramírez, R., Morrison, K.: Sustainability: systems thinking in complex situations. In: Education for Sustainable Human and Environmental Systems. Routledge, pp. 19–32 (2018)
36. Giordino, D., Crocco, E.: Sustainable development: a new frontier for SMEs. In: Management Strategies for Sustainability, New Knowledge Innovation, and Personalized Products and Services, IGI Global, pp. 68–93 (2022)
37. Saleh, H.M., Hassan, A.I.: Sustainability: a comprehensive approach to developing environmental technologies and conserving natural resources. In: Risk, Reliability and Sustainable Remediation in the Field of Civil and Environmental Engineering, Elsevier, pp. 437–48 (2022)
38. Mancini, M.: Sustainability. In: The Basalt Fiber—Material Design Art, Springer, pp. 29–34 (2023)
39. Campbell, S., Greenwood, M., Prior, S., Shearer, T., Walkem, K., Young, S., Bywaters, D., Walker, K.: Purposive sampling: complex or simple? Research case examples. J. Res. Nurs. **25**(8), 652–661 (2020)

40. Braun, V., Clarke, V.: Using thematic analysis in psychology. Qual. Res. Psychol. **3**(2), 77–101 (2006)
41. Taka, M.E., Bayarçelik, E.B.: Sustainable digital transformation of financial institutions. Bus. Manag. Stud. Int. J. **11**(1), 253–269 (2023)
42. Muthuraman, S.: Digital business models for sustainability. Gedrag Organ. Rev. **33**(2), 1095–1102 (2020)
43. Daver, G.: Dijitalleşirken Tasarruf Edilebilir Mi? Çalışan Sayısı Mı, Çalışanların Niteliği Mi? Yönetim Bilimleri Dergisi **19**(Özel Sayı):241–67 (2021)
44. Short, L.G.: Measuring trust for accountability in an untrusted digital world. IEEE Technol. Policy Ethics **6**(6), 1–5 (2021)
45. Tsaples, G., Papathanasiou, J., Manou, D.: Synergies and challenges: exploring organizational perspectives on digital transformation and sustainable development in the context of skills and education. Buildings **14**(2), 395 (2024)
46. Loukaka, A., Rahman, S.S.M.: security professionals must reinforce detect attacks to avoid unauthorized data exposure. Inf. Technol. Ind. **8**(1) (2020)
47. Ghosh, B.K., Mekhilef, S., Ahmad, S., Ghosh, S.K.: A review on global emissions by e-products based waste: technical management for reduced effects and achieving sustainable development goals. Sustainability **14**(7), 4036 (2022)
48. Puschmann, T., Leifer, L.: Sustainable Digital Finance: The Role of FinTech, InsurTech and Blockchain for Shaping the World for the Better (2020)
49. Ionescu, S.A., Diaconita, V.: Transforming financial decision-making: the interplay of AI, cloud computing and advanced data management technologies. Int. J. Comput. Commun. Control. **18**(6) (2023)
50. Ajayi-Nifise, A.O., Odeyemi, O., Mhlongo, N.Z., Ibeh, C.V., Elufioye, O.A., Falaiye, T.: Digital transformation in banking: the HR perspective on managing change and cultivating digital talent. Int. J. Sci. Res. Arch. **11**(1), 1452–1459 (2024)
51. Kemper, J.A., Michael Hall, C., Ballantine, P.W.: Marketing and sustainability: business as usual or changing worldviews? Sustainability **11**(3), 780 (2019)

# Deciphering Workplace Tech Adoption: Investigating Influential Factors in Shaping Employee Behaviour Within the Healthcare Sector

**Anisha P. Gopal and V. Mohana Sundari**

**Abstract** The rapid development of technology has pointedly impacted various industries, including healthcare sector. The adoption of workplace technologies in various healthcare settings has the potential to improve efficiency, enhance patient care, and streamline processes. The purpose of the study is to explore the influential factors that shape employee behaviour regarding the adoption of workplace technologies within the healthcare sector. This study employs a quantitative research design, utilizing questionnaire methodology to collect data from healthcare professionals. The sample size was determined using power analysis and the sampling technique will involve purposive sampling, targeting healthcare professionals across diverse roles within healthcare organizations. Statistical analysis conducted using structural equation modelling (SEM) techniques, employing software such as Analysis of Moment Structures (AMOS) and Statistical Package for the Social Sciences (SPSS). These tools will facilitate the examination of the proposed model, testing the relationships between, information sensitivity, social influence perceived usefulness, perceived ease of use, perceived risk (independent variables), attitude (the mediating variable), and the behavioural intention to use technology (dependent variable). The verdicts of this study embrace several inferences for healthcare establishments aiming to augment technology adoption initiatives aimed at promoting technology adoption and innovation, ultimately contributing to improved healthcare delivery and patient outcomes.

**Keywords** Artificial intelligence · Data security · Perceived privacy · Behavioural intention · Tech adoption

---

A. P. Gopal (✉) · V. Mohana Sundari
Faculty of Management, SRM Institute of Science and Technology, Kattankulathur, Tamil Nadu, India
e-mail: ap1083@srmist.edu.in

V. Mohana Sundari
e-mail: mohanas1@srmist.edu.in



N. D. K. Nguyen et al. (eds.), *Digital Transformation and Sustainability: Methods and Applications*, Information Systems Engineering and Management 47,
https://doi.org/10.1007/978-3-031-91278-8_2

# 1 Introduction

## *1.1 Evolution of Technology Adoption*

The technology adoption offers a vital part in assessing the efficacy and efficiency of health maintenance delivery system in ever-changing healthcare industry with more advanced facilities. The elements that influence employee behaviour towards these advancements are purely based on the new technologies which are prevailing [1]. The new technologies can be adopted for insisting on complying strict regulatory guidelines, the need for easy integration with current systems and the urgency to safeguard the standards of patient care are some of the hurdles which encounter the health care business industry [2]. This study addresses the complex patterns of technology espousal in the healthcare workplace. The objective is to identify, and substantiate the primary elements which facilitates the incorporation of technology advancements and innovations among healthcare practitioners [3].

Computer and information technologies are extensively utilized in modern enterprises including healthcare sector and have significantly grown in their range and usage also. The factors that hinder the adoption of technology have earned significant attention from various experts in this field [4]. The technological advancement only occurs when technologies become more precise, accurate, efficient, effective and more capable of providing standard outputs. The individual who possesses different degrees of technological proficiency and exhibit variable dispositions towards change includes mostly administrative staff and frontline medical personnel, are those issues prevailing has asserted by the divergent set of employees [5].

When it derives to adopting expertise in healthcare industry, the study identifies the most prominent elements that sway the behaviour of employees. The gaining acceptance of innovation is a lengthy process, and the swiftness at which the spread of innovation ensues might differ. Innovation becomes significant for individuals or organizations who are interested in incorporating new ideas or technologies [6].

Rogers states diffusion as the broadcast of an innovation over specific channels concluded a period of time amid the personalities in a society. In his work, he defines innovation as an idea, activity, or thing that is considered novel by a person or group that adopts it [7]. The research aspires to impart pragmatic insights and guidelines which can oblige healthcare industry for ensuring a flattering domain for the integration of technology and optimum utilisation of resources. This benefits in the improvement in healthcare delivery by developing more efficient and effective strategies for adopting technology and innovation [8].

### 1.2 Significance of Technology Adoption in Shaping Employee Behaviour

The most prominent alteration has witnessed in technologies in the era of progression. It includes the emergence of new creations, practices and innovations. Nowadays each and everything is intertwined with technology, ranging from the advanced medical equipment utilized in hospitals to the smartwatches we wear in our daily life [27]. The high acceptance of technologies in a business is mostly due to the fact that technology is one of the most prominent factors that initiates efficient and effective administration of operations inside a business. Due to the rapid and progressive expansion of technology in an organization leads to a rise in corporate transparency and efficiency [9]. Importance of incorporating technology in the healthcare sector has a prominent effect on multiple elements of patient care and operational effectiveness in various hospitals. While adopting the technology, it improves the quality of health care providing towards the patients by diagnostic instruments, treatment alternatives and monitoring systems [10].

This initiates to provide more precise diagnoses, and individualized treatment programs also for each and every patient. It optimizes the healthcare procedures, practises, diminishing administrative encumbrances, potential risks and enabling healthcare practitioners to prioritize the patient care over administrative tasks other than the traditional practises in the healthcare sector. Currently, the advanced technologies facilitate the enhanced data management and analytics, leading to improved decision-making [28]. The process of monitoring patient health trends the Electronic Health Records and the data analytics can effectively utilised through adopting the technology in this sector. The population residing in rural or underserved regions can have benefit of telemedicine which has augmented accessibility. This enables the patients to obtain medical attention at a distance [11].

The technology plays a significant role in reducing costs by avoiding redundant tests, practices and procedures, and also ensuring optimization of the allocation of resource and boosting operational efficiencies in an effective manner. The magnified coordination of treatment is the result of augmented communication among those healthcare practitioners, between the various clinicians and patients [29]. The electronic health records (EHRs) enable the effective storage, communication, processing and maintaining of medical information among different parties involved which leads to improved standard of care, improving the patient results and also in making comprehensible administrative procedures and practises. The practitioners in the healthcare sector can promote trust and can increase patients' confidence and the engagement in EHR systems by comprehending and resolving the various concerns arising. This will ultimately result in a more cohesive and efficient healthcare delivery paradigm with increased advancements in this field in the specified time duration [12].

## 1.3 Problem Statement

Adopting technology in the healthcare sector is a tough task as employee behaviour towards the usage of technology can be considered as one of the most significant tasks in the modern world. The research article seeks to embark on a study on how technology adoption techniques can greatly enhance the work of employees. This paper gives a point-by point examination of the complicated cycles of innovation reception with regards to worker conduct, underlining the meaning of the strong administration in the emergency clinic and the foundation that advances the vocation and self-awareness of representatives. Thus, by focusing attitude of employees towards the technology acceptance, research intends to identify the attitude and behaviour of employees towards the adoption of various advanced technologies and sophisticated equipment within the healthcare sector. In this context, the article aims to present original and profound findings that reveal the opportunities and threats faced during the technology adoption in the various hospitals. Finally, it is the hope of these findings to offer practical solutions to the healthcare professionals on how to maintain positive attitude towards technology adoption and to provide better services in the digital era.

## 1.4 Research Gap

Several research gaps are identified in relation to the technology adoption and investigated influential factors in shaping employee behaviour within the healthcare sector. Firstly, there is a need for more comprehensive researches in the health care sector that would focus on the effectiveness of the attitude, including their perception and evaluation on certain persons, ideas and the way of dealing with others individuals. Secondly, there is lack of numerical research for explicitly examining the effect of perceived ease of use and perceived usefulness on patient portal utilization regarding the patients' health condition. The existing literature primarily focuses on the qualitative aspects only, thereby leaving a need for empirical data to support these constructs in milieu of patient portals in healthcare division. Thirdly, certain studies pertain to the limited understanding of how social factors particularly within the framework of institutional theory, specifically influences the healthcare technology adoption among elderly patients in the healthcare sector.

Additionally, there is essential for empirical indication to support the applicability of these social influences in diverse healthcare settings. Most of the studies have explored general technology acceptance, research concentrates on unique aspects of the technology adoption in healthcare sector. Cross-sectionally, there is a lack of knowledge on how extended technology adoption initiatives or factors affects the privacy and information sensitivity of patients and the social influence and attitude towards the adoption of technology within the healthcare sector. Furthermore, there is research opportunity to understand how technology adoption and Artificial Intelligence complement the attitude of healthcare professionals in the various health upkeeping industry. More cross-cultural research is required to comprehend how the specific facets influencing the behavioral intentions to use technology in the healthcare settings. Filling these gaps will help healthcare professionals to understand the best ways of implementing the patient healthcare, diagnosis using technology creates earlier detection of various diseases to enhance the patient healthcare in the healthcare sector.

### *1.5 Research Objectives*

- To examine factors pertaining to perceived usefulness and the perceived risk in relation to attitude of healthcare professionals.
- To explore how information sensitivity and the social influence impact the attitude towards technology adoption.
- To examine the mediating role of attitude in association between perceived ease of use and the information sensitivity.
- To determine the best practices of promoting technology adoption for healthcare professionals' attitude towards the behavioural intention to use technology.
- To evaluate differences of attitude towards technological adoption with regard to their gender and age in the organization.

## 2 Review of Literature

### *2.1 Perceived Usefulness*

The perceived usefulness refers to the particular likelihood of a potential user has of an application system improving their job performance inside an organizational setting. Based on recent analysis it is founded that use of Technology Acceptance Model in healthcare consistently found that perceived usefulness is a key element of the desire to use information and communications technology [13]. Perceived usefulness can be referred as an extent to which individual believes that utilizing a precise technology would advance their job performance at the workplace. It is the concept that describes about health IT use which results to an improved job performance or benefits for its employees [14]. Perceived usefulness is the extent to which an individual admires that engaging a specific system would progress their job performance and leads to less time consuming. When the users perceive it as a means to enhance their productivity and efficiency in their professional tasks in an optimal manner but the users are more persuaded to enfold a technology.

## 2.2 Perceived Ease of Use

Perceived ease of use is the level to which an individual trusts that utilizing a specific system would require slight effort. It simply means the lack of physical or mental effort [15]. The perceived ease of use significantly affects personnel behavioural intention of the employees working in the hospitals. In the healthcare sector, through the perceived ease of use the training required a strong indirect effect on behavioural intention [16]. In the technology acceptance one's attitude towards utilizing is predisposed by their perception of the system's ease of use. It also affects their intention to take advantage of system and that ultimately impacts the actual usage of it [30].

Perceived ease of use (PEU) refers to the degree to which a person accepts as true that using a specific system or technology will be free of exertion. In context of the study on nursing students' perceptions of the simulated Electronic Medical Records (EMR), PEU indicates how easy the students found the Interactive Simulated Electronic Medical Record (ISEMR) systems to navigate and can utilize for their learning purposes.

## 2.3 Perceived Risk

The individuals have stated perceived risk as the personal perception and thought of the probability of an adverse result for them. It can be referred as the personal evaluation that an individual has concerning the frequency and seriousness of a distinct insecurity and others may discern it in stint of the probability and seriousness of the outcomes [17]. The perceive risk is influenced by the variable perception of risk which grasping the community's discernment of insecurity and trust allied to a certain subject. For establishing public confidence and to the earn social approvals towards prevailing the specific resources it is important that the patients should be willing to reveal personal health information. To concede the privacy fortification and risk extenuation alone may not sufficient to get the social authorization and it may be resulted because of the lack of knowledge of the social license and by refusals also [18].

In certain situations which is categorized by a lack of obscurity and control, evaluation of the possible adverse repercussion related to the utilization of a specific technology or service prevalent to perceived risk. The perceived risk refers to having uneasiness over the security and confidentiality of personal health information, dependability of the technology, possibility of data misuse, manipulation and exercising mal-practises [31]. When the users perceive a risk of their sensitive information being disclosed or mistreated or neglected the users' willingness to interrelate with technology decreases. This states that it may have a detrimental, results on their willingness to utilize such applications because individuals are having greater awareness of potential dangers. So, in some cases they may be reluctant to share their personal information or to trust upon the various digital health solutions without sufficient guarantees of confidentiality, security and privacy [19].

## 2.4 *Information Sensitivity*

Information sensitivity can be stated as the extent to which an information is kept as private, confidential or secret and the potential consequences of its unauthorized exposure on individuals, organizations and various sectors. The term "data" simply refers to collection of raw facts and various information. The sensitivity of information can be influenced according to the prevailing context in which it is used or shared. Moreover, when reviewing the sensitivity of the information shared in private may be considered more significant than the same information shared in a public setting to various people [20]. The possibility for getting harm is one of the main factors finding out the sensitivity of illegal access, identity theft, privacy invasion, operational disruption, financial loss, and reputational damage. For implementing effective and efficient privacy safeguards and to ensure compliance these elements want to be analysed in a crucial and critical manner [21].

Information sensitivity pertains to the degree of privacy concerns linked to particular categories of personal information within a specific setting in various industries. This explains the correlation between an individual's perception of the worth and most possible dangers and problems associated while disclosing their personal data and information with third parties [22]. While surrounding the request in connection with the specific circumstances, this perception in accordance with the information being asked to them. There are certain confidential data like medical records or financial information of any organisation which kept more private and hesitant to share it with others. The factors including the type of information being asked to shared, an individuals' concerns on his or her privacy and security shape an individual's attitudes and perceptions towards sharing the data and ensuring the privacy [23].

## 2.5 *Social Influence*

Social influence is the manner in which a persons' attitudes, beliefs, perceptions and behaviours are influenced by other individuals. It has a have a substantial impact on physicians' acceptance of new systems in the adoption of Healthcare Information Technology (HIT). There are four categories of social influence such as rewards, sanction identification and internalization [24]. The social variables influence the elderly's decision to embrace and utilize patient portals. It focuses on normative, mimetic, and coercive influences. The normative pressure had a negligible impact. In a social network which has a higher social standing the mimetic pressure refers to the deliberate imitation of actions exhibited by the esteemed peers [25]. Coercive pressure pertains to the persuasive impact exercised by healthcare practitioners. It

indicating that older persons are more inclined to utilize patient portals when specifically recommended by their medical providers. The elderly are most motivated to utilize patient portals due to coercive and mimetic pressures, but normative pressure does not have a significant impact on them while they are using the various portals in healthcare sector [26].

## 2.6 Attitude

The attitude refers to the ideas, values, and behaviours which are shared by the healthcare providers. It demonstrates their dedication to the safety management in various healthcare organizations expanded all over the world. The attitude can be examined by analysing different facets including communication, teamwork, error response and cognition. The word "attitude" refers to emotions, convictions, perceptions and behavioural tendencies of healthcare workers towards the employment of pharmacovigilance (PV) in the health care industry. The attitudes are mainly categorized as positive, negative, or neutral. It is also influenced by the blend of cognitive, emotional, and behavioural elements which an individual holds. The attitude of healthcare professionals includes their readiness to participate in pharmacovigilance activities, their sense of duty in reporting adverse drug reactions and their support for systems, development and education [32].

## 2.7 Behavioural Intention to Use

In health upkeep sector, the behavioural intention to use is extent to which an individual planned to either accomplish or not execute a detailed future activity or program by adopting information technology (IT) to perform it. The handler's level of engagement with technology in the healthcare sector is determined by evaluating the various aspects such as the performance expectancy, conducive conditions, effort expectancy, personal innovativeness, privacy concerns and trust in technology. These comprehending things are intended in order to enhance the acceptance, effective and efficient utilization of information technology in the management of humanitarian supply chains and optimum utilisation of resources [33].

Behavioural intention to use can be referred to as the notch to which a healthcare professional takes consciously developed plans to abstain from application of information technologies in the healthcare sector. The factors like perceived usefulness, privacy concerns, perceived ease of use, trust issues and perceived cost can influence its user's inclination to receive and use the technology at workplace. BI represents the user's preparedness to interact with the Internet of Things (IoT) services provided in healthcare settings [34].

# 3 Research Methodology

The study adopted descriptive method of research and primary data is used which is collected through structured questionnaire. The power analysis is used to identify the sample size and one hundred and ninety -two were collected. We processed the data and rejected eight responses and finally arrived to one hundred and eighty-four. The purpose of selection of the sample will be taking into account healthcare providers engaged in various activities in healthcare facilities. This study adopted Technology Acceptance Model (TAM) proposed by Fred Davis 1985 and this study included information sensitivity and social influence as two other independent variables along with the existing model. The questionnaire was designed in such a way that power analysis will guide the sample size. It will contain questions for independent variables like perceived usefulness, information sensitivity, perceived risk, social influence and perceived ease of use where applicable that have been validated by research. Inferential statistic such as SEM (Structural Equation Modeling) will be done using SPSS and AMOS applications to explore and test the relations inter amongst and across the mentioned constructs.

## *3.1 Conceptual Research Model*

See Fig. 1.

## *3.2 Research Hypothesis*

H1—The impact of perceived usefulness on behavioral intention to use technology.

H2—The impact of perceived ease of use on behavioral intention to use technology.

H3—The impact of perceived risk on behavioral intention to use technology.

H4—The impact of information sensitivity on behavioral intention to use technology.

H5—The impact of social influence on behavioral intention to use technology.

H6—The impact attitude on behavioral intention to use technology.

H7—The impact of perceived usefulness on attitude.

H8—The impact of perceived ease of use on attitude.

H9—The impact of perceived risk on attitude.

H10—The impact of information sensitivity on attitude.

H11—The impact of social influence on attitude.

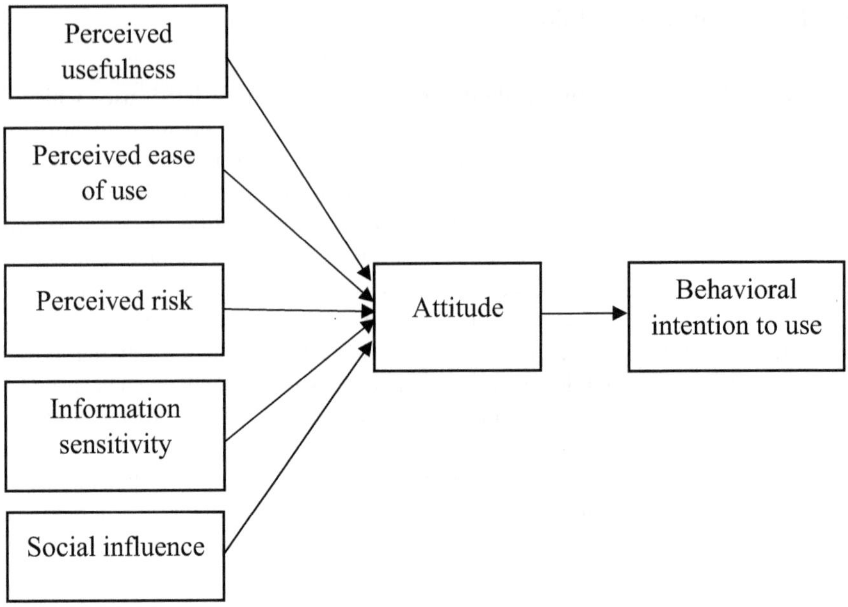


**Fig. 1** ...*Source* Author's work

## 4 Results

The below are the results obtained from the data analysis.

From Table 1, the gender distribution in the demographic data shows that the organisation employs more female employees than male employees (57.6% female employee). Most of the employees are in the age of 25–34 years (42.4%). The majority of the employees are either married. Most participants are enrolled as (79.3%) degree holders, (15.8%) or Master's, (4.9%) Doctorate and are either full-time (46.7%) or contract (33.2%) workers. The highest representation is the Nurses (55.4%) followed by Medical Coders and Billers (15.2%) of which 47.3% has a work experience of 1–3 years. The largest number of participants is employed in companies with 51–200 employees (39.7%).

From Table 2, both variables have positive mean values, and more than half of the respondents' ranked items at 4 or 5, so descriptive statistics would be positive about respondents' opinion on usefulness and simplicity of use by system. A lower mean and greater standard deviation in perceived risk will indicate the heterogeneity of opinions about the risks brought by the system. But apart from that, moderate perceptions could also be represented by social influence and information sensitivity, and the answers spread pretty uniformly across the scale.

From Table 3, the outcomes of the reliability analysis show that all the seven dimensions have reasonable levels of internal consistency, indicated by the Cronbach's alpha coefficients of 0. 737 to 0. 810. It implies that items within each

**Table 1** Demographic profile

| Gender | Frequency | Percentage |
|---|---|---|
| Female | 106 | 57.6 |
| Male | 78 | 42.4 |
| Total | 184 | 100 |
| *Age* | | |
| Under 25 | 62 | 33.7 |
| 25–34 | 78 | 42.4 |
| 35–44 | 33 | 17.9 |
| 45–54 | 11 | 6.0 |
| Total | 184 | 100 |
| *Marital status* | | |
| Married | 127 | 69.0 |
| Single | 57 | 31.0 |
| Total | 184 | 100 |
| *Highest level of education* | | |
| Bachelor's degree | 146 | 79.3 |
| Master's degree | 29 | 15.8 |
| Doctorate (PhD) | 9 | 4.9 |
| Total | 184 | 100 |
| *Current employment status* | | |
| Full time | 86 | 46.7 |
| Part time | 12 | 6.5 |
| Contract | 61 | 33.2 |
| Student | 25 | 13.6 |
| Total | 184 | 100 |
| *Job role/title* | | |
| Physicians and surgeons | 10 | 5.4 |
| Nurses | 102 | 55.4 |
| Therapists | 23 | 12.5 |
| Dentists | 5 | 2.7 |
| Medical coders and billers | 28 | 15.2 |
| Healthcare administrators | 9 | 4.9 |
| Others | 7 | 3.8 |
| Total | 184 | 100 |
| *Years of experience in healthcare sector* | | |
| Less than 1 year | 28 | 15.2 |
| 1–3 years | 87 | 47.3 |
| 4–6 years | 46 | 25.0 |
| 7–10 years | 23 | 12.5 |

(continued)

**Table 1** (continued)

| Gender | Frequency | Percentage |
|---|---|---|
| More than 10 years | 19 | 10.3 |
| Total | 184 | 100 |
| *Hospital size* | | |
| 1–50 employees | 32 | 17.4 |
| 51–200 employees | 73 | 39.7 |
| 201–500 employees | 29 | 15.8 |
| 501–1000 employees | 35 | 19.0 |
| More than 1000 employees | 15 | 8.2 |
| Total | 184 | 100 |

**Table 2** Descriptive statistics

| Variable | Mean | Median | Mode | Std. deviation | Range | 1 (%) | 2 (%) | 3 (%) | 4 (%) | 5 (%) |
|---|---|---|---|---|---|---|---|---|---|---|
| Perceived usefulness | 3.85 | 4 | 4 | 0.92 | 4 | 3.5 | 7.5 | 18 | 42.5 | 28.5 |
| Perceived ease of use | 3.72 | 4 | 4 | 1.01 | 4 | 4 | 10 | 22 | 38 | 26 |
| Perceived risk | 2.95 | 3 | 3 | 1.15 | 4 | 12 | 22.5 | 35.5 | 19 | 11 |
| Information sensitivity | 3.62 | 4 | 4 | 1.08 | 4 | 5.5 | 12 | 23.5 | 33 | 26 |
| Social influence | 3.41 | 3 | 3 | 1.06 | 4 | 6 | 14.5 | 29.5 | 32.5 | 17.5 |
| Attitude | 3.78 | 4 | 4 | 0.98 | 4 | 3 | 9.5 | 20.5 | 40.5 | 26.5 |
| Behavioural intention to use | 3.69 | 4 | 4 | 1.05 | 4 | 4.5 | 11 | 21.5 | 37 | 26 |

dimension are valid and reliable in terms of the intended constructs. Peculiarly, the reliability of "Perceived ease of use" is the highest (0. 810), whereas "Perceived usefulness and Attitude", is the lowest yet reasonable (0. 737). In summary, these findings deliver evidence for the dependability of instrument employed for assessing these dimensions in the research study.

From Table 4, the analysis of the results obtained with the support of the SPSS program indicates that all the factors are positively correlated. Most importantly, the correlation between perceived risk and is the highest ($r = 0.796$). Other factors such as training and development, work environment, job satisfaction, employee recognition, and employee retention had other moderate positive relationship (r values ranging from 0.718 to 0.778). These findings imply that these factors are interrelated and reciprocally affecting one another in the workplace.

**Table 3** Reliability test

| Variable | No of items | Cronbach's Alpha |
|---|---|---|
| Perceived usefulness | 5 | 0.737 |
| Perceived ease of use | 5 | 0.810 |
| Perceived risk | 5 | 0.757 |
| Information sensitivity | 5 | 0.805 |
| Social influence | 5 | 0.775 |
| Attitude | 5 | 0.737 |
| Behavioural intention to use | 5 | 0.756 |

**Table 4** Correlation test

| Constructs | Perceived usefulness | Perceived ease of use | Perceived risk | Information sensitivity | Social influence | Attitude | Behavioural intention to use |
|---|---|---|---|---|---|---|---|
| Perceived usefulness | 1 | – | – | – | – | – | – |
| Perceived ease of use | 0.773** | 1 | – | – | – | – | – |
| Perceived risk | 0.788** | 0.784** | 1 | – | – | – | – |
| Information sensitivity | 0.796** | 0.762** | 0.778** | 1 | – | – | – |
| Social influence | 0.780** | 0.786** | 0.747** | 0.783** | 1 | – | – |
| Attitude | 0.754** | 0.747** | 0.742** | 0.783** | 0.738** | 1 | – |
| Behavioural intention to use | 0.764** | 0.779** | 0.765** | 0.786** | 0.769** | 0.757** | 1 |

From Table 5, the regression model is statistically substantial for the product variable ($p < 0.001$). This indicates that the model has a predictive power of variance about 69.1% on the dependent variable (Adjusted R Square = 0.728). The variables SI, PR, IS, PEU, and PU are the predictors that have been used in this study and they are closely related to the outcome variable with the correlation coefficient of 0.853. The adjusted R Square was 0.720 indicates that the model is expected to do well on new data because the model is likely to be correct most of the time. The standard error of the evaluation (1.97877) is measure of the average difference among the actual and the predicted values.

From the Table 6, the ANOVA table shows the regression model to be statistically significant for ERN ($F(5, 178) = 372.875$, $p < 0.001$). The model with five predictors (SI, PR, IS, PEU, PU) accounts for a significant sum of variance in BI as evidenced by large F-value and small p-value. The regression sum of squares (1864.377) is

**Table 5** Regression table—model summary

| Model summary | | | | |
|---|---|---|---|---|
| Model | R | R square | Adjusted R square | Std. error of the estimate |
| 1 | 0.853[a] | 0.728 | 0.720 | 1.97877 |

[a] Predictors: (Constant), SI, PR, IS, PEU, PU

significantly larger than the residual sum of squares (696.965), which indicates that the predictors as a whole significantly influence BI.

From Table 7, the regression analysis shows that all the independent variables, namely Perceived Usefulness (PU), Perceived Risk (PR), Information Sensitivity (IS), Perceived Ease of Use (PEU) and Social Influence (SI) have a positive and a significant relationship with Behavioral Intention to Use (BI) at $p < 0.05$. Thus, the standardized coefficients (Beta) reveal that IS has the highest impact on the BI, with a coefficient of 0.251, PEU with a coefficient of 0.224, SI with a coefficient of 0.177, PR with a coefficient of 0.167, and PU with a coefficient of 0.121. The model accounts for an important part of the total variability of BI signifying that these antecedents are essential for meeting the retention needs of employees.

**Table 6** Regression—ANOVA

| ANOVA[a] | | | | | | |
|---|---|---|---|---|---|---|
| Model | | Sum of squares | df | Mean square | F | Sig |
| 1 | Regression | 1864.377 | 5 | 372.875 | 95.230 | 0.000[b] |
| | Residual | 696.965 | 178 | 3.916 | | |
| | Total | 2561.342 | 183 | | | |

[a] Dependent Variable: BI
[b] Predictors: (Constant), SI, PR, IS, PEU, PU

**Table 7** Regression table—coefficients

| Coefficients[a] | | | | | | | | |
|---|---|---|---|---|---|---|---|---|
| Model | | Unstandardized coefficients | | Standardized coefficients | t | Sig | Collinearity statistics | |
| | | B | Std. error | Beta | | | Tolerance | VIF |
| 1 | (Constant) | 1.635 | 0.829 | | 1.972 | 0.050 | | |
| | PU | 0.128 | 0.081 | 0.121 | 1.571 | 0.118 | 0.257 | 3.887 |
| | PEU | 0.207 | 0.069 | 0.224 | 2.995 | 0.003 | 0.273 | 3.658 |
| | PR | 0.171 | 0.076 | 0.167 | 2.244 | 0.026 | 0.276 | 3.621 |
| | IS | 0.229 | 0.069 | 0.251 | 3.310 | 0.001 | 0.267 | 3.747 |
| | SI | 0.174 | 0.073 | 0.177 | 2.385 | 0.018 | 0.277 | 3.610 |

[a]Dependent Variable: BI

From Table 8, the percentages of respondents who rated them at 4 or 5 are more than half, descriptive statistics reveal that respondents were generally positive regarding the ease of use and usefulness. Having a lower mean and a higher standard deviation, perceived risk depicts that opinions on the risks of the system vary more. The social influence and information sensitivity perception levels are moderate, as answers have a uniform distribution across the scale.

From Table 9, the percentages of respondents who rated them at 4 or 5 are more than half, descriptive statistics reveal that respondents were generally positive regarding the usefulness and ease of use. Having a lower mean and a higher standard deviation, perceived risk depicts that opinions on the risks of the system vary more. Other evidence, also coming from the field of social influence and information sensitivity, is that perception levels are moderate, as answers have a rather uniform distribution across the scale.

From Table 10, the results of the hypothesis test reveal that most routes are supported and predictors of behavioral intention to use are substantially correlated with attitude, especially including perceived ease of use, information sensitivity, perceived risk and social influence. On the other hand, there was no correlation between perceiving use and BI. Attitude was also significantly affected by all the factors and contributed mightily to the formation of BI.

**Path Model**

The figure depicts a model demonstrating influences on people's intention to use a good or service. Among the factors that influence attitudes toward its use are risk, sensitivity of the information, social influence, usefulness, and how convenient it is to use. Of these, attitude has the greatest direct influence on behavioral intention, followed by perceived risk and usefulness (Fig. 2).

- IV = Independent Variables;
- CFI = Comparative Fit Index;
- TLI = Tucker-Lewis Index;
- RMSEA = Root Mean Square Error of Approximation;
- SRMR = Standardized Root Mean Square Residual;
- AIC = Akaike Information Criterion;
- BIC = Bayesian Information Criterion

From Table 11, the original model is best fitted to the data. It encompasses the lowest AIC and BIC values coupled with high indices (CFI = 0.956, RMSEA = 0.054). Both the full mediation model and the direct effects only model with no mediation had a slightly worse fit but within tolerable bounds. In comparison, the models by reverse causality and attitude as moderator are the worst performers with larger $\chi 2$/df ratios and worse fit indices: they are likely to fail to capture the true connections within variables in question.

From Table 12, the outcome of the sensitivity analysis is that significant cases did not influence the fit of the model as the value of $\chi 2$/df, CFI, TLI, RMSEA, and SRMR does not change at any level of outlier removal. Removing outlier's one percent at a time (1%, 5%, and 10%) for each of the above cases, the coefficient between

**Table 8** Reliability and validity analysis

| Construct | Item | Factor loading | CR | AVE | MSV | ASV |
|---|---|---|---|---|---|---|
| Perceived usefulness (PU) | PU1 | 0.82 | 0.891 | 0.621 | 0.564 | 0.498 |
| | PU2 | 0.79 | | | | |
| | PU3 | 0.85 | | | | |
| | PU4 | 0.76 | | | | |
| | PU5 | 0.72 | | | | |
| Perceived ease of use (PEOU) | PEOU1 | 0.84 | 0.902 | 0.649 | 0.529 | 0.476 |
| | PEOU2 | 0.81 | | | | |
| | PEOU3 | 0.87 | | | | |
| | PEOU4 | 0.78 | | | | |
| | PEOU5 | 0.73 | | | | |
| Perceived risk (PR) | PR1 | 0.76 | 0.876 | 0.585 | 0.481 | 0.423 |
| | PR2 | 0.81 | | | | |
| | PR3 | 0.79 | | | | |
| | PR4 | 0.74 | | | | |
| | PR5 | 0.7 | | | | |
| Information sensitivity (IS) | IS1 | 0.83 | 0.895 | 0.631 | 0.551 | 0.487 |
| | IS2 | 0.8 | | | | |
| | IS3 | 0.85 | | | | |
| | IS4 | 0.77 | | | | |
| | IS5 | 0.72 | | | | |
| Social influence (SI) | SI1 | 0.79 | 0.885 | 0.605 | 0.509 | 0.452 |
| | SI2 | 0.82 | | | | |
| | SI3 | 0.78 | | | | |
| | SI4 | 0.75 | | | | |
| | SI5 | 0.73 | | | | |
| Attitude (AT) | A1 | 0.85 | 0.907 | 0.661 | 0.576 | 0.511 |
| | A2 | 0.82 | | | | |
| | A3 | 0.87 | | | | |
| | A4 | 0.79 | | | | |
| | A5 | 0.74 | | | | |
| Behavioural intention to use (BI) | BI1 | 0.86 | 0.915 | 0.683 | 0.576 | 0.518 |
| | BI2 | 0.83 | | | | |
| | BI3 | 0.88 | | | | |
| | BI4 | 0.81 | | | | |
| | BI5 | 0.76 | | | | |

**Table 9** Mediation analysis with bootstrap results

| Path | Direct effect | Indirect effect | Total effect | 95% CI (Indirect) |
|---|---|---|---|---|
| PU → A → BI | 0.12** | 0.09** | 0.21** | [0.04, 0.15] |
| PEOU → A → BI | 0.22** | 0.13** | 0.35** | [0.07, 0.20] |
| PR → A → BI | 0.17** | 0.08** | 0.25** | [0.03, 0.14] |
| IS → A → BI | 0.25** | 0.15** | 0.40** | [0.09, 0.22] |
| SI → A → BI | 0.18** | 0.11** | 0.29** | [0.05, 0.17] |

**Table 10** Path coefficients and hypothesis testing

| Hypothesis | Path | Std. coefficient | t-value | p-value | Result |
|---|---|---|---|---|---|
| H1 | PU → BI | 0.12 | 1.57 | 0.118 | Not supported |
| H2 | PEOU → BI | 0.22 | 3 | 0.003 | Supported |
| H3 | PR → BI | 0.17 | 2.24 | 0.026 | Supported |
| H4 | IS → BI | 0.25 | 3.31 | 0.001 | Supported |
| H5 | SI → BI | 0.18 | 2.39 | 0.018 | Supported |
| H6 | A → BI | 0.53 | 11.24 | < 0.001 | Supported |
| H7 | PU → A | 0.23 | 4.76 | < 0.001 | Supported |
| H8 | PEOU → A | 0.28 | 5.87 | < 0.001 | Supported |
| H9 | PR → A | 0.18 | 3.89 | < 0.001 | Supported |
| H10 | IS → A | 0.25 | 5.12 | < 0.001 | Supported |
| H11 | SI → A | 0.14 | 3.21 | < 0.01 | Supported |

Attitude and Behavioral Intention to Use remains significant. But its magnitude reduces substantially from 0.53 to 0.50. These results indicate the stability of the model and its non-gross vulnerability to anomalies (Fig. 3).

The perceived usefulness, perceived ease of use, perceived risk, information security, and social influence that can result in affecting the attitude of an individual toward the specified behavior or technology. All these factors have a direct, positive relation with attitude. Attitude is found to be the factor that strongly influences behavioral intention and with standardized path coefficients of 1.08, indicates to a greater extent that it takes one's intention to carry out this behavior to a higher level when one's attitude has a more positive trend. The deviations of the components e1-attitude, AT and e2:BI are marking negative connections between them. This indicates that the model does not show accounting for some variance both in affecting AT and BI. This therefore implies that latent variables or other unobserved factors could influence both the constructs under investigation, thereby creating shared variance.

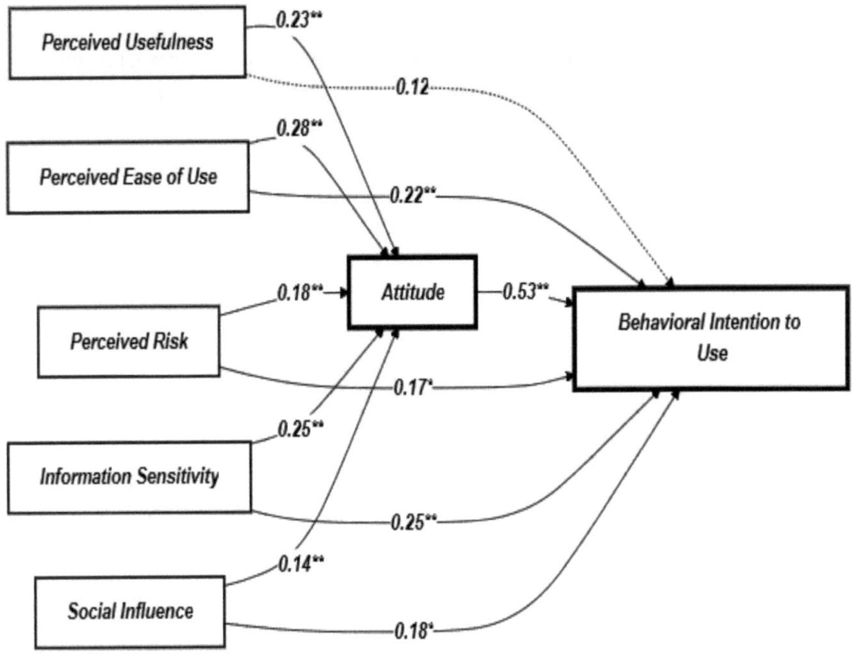


**Fig. 2** Path diagram. *Source* Author's work

**Table 11** Alternative model specifications

| Model | Description | χ2/df | CFI | TLI | RMSEA | SRMR | AIC | BIC |
|---|---|---|---|---|---|---|---|---|
| 1. Original model | As hypothesized | 2.41 | 0.956 | 0.951 | 0.054 | 0.038 | 2145.32 | 2398.76 |
| 2. Direct effects only | No mediation through attitude | 2.87 | 0.932 | 0.925 | 0.062 | 0.056 | 2312.45 | 2542.89 |
| 3. Full mediation | No direct paths to BI | 2.63 | 0.944 | 0.938 | 0.058 | 0.047 | 2231.87 | 2473.21 |
| 4. Reverse causality | BI → A → IVs | 3.12 | 0.921 | 0.914 | 0.066 | 0.061 | 2398.63 | 2629.07 |
| 5. AT as moderator | IVs x A → BI | 2.75 | 0.938 | 0.93 | 0.060 | 0.052 | 2278.91 | 2521.35 |

**Table 12** Sensitivity analysis—removal of influential cases

| Model | N | χ2/df | CFI | TLI | RMSEA | SRMR | Key path coefficients |
|---|---|---|---|---|---|---|---|
| Original | 184 | 2.41 | 0.956 | 0.951 | 0.054 | 0.038 | A → BI: 0.53** |
| −1% Outliers | 182 | 2.39 | 0.957 | 0.952 | 0.053 | 0.037 | A → BI: 0.52** |
| −5% Outliers | 175 | 2.36 | 0.959 | 0.954 | 0.052 | 0.036 | A → BI: 0.51** |
| −10% Outliers | 166 | 2.33 | 0.961 | 0.961 | 0.051 | 0.035 | A → BI: 0.50** |

*Note* ** $p < 0.01$; Outliers identified using Mahalanobis distance

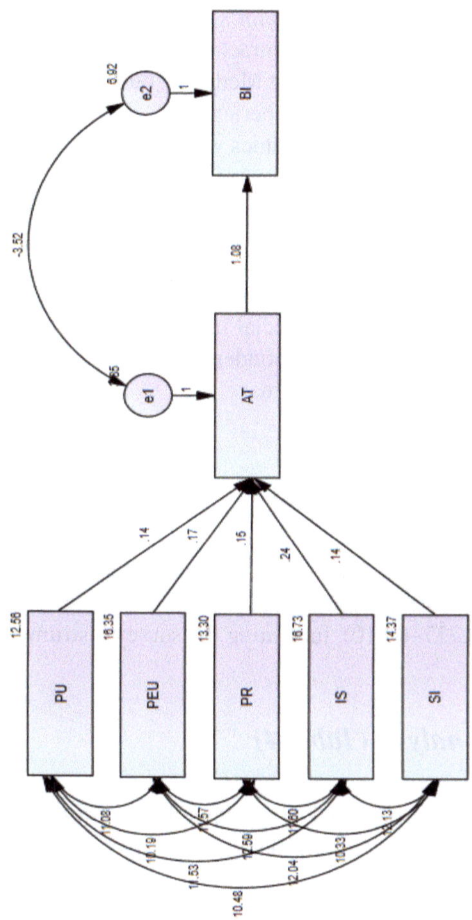


**Fig. 3** SEM model. *Source* Author's work

# 5 Summary of the Key Findings

## 5.1 Demographics (Table 1)

- Gender: More females (57.6%) than males (42.4%) in the sample.
- Age: Majority of employees are 25–34 years old (42.4%).
- Marital Status: Majority of employees are married.
- Education: Most employees hold Bachelor's or Master's degrees.
- Employment: Primarily full-time or contract workers are in the healthcare sector, with a significant portion in Nurses and Medical Coders and Billers roles.
- Experience: Most have 1–3 years of experience.
- Company Size: Concentrated in companies with 51–200 employees.

## 5.2 Descriptive Statistics (Table 2)

- The descriptive statistics reveal that social influence and information sensitivity reflect mid-level perceptions and respondents had positive opinions regarding usefulness and ease of use of the system.

## 5.3 Reliability Analysis (Table 3)

- The dimensions (Perceived Usefulness, Social Influence, Perceived Ease of Use, Perceived Risk, Information Sensitivity,) shows good internal consistency (Cronbach's alpha 0.737–0.810), indicating the survey instrument is reliable.

## 5.4 Correlation Analysis (Table 4)

- All factors are positively correlated, suggesting they reinforce each other.
- Strongest correlation is between Information Sensitivity and Attitude ($r = 0.796$).

## 5.5 Regression Analysis (Table 5)

- The model is statistically significant ($p < 0.001$) and explains 72.8% of the change in Behavioral Intention to Use (BI).
- All five predictors (SI, PR, IS, PEU, PU) have a positive and substantial relationship with BI.

- Perceived Ease of Use (PEU) has strongest impact on BI, followed by Information Sensitivity.

## 5.6 Reliability and Validity Analysis (Table 8)

- The mean was lower along with higher standard deviation for perceived risk, although the people do mention it more with varying opinions.
- Social influence and information sensitivity perceived the system moderately with uniform responses along the scale.
- The majority of the respondents rated the structure positively in terms of ease of use and usefulness.

## 5.7 Mediation Analysis (Table 9)

- The most respondents gave the system high marks with respect to its helpfulness and usability; responses were fairly evenly distributed on the scale but were moderate with respect to social influence and information sensitivity, whereas perceived risk was much more variable because it was recorded with a lower mean and greater standard deviation.

## 5.8 Path Coefficients and Hypothesis Testing (Table 10)

- The predictors including perceived risk, information sensitivity, social influence and perceived ease of use correlated with behavioural intention to use while the other perceived usefulness did not have any relationship besides, all factors highly influence attitude, which is a main factor for the Behavioral Intention.

## 5.9 Alternative Model (Table 11)

- The full mediation and direct effects-only models fit slightly worse, the reverse causality and attitude as moderator models are poor, so they do not depict the relationships reasonably well, and the original model seems to be the best fit for the data since the lowest AIC and BIC values are accompanied by high indices.

### 5.10 Sensitivity Analysis (Table 12)

- The association between Attitude and Behavioral Intention to Use, reducing from 0.53 to 0.50 but still significant, thus indicates the stability and resilience of the model against anomalies.

### 5.11 Key Takeaways

- To enhance the behaviour intention to use technology, key areas that need to be worked on are—perceived usefulness, perceived ease of use, perceived risk, information sensitivity, social influence and attitude.
- One of the most important issues in the technology adoption in healthcare sector is the level of information sensitivity that is crucial for the technology adoption, especially if it is upheld by the attitude of healthcare professionals and the patients.

## 6 Suggestions and Recommendations

| Area of focus | Suggestions and recommendations | Innovative studies for reference |
|---|---|---|
| Perceived usefulness | Conduct training sessions to help users to understand and utilize the functionalities of the technology | Study on the adoption of the Electronic Health Records systems in hospitals and their impact on workflow efficiency and employee performance |
| | Demeanour testimonials or case studies from professionals who have experienced positive outcomes and emphasize the tangible benefits of the technology in day today tasks | Study about the telemedicine during the COVID-19 pandemic and how it enhanced health care accessibility and the advancement |
| Perceived ease of use | Perform regular experience testing in order to find and address problem areas | Study on the analysis of Mobile Health Application Designs and How Simplified Interfaces Increase Adoption among Older Health Care Providers |
| | Provide continuous support for troubleshooting and problem resolution | Study on the voice command integration in clinical software for hand-free operations |
| Perceived risk | The healthcare system should deal with patients concerns on security through encryption and multi-factor authentication | Study on the Cybersecurity Risks in Healthcare and the various Measures Used by Healthcare Institutions for Risk Mitigation |

(continued)

(continued)

| Area of focus | Suggestions and recommendations | Innovative studies for reference |
|---|---|---|
| | Clear handling and storage of personal and professional data following guidelines and security measures | Conduct research on the implementation of blockchain technology in securing patient records |
| Information sensitivity | The system should have strict policies in place relating to the availability of confidential information | Study on the breaches of patients' data information and also the role that encryption plays in preventing information leaks |
| | With the ongoing evolution in health care field, regular privacy impact assessments would be a way of conforming with data protection procedures | Study on General Data Protection Regulation compliance in healthcare IT systems and implications for global data exchange |
| Social influence | Engage the people who are influential enough to have them become supportive of the tool implementation in the | Study the critical role of peer influence in technology adoption in healthcare |
| | The healthcare sector should foster a friendly environment where early adopters will have many options to share their experiences with others | Study on different social networks within hospitals and how they influence the diffusion of innovations |

## 7 Future Directions

The study on the influential factors affecting employee behavior in the implementation of workplace technologies into healthcare has quite a number of potential avenues for future study. First, generalizing the sample size across various healthcare sceneries, such as hospitals, long-term care institutions or clinics could better illustrate the variation in such factors by organizational context. Future research should investigate the difference in the adoption of technology behaviors across cultures since the attitudes of health care institutions across regions are quite different and influenced by their local regulations as well as norms and the type of technological infrastructures adopted. It could ultimately enable one to identify variables that are relevant at different points in the phase of adoption of a new technology, from introduction to habitual use.

Further researches will cover other factors including perceived cost of using technology, organizational support, availability of training and so on. It will then be able to give an even more holistic view of what really drives or hinders the employees from forming the corresponding behavioral intention in adoption of technology. The mediating effect of attitude can be extended to include independent variables, such as perceived usefulness or ease of use, and a dependent variable, such as that described above, in order to determine if there are additional constraints on the effect of other subsets of variables, such as job role, age, or experience demographics, that could affect the strength of a relationship. More importantly, most of the modern high-tech innovations that health care organizations are nowadays embracing, including wearable health devices, artificial intelligence, and machine learning, will greatly leverage future research on the way employees perceive those particular technologies and how their perceptions influence their behavioral intentions. This would help health care organizations understand the problems that begin with innovative adoption and the targeted, strategic interventions meant to overcome opposition.

## 8 Conclusion

This study offers fruitful perceptions into variables affecting the attitudes of workers in the healthcare industry toward implementing the newer technologies. From this relation analysis between the independent variables social influence, perceived ease of use, perceived risk, information sensitivity, and perceived usefulness, how these influence employees' attitudes in regard to the behavior intended for using technology there are various key conclusions. First, the results underscore how perceived utility has an impact on affecting workers' attitude towards new technologies. There is a higher chance to be more favourably accepted if it is to increase productivity in any way or to make things better for patients and smooth out administrative roles for healthcare professionals. Subsequently, their behavioral intention toward adopting the use of technology in their daily lives will also be increased. Hence, healthcare organizations can use such knowledge in practice by making staff members aware of the material benefits of a technology and how the technology applies through examples that are there in practical situations.

The employees' attitudes toward a technology were directly and positively affected by perceived ease of use. The review indicates information sensitivity and perceived risk could prove to be critical barriers to the adoption of technology. It is only logical why physicians were hesitant to adopt new data management systems in healthcare settings as some security lapses can have severe effects on the operation, ethics, and the law. A worse attitude toward technology can spring based on the perceived risks associated with personal data, the reliability of the system, and even security breaches; in summary, it diminishes the adoption likelihood. To alleviate such concerns, healthcare organizations must invest greatly in data security measures and educate staff about the protections available. Patient data security can be communicated transparently, meeting legal standards and keep individuals with positive attitudes toward technology.

Social influence is another factor that affects the adoption of technology. Workers would be more receptive to new technology if they are finding respected peers, managers, or coworkers using and promoting it. An important finding of the study is attitude played mediating role. Employees' technological views then served as a binding between their behavioral intention to adopt the technology and independent factors. It has significant implications for healthcare institutions. Awareness of which factors drive the attitudes and intentions can provide organizations with more effective plans to further promote the use of the technology. In addition to making decisions on useful technology that is friendly to users, this will control risk and sensitive information issues, apply social influence, and ensure staff adopt a good attitude.

In the long term, successful execution of technology will lead workflow to be more effectual, increase organizational productivity, most importantly, show improved quality of care for their patients. The healthcare organization should have an open environment toward technological change in order to better assist their workforce in dealing with the challenges in modern delivery in healthcare. This fits within the overall patient results improvement and experience, yet promotes operational results. This, in turn, leads to improved compliance with treatment regimens and the adoption of healthier lifestyle decisions. In summary, reflective influence of technology on healthcare sector highlights crucial function in increased outcomes, productivity and patient contentment.

## References


1. Edo, O.C., Ang, D., Etu, E.E., Tenebe, I., Edo, S., Diekola, O.A.: Why do healthcare workers adopt digital health technologies-a cross-sectional study integrating the TAM and UTAUT model in a developing economy. Int. J. Inf. Manag. Data Insights **3**(2), 100186 (2023)
2. Liu, L., Miguel-Cruz, A.: Technology adoption and diffusion in healthcare at onset of COVID-19 and beyond. In: Healthc. Manag. Forum 35(3), 161–167. SAGE Publications, Sage CA, Los Angeles, CA (2022)
3. Phichitchaisopa, N., Naenna, T.: Factors affecting the adoption of healthcare information technology. EXCLI J. **12**, 413 (2013)
4. Patel, H., Connolly, R.: Factors influencing technology adoption: a review. Inf. Manag. Netw. Econ.: Issues Solut. pp. 416–431 (2007)
5. Stoumpos, A.I., Kitsios, F., Talias, M.A.: Digital transformation in healthcare: technology acceptance and its applications. Int. J. Environ. Res. Public Health **20**(4), 3407 (2023)
6. Kesse-Tachi, A., Asmah, A.E., Agbozo, E.: Factors influencing adoption of eHealth technologies in Ghana. Digit. Health **5**, 2055207619871425 (2019)
7. Rogers, E.M., Singhal, A., Quinlan, M.M.: Diffusion of innovations. In: An integrated approach to communication theory and research, Routledge, pp. 432–448 (2014)
8. Guisado-Fernández, E., Giunti, G., Mackey, L.M., Blake, C., Caulfield, B.M.: Factors influencing the adoption of smart health technologies for people with dementia and their informal caregivers: scoping review and design framework. JMIR aging **2**(1), e12192 (2019)
9. Baskaran, S., Lay, H.S., Ming, B.S., Mahadi, N.: Technology adoption and employee's job performance: an empirical investigation. Int. J. Acad. Res. Econ. Manag. Sci. **9**(1), 78–105 (2020)
10. Alotaibi, Y.K., Federico, F.: The impact of health information technology on patient safety. Saudi Med. J. **38**(12), 1173 (2017)
11. Buntin, M.B., Burke, M.F., Hoaglin, M.C., Blumenthal, D.: The benefits of health information technology: a review of the recent literature shows predominantly positive results. Health Aff. **30**(3), 464–471 (2011)
12. Angst, C.M., Agarwal, R.: Adoption of electronic health records in the presence of privacy concerns: the elaboration likelihood model and individual persuasion. MIS quarterly, pp. 339–370 (2009)

13. Gajanayake, R., Sahama, T., Iannella, R.: (2013). The role of perceived usefulness and attitude on electronic health record acceptance. In: 2013 IEEE 15th international conference on e-health networking, applications and services (Healthcom 2013), IEEE, pp. 388–393.
14. Holden, R.J., Karsh, B.T.: The technology acceptance model: its past and its future in health care. J. Biomed. Inform. **43**(1), 159–172 (2010)
15. Ologeanu-Taddei, R., Morquin, D., Bourret, R.: Understanding the perceived usefulness and the ease of use of a hospital information system: the case of a French university hospital. In: Digital healthcare empowering Europeans, IOS Press, pp 531–535 (2015)
16. Chirchir, L.K., Aruasa, W.K., Chebon, S.K.: Perceived usefulness and ease of use as mediators of the effect of health information systems on user performance. Eur. J. Comput. Sci. Inf. Technol. **7**(1), 22–37 (2019)
17. Krahe, M., Milligan, E., Reilly, S.: Personal health information in research: perceived risk, trustworthiness and opinions from patients attending a tertiary healthcare facility. J. Biomed. Inform. **95**, 103222 (2019)
18. Platt, J.E., Jacobson, P.D., Kardia, S.L.: Public trust in health information sharing: a measure of system trust. Health Serv. Res. **53**(2), 824–845 (2018)
19. Choudhury, A.: Factors influencing clinicians' willingness to use an AI-based clinical decision support system. Front. Digit. Health **4**, 920662 (2022)
20. Bansal, G., Gefen, D.: The impact of personal dispositions on information sensitivity, privacy concern and trust in disclosing health information online. Decis. Support. Syst. **49**(2), 138–150 (2010)
21. Dang, Y., Guo, S., Guo, X., Wang, M., Xie, K.: Privacy concerns about health information disclosure in mobile health: questionnaire study investigating the moderation effect of social support. JMIR Mhealth Uhealth **9**(2), e19594 (2021)
22. Kang, J., Lan, J., Yan, H., Li, W., Shi, X.: Antecedents of information sensitivity and willingness to provide. Mark. Intell. Plan. **40**(6), 787–803 (2022)
23. Karampela, M., Ouhbi, S., Isomursu, M.: Connected health user willingness to share personal health data: questionnaire study. J. Med. Internet Res. **21**(11), e14537 (2019)
24. Lu, Z., Cui, T., Tong, Y., Wang, W.: Examining the effects of social influence in pre-adoption phase and initial post-adoption phase in the healthcare context. Inf. Manag. **57**(3), 103195 (2020)
25. Bozan, K., Davey, B., Parker, K.: Social influence on health IT adoption patterns of the elderly: an institutional theory based use behaviour approach. Procedia Computer Science **63**, 517–523 (2015)
26. Zhu, P., Shen, J., Xu, M.: Patients' willingness to share information in online patient communities: questionnaire study. J. Med. Internet Res. **22**(4), e16546 (2020)
27. Thimbleby, H.: Technology and the future of healthcare. J. Public Health Res. **2**(3), e28 (2013). https://doi.org/10.4081/jphr.2013.e28
28. Corridon, P.R., Wang, X., Shakeel, A., Chan, V.: Digital technologies: advancing individualized treatments through gene and cell therapies, pharmacogenetics and disease detection and diagnostics. Biomedicines **10**(10), 2445 (2022). https://doi.org/10.3390/biomediscines10102445
29. Safi, S., Thiessen, T., Schmailzl, K.J.: Acceptance and resistance of new digital technologies in medicine: qualitative study. JMIR Res. Protoc. **7**(12), e11072 (2018). https://doi.org/10.2196/11072
30. Sherifi, D.: Perceived usefulness and perceived ease of use impact on patient portal use (2018). https://scholarworks.waldenu.edu/dissertations/4756
31. Schnall, R., Higgins, T., Brown, W., Carballo-Dieguez, A., Bakken, S.: Trust, perceived risk, perceived ease of use and perceived usefulness as factors related to mHealth technology use. Stud. Health Technol. Inform. **216**, 467–471 (2015)
32. Alshammari, T.M., Alamri, K.K., Ghawa, Y.A., Alohali, N.F., Abualkol, S.A., Aljadhey, H.S.: Knowledge and attitude of health-care professionals in hospitals towards pharmacovigilance in Saudi Arabia. Int. J. Clin. Pharm. **37**(6), 1104–1110 (2015). https://doi.org/10.1007/s11096-015-0165-5

33. Kabra, G., Ramesh, A., Akhtar, P., Dash, M.K.: Understanding behavioural intention to use information technology: insights from humanitarian practitioners. Telematics Inform. **34**(7), 1250–1261 (2017). https://doi.org/10.1016/j.tele.2017.05.010
34. Alanazi, M.A., Tully, M.P., Lewis, P.J.: Prescribing errors by junior doctors – a comparison of errors with high risk medicines and non-high risk medicines. PloS One **14**(1), e0211270 (2019). https://doi.org/10.1371/journal.pone.0211270

# Quantifying the Social Impact of Carbon Projects and Generative AI Tool for United Nations Social Development Goals Claims Verification

**Drishtant Leuva, Farookh Hussain, and Morteza Saberi**

**Abstract** Climate change is a global challenge that shas accelerated the adoption of carbon credit mechanisms to reduce emissions. However, these projects' impacts extend beyond the environment, encompassing significant social outcomes in local communities. This research introduces an innovative framework to quantify the social impacts of carbon credit projects by integrating the United Nations Sustainable Development Goals (UN SDGs). The study addresses three core objectives. First, it incorporates UN SDG indicators like poverty reduction, health improvement, and education into the evaluation, creating a nuanced valuation model for comprehensive social impact assessment. Second, it validates statistical models that predict social outcomes of carbon credit projects by analysing variables such as job creation, methodology, and UN SDG claims. Third, it develops a rule-based verification system, leveraging AI to ensure project alignment with specific UN SDGs, promoting transparency and accountability. By advancing these objectives, the research provides a robust methodology for assessing UN SDG claims' social impact and credibility in carbon credit projects. The framework demonstrates the potential to predict and quantify diverse social implications through literature review and real-world data analysis, establishing a new benchmark for evaluating carbon credit projects.

**Keywords** Carbon credit project · UN SDG (United Nations Social Development Goals) · Statistical modelling

D. Leuva (✉) · F. Hussain · M. Saberi
School of Computer Science, University of Technology Sydney, Ultimo, NSW, Australia
e-mail: Drishtant.Leuva@uts.edu.au

F. Hussain
e-mail: farookh.hussain@uts.edu.au

M. Saberi
e-mail: morteza.saberi@uts.edu.au



N. D. K. Nguyen et al. (eds.), *Digital Transformation and Sustainability: Methods and Applications*, Information Systems Engineering and Management 47,
https://doi.org/10.1007/978-3-031-91278-8_3

# 1 Introduction

The ongoing battle against climate change has amplified the importance of carbon credit mechanisms, which incentivise emission reduction initiatives through financial rewards. Traditionally, the evaluation of carbon credit projects has centred around their environmental outcomes, particularly their contribution to reducing greenhouse gas emissions. However, these projects also generate significant social impacts, influencing local communities in areas such as poverty alleviation, healthcare access, and education. Despite this, the social dimensions of these projects are often overlooked in conventional evaluations [1].

This research aims to bridge this gap by developing a comprehensive framework that quantifies the social impact of carbon credit projects, explicitly aligning the assessment process with the United Nations Sustainable Development Goals (UN SDGs). Integrating UN SDG indicators into carbon credit evaluations allows for a more holistic understanding of these projects, moving beyond a purely environmental focus to consider their broader societal contributions [2].

The chapter presents a multi-faceted approach that combines statistical modelling and artificial intelligence (AI) techniques to quantify, predict, and verify the social impacts of carbon credit projects. By integrating AI-based verification, the proposed framework also ensures transparency and credibility, addressing critical challenges in sustainability assessments. This innovative approach improves the accuracy of social impact evaluations and aligns the carbon credit sector with global sustainability objectives, ultimately fostering more informed decision-making among stakeholders [3, 4].

# 2 Literature Review

## 2.1 Introduction

The evaluation of carbon credit projects traditionally focuses on environmental outcomes, such as greenhouse gas (GHG) emissions reduction. However, an increasing body of research highlights the importance of integrating socio-economic dimensions, particularly in alignment with the United Nations Sustainable Development Goals (UN SDGs). This literature review synthesizes the current state of research on carbon credit projects, addressing their social impact, carbon pricing mechanisms, and the potential for AI-driven evaluations.

**Table 1** Search categories and keywords

| Search category | Keywords |
|---|---|
| Social Impact of carbon credit or social impact caused by carbon reduction or carbon credit projects | ((social impact) OR (social development)) AND ((carbon credit project) OR (carbon emission project) OR (carbon reduction project)) |
| Calculate social cost | (Calculate social cost) and (carbon credit project) |

## 2.2 *Systematic Literature Review*

A systematic literature review (SLR) was conducted to comprehensively assess the current methodologies, identify gaps, and provide a foundation for this study. Following the protocol proposed by Kitchenham and Charters [5], the SLR involved a structured process of search, inclusion/exclusion criteria, study selection, and data extraction [5].

### 2.2.1 Search Protocol

Relevant studies were sourced from prominent databases, including Scopus, IEEE Xplore, SpringerLink, and ACM Digital Library. Boolean operators and targeted keywords, such as "social impact of carbon credit," "carbon pricing," and "sustainability frameworks," were employed to maximize search results. Table 1 in the document details the search categories and keywords.

### 2.2.2 Inclusion and Exclusion Criteria

The review included peer-reviewed, full-text articles in English that proposed solutions for calculating the social impact or cost of carbon credit projects or discussed frameworks aligning carbon credits with UN SDGs. Duplicate, non-English, or inaccessible studies were excluded to ensure the review's quality and robustness.

### 2.2.3 Study Selection

The study selection process identified 267 articles, with 15 ultimately meeting the inclusion criteria. Figure 3 in the document visually represents the filtration process, showing the reduction of studies at each stage (Figs. 1 and 2).

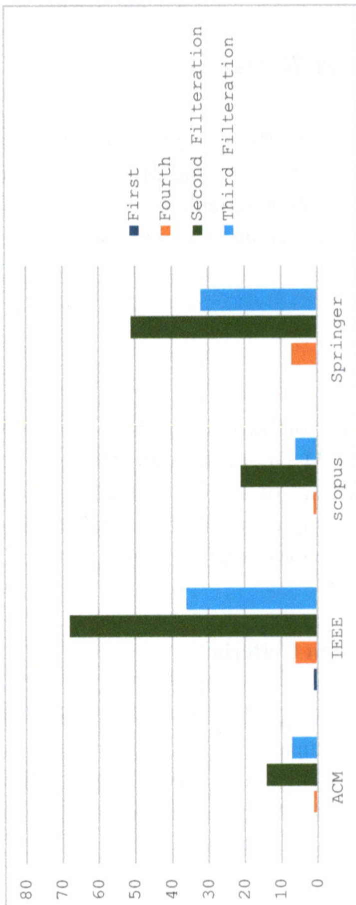


**Fig. 1** Studies filtration process (Author's work)

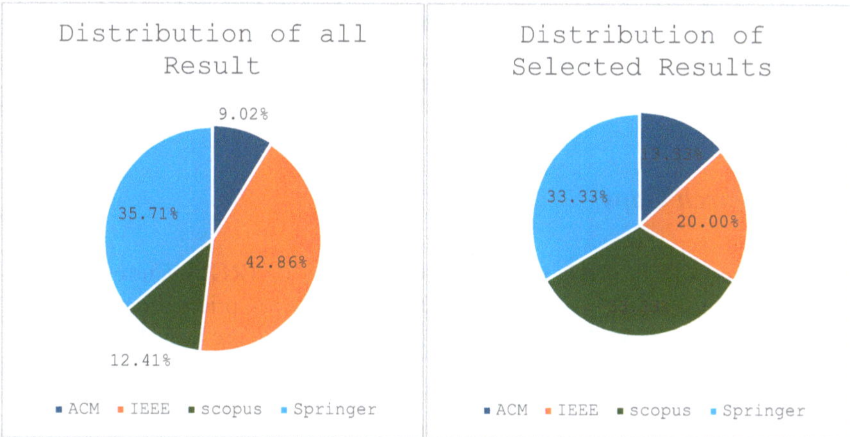


**Fig. 2** Distribution of results across search engines (Author's work)

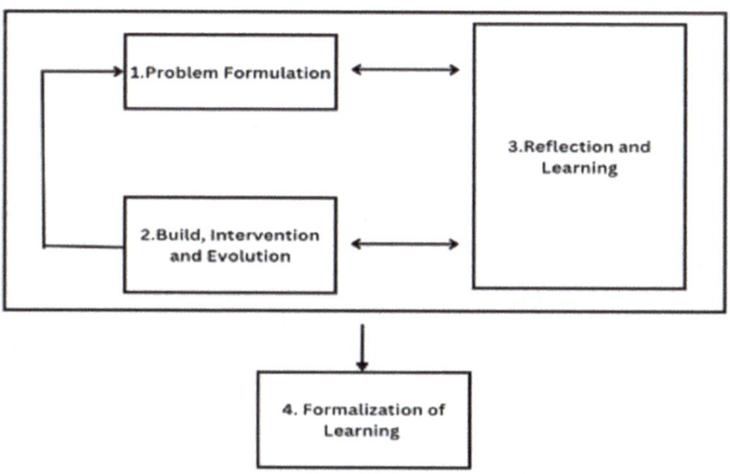


**Fig. 3** DSRM framework (Author's work)

## *2.3 Key Themes in Existing Literature*

### 2.3.1 Social Impact of Carbon Credit Projects and Social Cost

Many studies highlight the co-benefits of carbon credit projects, such as job creation, improved air quality, and enhanced community well-being. However, methodologies for quantifying these benefits remain underdeveloped. For instance, Lee et al. [6]

emphasized the role of additional standards, such as Climate, Community, and Biodiversity (CCB), in enhancing market competitiveness through co-benefit labelling. However, their approach lacked integration with SDG indicators [6].

### 2.4 Research Gaps

- **Integration of UN SDGs in Carbon Credit Pricing (RG1):** Current methodologies fail to explicitly incorporate SDG indicators into pricing frameworks.
- **Quantification of Social Impact (RG2):** Limited tools are available to systematically assess both direct and indirect socio-economic benefits of carbon credit projects.
- **Integrated Assessment Models (RG3):** A lack of comprehensive models that capture environmental, social, and economic impacts in a unified framework.
- **AI-Driven SDG Verification (RG4):** Absence of systematic methodologies to verify project claims against SDG contributions using AI and machine learning.

## 3 Research Objectives

This chapter addresses three core research objectives to enhance the evaluation of carbon credit projects' social impacts. Each objective contributes to a comprehensive assessment framework that measures social outcomes and ensures the credibility of UN SDG claims made by project developers.

### 3.1 Integration of UN SDGs

The first objective centres on establishing a robust framework incorporating the United Nations Sustainable Development Goals (UN SDGs) as critical indicators for assessing the social impact of carbon credit projects. Traditionally, the effect of these projects has been measured in terms of emission reductions alone, overlooking broader socio-economic benefits. However, by integrating UN SDG indicators such as poverty reduction, health improvements, and education, the framework provides a more holistic view of project outcomes [7].

**Key aspects of this objective include**

- **Indicator Selection**: Choosing relevant UN SDG indicators that reflect the potential social impacts of carbon credit projects. For example, projects promoting renewable energy may contribute to SDG 7 (Affordable and Clean Energy) and SDG 1 (No Poverty) by providing jobs and affordable energy access in underserved areas.

- **Quantification Methodology**: Transforming these indicators into measurable variables that can be included in statistical models. This enables a structured and data-driven approach to evaluating how each project aligns with specific SDGs, thereby supporting a comprehensive understanding of social outcomes [3, 7].

## *3.2 Predictive Statistical Modeling*

The second objective focuses on developing statistical models to quantify and predict the social impacts of carbon credit projects. By analysing key variables—such as job creation, project methodology, and claimed UN SDG contributions—these models offer insights into such projects' current and future social implications.

**Key aspects of this objective include**

- **Model Development**: Implementing machine learning models like Random Forest, Gradient Boosting, and XGBoost to handle complex, non-linear relationships within the data. These models are chosen for their effectiveness in managing diverse datasets and their ability to predict social outcomes based on historical data [8].
- **Variable Analysis**: Identifying and creating variables that significantly influence social impact. For instance, project size, geographic location, and stakeholder engagement are critical in determining outcomes such as employment generation and community well-being [9].
- **Validation and Performance Metrics**: Using evaluation metrics like Mean Squared Error (MSE) and R2 scores to validate model performance. This ensures that the models provide accurate predictions of social outcomes, which can guide decision-making for both project developers and stakeholders.

## *3.3 AI-Based Verification System*

The third objective aims to develop an AI-based rule-driven system to verify the UN SDG claims made by carbon credit projects. While statistical models can predict social outcomes, the accuracy and credibility of claimed contributions to UN SDGs require an additional layer of verification.

**Key aspects of this objective include**

- **Rule-Based AI System**: Designing an AI system that uses a set of structured prompts to analyse project descriptions against specific UN SDG criteria. This involves simple, chain-of-thought, and contrastive prompts to assess the alignment of projects with claimed SDGs [10].
- **Verification Accuracy**: The AI system evaluates each project's stated goals, outcomes, and methodologies to confirm whether they align with the claimed

UN SDGs. The system also assigns a confidence score to each verification result, indicating the degree of reliability in the alignment assessment.

- **Transparency and Accountability**: By automating the verification process, this system ensures greater transparency in reporting and accountability for claimed social impacts. It provides stakeholders with a clear, evidence-based understanding of how projects contribute to broader sustainability goals.

# 4 Methodology

This chapter employs a rigorous Design Science Research Methodology (DSRM) to develop and validate a novel framework for quantifying and verifying the social impacts of carbon credit projects. The methodology is aligned with the three core research objectives: integrating UN SDG indicators, developing predictive statistical models, and implementing an AI-based verification system. It is structured to ensure both theoretical robustness and practical applicability [11].

## *4.1 Design Science Research Methodology (DSRM)*

The DSRM framework was chosen for its iterative nature, which allows for creating and refining new systems within real-world contexts. It comprises four key phases (Fig. 3):

1. **Problem Formulation**
    - The first phase involves identifying limitations in current carbon credit assessments, which often need to pay more attention to social impacts beyond environmental outcomes. This phase emphasises the importance of incorporating the United Nations Sustainable Development Goals (UN SDGs) to evaluate carbon credit projects comprehensively.
    - The problem formulation establishes the need for a framework that quantifies the broader social implications of projects, guiding the development of subsequent models and verification tools.
2. **Build, Intervention, and Evaluation**
    - In this phase, the research focuses on building statistical models and an AI-based verification system.
    - Statistical models are constructed using data from Core Markets, analysing variables such as job creation, project type, and alignment with specific UN SDGs.
    - The evaluation process involves assessing model accuracy through metrics like Mean Squared Error (MSE), R2 scores, and precision in identifying project

impacts, ensuring that the models can provide reliable predictions of social outcomes [12].

3. **Reflection and Learning**

   - The iterative nature of DSRM allows for continuous refinement. Insights from model testing and AI verification inform further improvements, addressing challenges such as data variability and AI performance in accurately interpreting complex project descriptions.
   - Adjustments to the models and the AI system are made based on feedback from initial iterations, ensuring that the framework remains adaptive and capable of capturing diverse project characteristics.

4. **Formalisation of Learning**

   - The final phase involves consolidating the developed framework into a comprehensive stakeholder tool. This tool can be used for decision-making, policy formulation, and strategic planning, aligning carbon credit projects with global sustainability goals.

Recent advancements in machine learning, such as Gradient Boosting and Random Forest models, have proven effective in capturing complex non-linear relationships in social impact datasets. These models address challenges in measuring indirect impacts, such as job creation and health improvements [13].

## *4.2 Data Collection and Pre-Processing*

The dataset used for this research includes information from 161 carbon credit projects, detailing attributes like project type, scale, geographic location, and UN SDG claims. The pre-processing phase is critical for ensuring data quality and reliability:

To ensure accurate and reliable results in our statistical modelling, a rigorous data pre-processing phase is necessary. This phase prepares the dataset for analysis by cleaning, transforming, and engineering variables that reflect the social impact of carbon credit projects. The pre-processing steps are outlined below (Fig. 4):

1. **Clean Dataset**

   - The initial step involves reviewing the dataset to ensure that all entries are correctly formatted, especially those aligned with the UN SDG indicators. This step helps in standardizing the dataset for further analysis.

2. **Variable Selection**

   - In this step, we select variables that have a significant impact on the social outcomes of carbon credit projects. Important variables include project type, geographic location, scale, and job creation metrics.

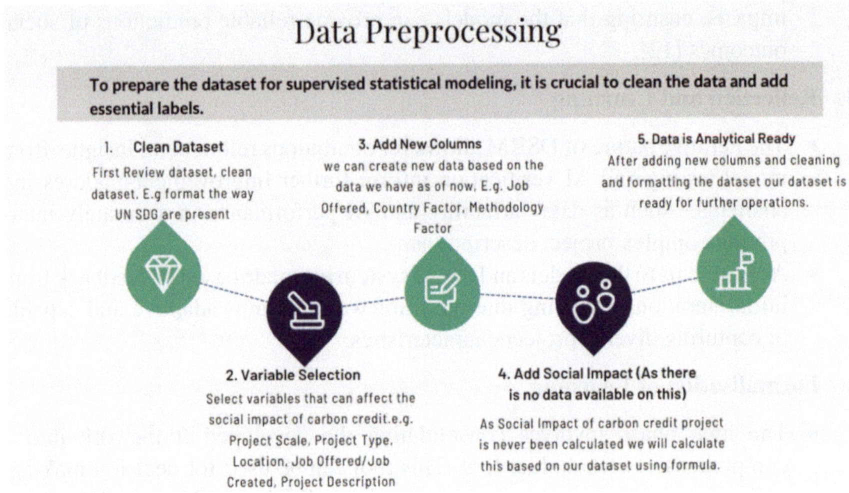


**Fig. 4** Data pre-processing (Author's work)

3. **Add New Columns**
   - New variables are engineered based on the existing dataset to enhance the model's predictive capacity. For example, additional columns like 'Job Offered,' 'Country Factor,' and 'Methodology Factor' are added to provide more insights into potential social impacts.
4. **Add Social Impact (Calculated)**
   - Since direct data on social impact is often unavailable, we use a formula-based approach to estimate social impact based on existing variables. This calculated value serves as a target variable for supervised learning models.
5. **Data is Analytical Ready**
   - After cleaning, adding new variables, and estimating social impact, the dataset is now ready for further analysis, including statistical modeling and AI-based verification (Fig. 5).

Implementing the methodology to assess and verify the social impacts of carbon credit projects involves a comprehensive framework that integrates UN SDGs using statistical modelling and AI-based verification systems. The solution is structured around four key components, each addressing different aspects of social impact evaluation:

1. **Carbon Credit Project's Social Impact Classification Using Only SDG UN Data**
   - This component focuses on evaluating social impacts based solely on UN SDG indicators. It uses statistical models trained exclusively with SDG-specific

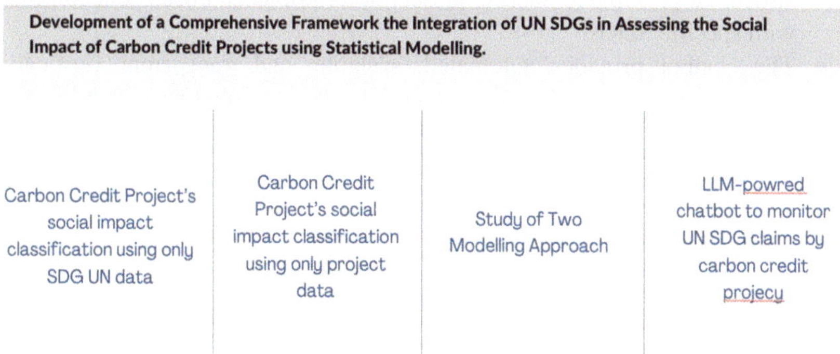


**Fig. 5** Methodology overview (Author's work)

data to determine whether projects align with global sustainability objectives [14].

- The models assess direct contributions to SDGs, such as poverty reduction, education improvement, and healthcare access, providing a baseline understanding of the social outcomes derived from SDG-related indicators [15].

2. **Carbon Credit Project's Social Impact Classification Using Only Project Data**

   - This component develops statistical models using project-specific attributes such as geographic location, scale, duration, and type. These models predict direct social impacts without integrating SDG data, offering a bottom-up analysis.
   - This approach aims to identify how core project attributes influence direct social outcomes independent of broader global indicators.

3. **Study of the Two Modelling Approaches**

   - This phase involves a comparative analysis of the two statistical modeling approaches—using only SDG UN data versus project data. The study aims to determine the effectiveness and accuracy of each approach in predicting social outcomes.
   - Examining how each model performs with different data inputs helps identify which approach is more reliable and scalable for different stages of carbon credit project evaluation, ensuring accurate assessments of social impacts.

4. **LLM-Powered Chatbot to Monitor UN SDG Claims by Carbon Credit Projects**

   - The final component leverages a large language model (LLM)-powered chatbot to verify UN SDG claims made by carbon credit projects. It uses structured prompts to analyse project descriptions and determine alignment with specific SDGs.

- The chatbot provides automated, scalable, and transparent verification, ensuring that projects' claimed contributions to global sustainability goals are credible and accurately reported [13, 16].

## *4.3 Statistical Modelling for Social Impact Prediction*

To predict and quantify the social impacts of carbon credit projects, various statistical models are developed using the following approach:

1. **Variable Analysis**
    - The analysis begins with selecting variables that are likely to influence social outcomes. Variables include project scale, geographic location, claimed UN SDGs, and project methodologies, each having a unique impact on social indicators like employment, economic growth, and community well-being.
2. **Model Development**
    - Machine learning models, such as Random Forest, Gradient Boosting, and XGBoost, are employed for their ability to manage non-linear relationships and handle complex interactions within the data.
    - These models are trained to predict social impacts based on historical project data, focusing on job creation, poverty reduction, and health improvements.
3. **Labelling Formula for Social Impact**
    - To label data for model training, a comprehensive formula is used to estimate the social impact of each project. The formula combines multiple factors, each representing a different dimension of impact:

$$\text{Social Impact} = \frac{\left(1 - \frac{1}{DCR}\right) * SDGA * CI * ME}{Max.DCR * Max.SDGA * Max.CI * Max.ME} \tag{1}$$

***Social Impact Calculation (Author's work)***

where:

- **DCR (Direct Carbon Reduction)**: Quantifies the carbon reduction achieved by the project, validated through third-party audit reports.
- **SDGA (SDG Alignment)**: A binary or scaled score indicating the project's alignment with specific UN SDGs, with a higher score representing a more robust alignment.
- **CI (Community Impact)**: Captures direct benefits to the community, such as job creation, healthcare improvements, or educational advancements.
- **ME (Multiplier Effect)**: Accounts for cascading impacts, like the ripple effects of job creation or improved infrastructure on the local economy.

   - This labelling process provides a normalised score that serves as the target variable for supervised learning models, enabling precise predictions of social impact based on project attributes.

4. **Evaluation Metrics**

   - The predictive models are evaluated using metrics like Mean Squared Error (MSE) for continuous variables and classification accuracy for categorical outcomes. The R2 score assesses the model's explanatory power, ensuring that the models offer reliable predictions.

## *4.4 AI-Based Rule-Driven Verification System*

An AI-based rule-driven verification system is developed to ensure transparency and credibility in UN SDG claims made by carbon credit projects. This system leverages Open AI models, integrating prompt engineering and rule-based reasoning for robust assessment:

1. **Prompt Engineering**

   - Carefully designed prompts guide the AI model in analysing project descriptions and verifying their alignment with claimed UN SDGs. These prompts include:

     - Project Impact Analysis: Evaluate how the project contributes to specific SDGs based on its goals, methodologies, and outcomes.
     - SDG Relevance Identification: Identify which SDGs are most relevant to the project and provide a rationale for each identified SDG [16, 17].

2. **Rule-Based System**

   - The AI operates within a rule-based framework that ensures alignment with predefined criteria for SDG verification. This approach reduces the likelihood of false positives, making the verification process more accurate and reliable.

3. **Confidence Scoring**

   - The AI assigns a confidence score to each verification result, indicating the degree of alignment between the project's stated claims and its actual contributions. This score helps stakeholders assess the credibility of the project's impact claims, fostering greater accountability [18].

4. **Implementation and Transparency**

   - The AI-based verification system is designed to offer transparency in reporting, providing clear insights into whether projects genuinely contribute to the claimed UN SDGs. This approach supports decision-making and enhances the accountability of carbon credit projects.

## *4.5 Summary*

The methodology outlined here combines statistical modelling with AI-driven verification to create a comprehensive framework for quantifying and assessing the social impacts of carbon credit projects. The iterative use of DSRM, combined with data-driven insights and AI verification, ensures that the framework is practical and aligned with global sustainability objectives.

# 5 Results and Analysis

This section presents the findings from the statistical models and the AI-based verification system, focusing on how effectively they quantify and verify the social impacts of carbon credit projects. The analysis is strictly based on the data and methodologies.

## *5.1 Integration of UN SDGs*

Integrating UN Sustainable Development Goals (SDGs) into evaluating carbon credit projects is crucial for understanding their broader socio-economic impacts. This section details how UN SDG data is incorporated into the modelling and verification process, providing a comprehensive assessment of projects' alignment with global sustainability targets.

Existing frameworks often lack multi-dimensional assessments, focusing narrowly on environmental metrics. There is a growing need to develop comprehensive models that integrate socio-economic dimensions alongside environmental benefits [14].

### 5.1.1 Overview of UN SDG Integration in Project Evaluation

Integrating UN SDGs involves aligning project attributes with specific SDG indicators, allowing for a more nuanced evaluation of social impacts. This approach quantifies the direct environmental benefits of carbon reduction and assesses contributions to broader socio-economic goals such as poverty reduction, education access, and clean energy deployment.

**Key Elements of UN SDG Integration**

- **SDG Relevance Mapping**: Each project is mapped to relevant SDGs based on its stated objectives, methodologies, and outcomes. For example, a renewable energy project may align with SDG 7 (Affordable and Clean Energy) and SDG 13 (Climate Action).

**Table 2** Example of data alignment with UN SDGs

| Attribute | Example values | Aligned SDGs |
|---|---|---|
| Project type | Renewable energy | 7, 13 |
| Region | India | 1, 7, 13 |
| Job creation | 200 jobs created | 8 |
| Improved water access | 5,000 households | 6 |

- **Indicator-Based Alignment**: Projects are assessed using specific SDG indicators that measure socio-economic outcomes. For instance, SDG 8 (Decent Work and Economic Growth) is evaluated based on job creation metrics, while SDG 6 (Clean Water and Sanitation) is measured through improvements in water access.
- **Normalisation of SDG Contributions**: The strength of a project's alignment with each SDG is normalised to a scale of 0 to 1, indicating the extent of its contribution to achieving the targeted SDG. This normalisation facilitates direct comparison across projects, irrespective of scale or location.

### 5.1.2 Data Preparation for SDG Integration

The dataset used for integration contains attributes specifically aligned with UN SDGs, including:

- **Project Type and Methodology**: Indicates the primary approach (e.g., renewables, afforestation) and its corresponding SDGs.
- **Geographic Focus**: Geographic data helps identify regional disparities and priorities, ensuring that SDG alignment reflects local needs.
- **SDG Indicators**: The dataset includes specific metrics associated with targeted SDGs, such as healthcare access, educational improvements, or employment generation (Table 2).

### 5.1.3 Model Preparation for SDG Integration

The integration of SDG data into statistical models requires pre-processing steps that ensure alignment with specific SDG indicators:

- **Feature Selection**: The dataset is refined to retain attributes most relevant to SDG targets, such as economic upliftment, educational outreach, and environmental restoration.
- **SDG Scoring System**: Each project receives an SDG score based on its alignment with relevant goals, facilitating direct evaluation within the models.
- **Normalisation and Scaling**: All SDG indicators are scaled to a 0–1 range, allowing the models to treat different SDGs uniformly, irrespective of metric types or units.

To classify the social impact of carbon credit projects using the UN SDG dataset, a structured approach is adopted. This approach ensures that projects are evaluated not only for environmental outcomes but also for their alignment with broader socio-economic goals defined by the UN Sustainable Development Goals (SDGs). The process involves several key steps:

1. **Gather UN SDG Baseline Data**
   - We begin by collecting baseline data from reputed sources that align with specific UN SDG indicators. This dataset serves as the foundation for further analysis and comparison.
   - Schema Matching
   - The collected dataset is reviewed to ensure compatibility with existing project data. Schema matching involves aligning the data structures to facilitate effective merging and analysis.
2. **Schema Matching**
   - The collected dataset is reviewed to ensure compatibility with existing project data. Schema matching involves aligning the data structures to facilitate effective merging and analysis.
3. **Merging: Match Project Details with Baseline Data**
   - Project-specific details are combined with baseline SDG data to create a comprehensive dataset. This involves computing baseline averages by country and project type, ensuring that the evaluation reflects the local context.
4. **Handle Missing Data**
   - Missing values in the dataset are handled using appropriate techniques, such as filling NaN values with the median of each numeric column. This ensures data consistency and accuracy.
5. **Social Impact Classification Training**
   - Once the dataset is prepared, various statistical models—like OLS, Ridge, Random Forest, XGBoost, Lasso, and Gradient Boosting—are applied to classify the social impact of projects. These models analyze how well the projects align with the targeted SDGs.
6. **Compare Models and Improve**
   - The performance of different models is compared, and the best-performing model is further improved to enhance accuracy. This iterative process ensures that the model reliably classifies social impacts based on UN SDG data.

Below is a diagram summarizing this SDG-based classification process (Fig. 6):

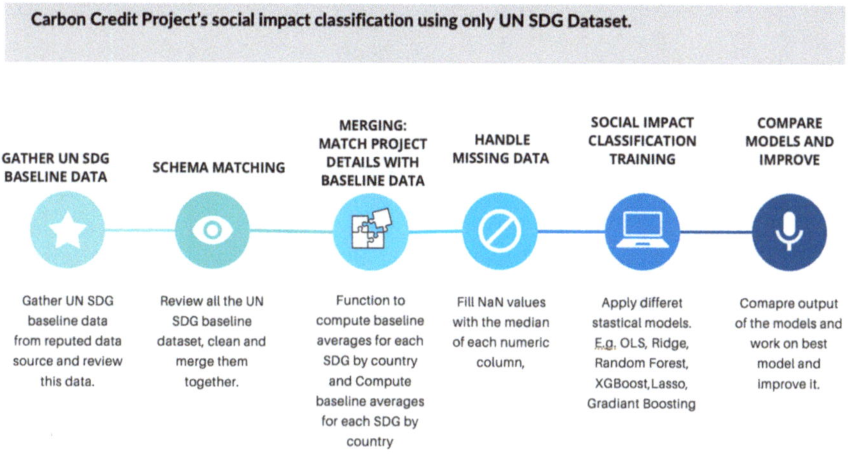


**Fig. 6** Carbon credit project's social impact classification using only UN SDG dataset (Author's work)

### 5.1.4 Initial Findings from SDG Integration

The initial analysis of SDG integration revealed several trends in the dataset:

- **High Alignment with SDG 7 and 13**: Renewable energy projects showed the highest alignment with SDG 7 (Affordable and Clean Energy) and SDG 13 (Climate Action), indicating a strong focus on clean energy deployment and climate action.
- **Significant Contributions to SDG 8**: Projects focused on economic growth and job creation demonstrated substantial alignment with SDG 8 (Decent Work and Economic Growth).
- **Regional Disparities in SDG Alignment**: The analysis revealed geographic disparities, with certain regions showing higher alignment with specific SDGs based on local priorities and project focus (Fig. 7).

*This figure illustrates the distribution of carbon credit projects across different UN SDGs, highlighting the most targeted goals and potential gaps in impact areas.*

### 5.1.5 Implications of SDG Integration in Project Evaluation

The integration of UN SDGs provides a holistic view of the socio-economic impacts of carbon credit projects. By incorporating SDG indicators into statistical modelling, this approach ensures that projects are evaluated based on carbon reduction and their broader contributions to global sustainability goals. This comprehensive assessment framework allows stakeholders to make more informed decisions, aligning project strategies with long-term socio-economic benefits [19, 20].

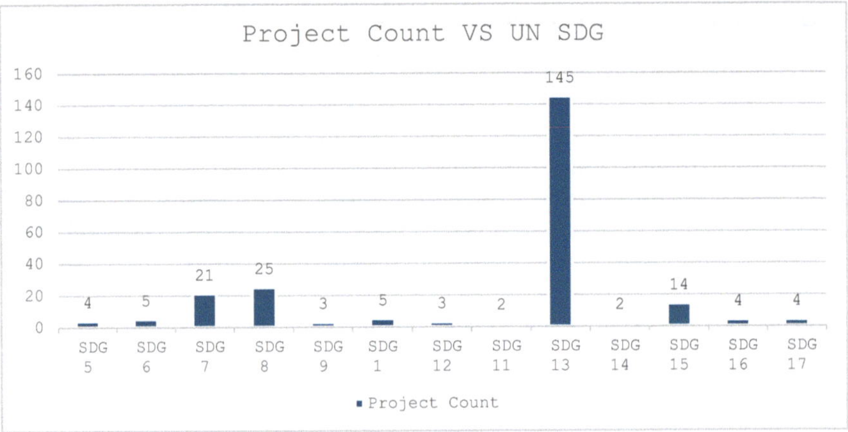


**Fig. 7** Distribution of projects (Author's work)

## 5.2 *Model Performance*

This section presents an in-depth analysis of the predictive capacity of statistical models developed to assess the social impacts of carbon credit projects. Two distinct modelling approaches were used. The choice of these two approaches reflects different stages of project evaluation and aims to provide a comprehensive understanding of how carbon credit projects generate social impacts [21].

### 5.2.1 Modelling with Project Details Only

This approach uses core project characteristics, such as project type, geographic location, scale, and duration, to predict direct social outcomes. It focuses on how these factors contribute to specific social impacts without considering broader contextual influences.

This approach involves using only project-specific attributes to predict and classify the social impacts of carbon credit projects. It provides a bottom-up analysis by focusing exclusively on the core features of each project, such as geographic location, scale, duration, and type. The goal is to understand how these inherent attributes contribute to social outcomes without incorporating broader indicators like UN SDGs.

The process includes the following key steps

1. **Generate New Dataset from Core Market API**
    - A new dataset is generated from the Core Market API, ensuring that the dataset is always updated with the latest project details. This dataset forms the foundation for analysis.

2. **Clean Data**
   - The dataset undergoes a cleaning process where blanks and missing values are managed. Proper formatting is applied to ensure data consistency and reliability.
3. **Add/Remove Columns**
   - Additional variables that could influence social impact—such as Region Factor, Country Factor, Methodology Factor, and Job Offered—are added to enhance predictive accuracy. Unnecessary columns are removed to focus on relevant data.
4. **Statistical Modeling**
   - Various statistical models, such as OLS, Ridge Regression, and Random Forest, are applied to the cleaned dataset. The models aim to predict social outcomes based on project attributes, using supervised learning techniques.
5. **Publish Results**
   - The results from the best-performing model are compared and published. This iterative approach ensures the most accurate classification of social impacts based on project data alone (Fig. 8).

- **Reasoning Behind the Approach**
  - The primary goal of this approach is to establish a baseline understanding of how intrinsic project features directly affect social outcomes. It follows

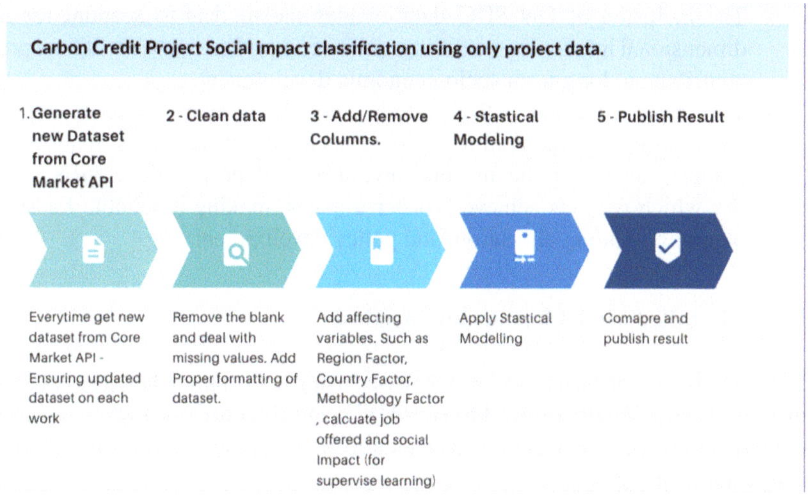


**Fig. 8** Carbon credit project social impact classification using only project data (Author's work)

a bottom-up evaluation framework, emphasising direct impacts without accounting for broader indicators such as UN SDGs.

- This approach is particularly useful in early-stage project evaluations, where projects are assessed based solely on their core attributes, with limited or no access to SDG-related data. It provides valuable insights into the potential impacts based on project type, location, and scale, making it suitable for projects in the planning or development phase.
- Additionally, in cases where UN SDG data is incomplete or unavailable, this approach allows stakeholders to evaluate projects' potential impacts based on fundamental characteristics alone. It is a practical tool for identifying which project features contribute the most to direct outcomes like job creation, poverty alleviation, and energy access.

### 5.2.2 Modelling with Project Details and UN SDG Data

This approach integrates UN SDG indicators with project-specific attributes, offering a broader and more nuanced analysis of social impacts. It captures both direct and indirect effects, providing insights into how projects align with global sustainability goals.

- **Reasoning Behind the Approach**
  - This approach aims to provide a top-down evaluation that aligns with global socio-economic objectives by including UN SDG indicators. It captures the direct impacts of project attributes and how these projects contribute to broader socio-economic goals such as healthcare improvements, education access, and poverty reduction.
  - The inclusion of UN SDG data enables the models to account for multi-dimensional interactions, offering a more comprehensive view of how projects contribute to long-term socio-economic development.
  - This approach aligns with the study's broader objective of assessing carbon credit projects' alignment with global sustainability standards. It allows stakeholders, such as policymakers, investors, and project developers, to identify which projects achieve broader impacts, making it suitable for strategic planning, funding allocation, and policy development.

### 5.2.3 Detailed Model Results and Analysis

The performance of each model was assessed using both approaches. The results were measured using **Mean Squared Error (MSE)** and **$R^2$ scores**, providing insights into the models' predictive accuracy and reliability in assessing social impacts (Figs. 9 and 10 and Table 3).

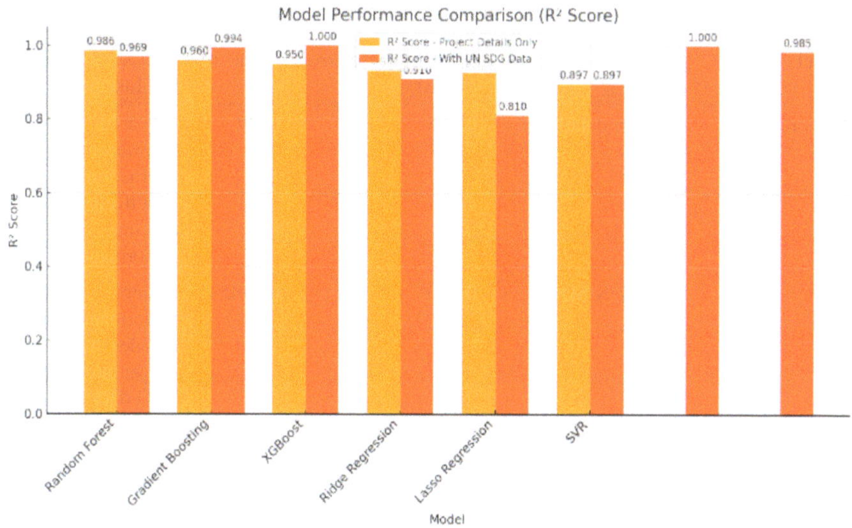


Fig. 9 Model performance comparison 1 (Author's work)

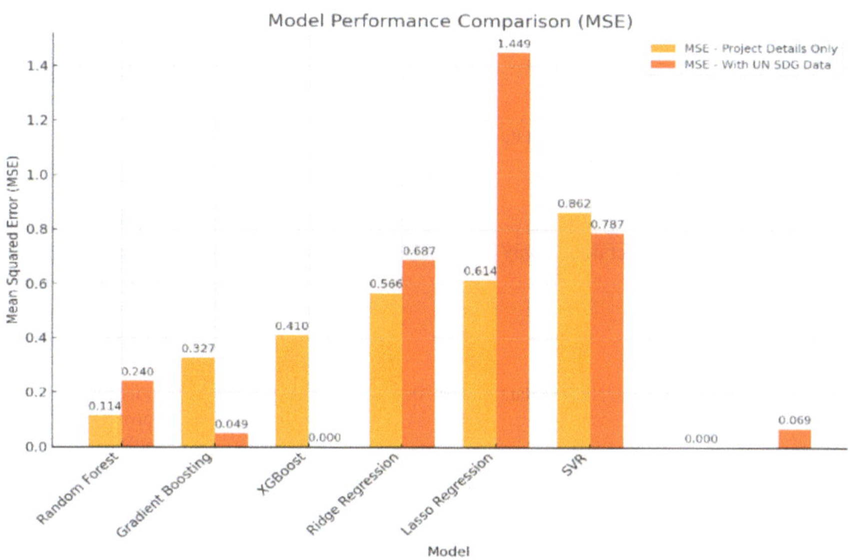


Fig. 10 Model performance comparison 2 (Author's work)

### 5.2.4 In-Depth Insights from Model Comparison

1. Random Forest
   - Performance with Project Details Only

**Table 3** Model result and analysis

| Model | Approach | Mean Squared error (MSE) | $R^2$ Score | Interpretation and key Insights |
|---|---|---|---|---|
| Random forest | Project details only | 0.114 | 0.986 | Captured non-linear interactions effectively; best baseline performance |
| | With UN SDG data | 0.2401 | 0.9685 | Strong performance with broader impact scope; integrates SDG data well |
| Gradient boosting | Project details only | 0.327 | 0.960 | Robust for handling complex relationships; strong baseline performance |
| | With UN SDG Data | 0.0488 | 0.9936 | High accuracy; best model with SDG integration |
| XGBoost | Project details only | 0.4104 | 0.950 | Good performance; limited by lack of broader context |
| | With UN SDG data | 0.0001 | 0.9999 | Near-perfect accuracy; excels with SDG integration |
| Bridge regression | Project details only | 0.5659 | 0.9321 | Limited in capturing complex relationships; linear model limitations |
| | With UN SDG data | 0.6870 | 0.9100 | Moderate improvement; handles multicollinearity better with SDG integration |
| Lasso regression | Project Details only | 0.6141 | 0.9264 | Effective in feature selection; limited by linear assumptions |

(continued)

**Table 3** (continued)

| Model | Approach | Mean Squared error (MSE) | $R^2$ Score | Interpretation and key Insights |
|---|---|---|---|---|
| | With UN SDG data | 1.4491 | 0.8102 | Low accuracy with SDG integration; extreme feature reduction affects performance |
| SVR | Project details only | 0.8623 | 0.8967 | Least effective without SDG data |
| | With UN SDG data | 0.7871 | 0.8969 | Moderate improvement with SDG data; still limited by non-linear complexity |
| AdaBoost | With UN SDG data | ~ 0 | 1.0 | Excellent performance; risk of overfitting |
| Stacking model | With UN SDG data | 0.0691 | 0.9850 | Combines strengths of multiple models; stable performance |

- Random Forest achieved the lowest MSE (0.114) and the highest R2 score (0.986) among all models in the baseline approach. It effectively captured the non-linear relationships between project characteristics, making it particularly useful for predicting direct social outcomes based on fundamental project attributes.
- The strong performance of Random Forest in this approach demonstrates its adaptability to varying types of data, making it suitable for early-stage evaluations where projects are assessed primarily based on their core characteristics.
- However, despite its success in the baseline approach, Random Forest's performance slightly diminished when broader contextual factors were added, as indicated by a higher MSE (0.2401) and a lower R2 score (0.9685) with SDG integration. This suggests that while the model is highly effective for predicting direct outcomes, it is less capable of capturing complex interactions with broader socio-economic indicators.

• Interpretation

  - The adaptability of Random Forest makes it valuable for both direct impact assessments and more comprehensive evaluations involving broader indicators. However, for projects aiming to achieve both direct and indirect social outcomes, it may be more effective to pair Random Forest with models that handle multi-dimensional data better.

2. Gradient Boosting

   - Performance with Project Details Only

     – Gradient Boosting demonstrated strong baseline performance with an MSE of 0.327 and an R2 score of 0.960. Its robustness in handling complex relationships among project attributes makes it highly effective for direct impact assessments. The model's strength lies in its ability to capture subtle variations in data, making it useful for identifying less obvious patterns in project characteristics that contribute to social outcomes.
     – In the baseline approach, Gradient Boosting was able to model non-linear relationships effectively, indicating its potential for capturing intricate interactions within project-specific data.

   - Performance with SDG Integration

     – When integrated with SDG data, Gradient Boosting's performance improved significantly, achieving an MSE of 0.0488 and an R2 score of 0.9936. The model's superior handling of broader data highlights its capacity to predict both direct and indirect social outcomes, making it one of the best-performing models in the SDG-integrated approach.
     – The ability of Gradient Boosting to leverage SDG data demonstrates its effectiveness in capturing multi-dimensional impacts, aligning well with the broader evaluation framework of this study.

   - Interpretation

     – Gradient Boosting's adaptability makes it ideal for both initial assessments and more comprehensive evaluations. It effectively handles complex interactions and multi-dimensional impacts, making it suitable for projects with well-defined sustainability targets.

3. XGBoost

   - Performance with SDG Integration

     – XGBoost achieved near-perfect accuracy with the inclusion of SDG data, recording an MSE of 0.0001 and an R2 score of 0.9999. This exceptional performance highlights its ability to model complex, non-linear relationships and multi-dimensional interactions effectively. It surpassed all other models when integrating broader indicators, demonstrating its ability to capture direct and indirect impacts.
     – The model's performance indicates that it effectively uses decision trees and gradient-boosting techniques to identify and leverage interactions between project attributes and SDG indicators. This makes it a powerful tool for comprehensive assessments where global sustainability alignment is crucial.

   - Interpretation

  - XGBoost's ability to handle large, complex datasets with diverse variables makes it highly suitable for modeling projects' broader socio-economic impacts. Its effectiveness in integrating SDG data aligns with the study's objective to provide holistic evaluations. It is the best choice for projects with well-defined sustainability goals, as it maximises predictive accuracy and captures multi-faceted impacts.

4. Ridge Regression

   - Performance with Project Details Only

     - Ridge Regression, a linear model, showed moderate performance with an MSE of 0.5659 and an R2 score of 0.9321 in the baseline approach. The model's L2 regularisation helps it manage multicollinearity among the predictors, making it relatively stable in the presence of correlated variables.
     - Despite its ability to handle linear relationships, Ridge Regression struggled with the non-linear interactions among project attributes. This limitation is evident from its higher MSE than non-linear models, indicating that it is less effective for direct social impact predictions when complex relationships are involved.

   - Performance with SDG Integration

     - With SDG data integration, Ridge Regression showed slight improvement, achieving an MSE of 0.6870 and an R2 score of 0.9100. Adding broader socio-economic indicators improved the model's ability to predict outcomes but remained limited due to its linear nature.
     - Ridge Regression's performance indicates that while it can handle a broader range of variables with regularisation, it lacks the capacity to capture deeper, non-linear relationships crucial for accurately predicting comprehensive social impacts.

   - Interpretation

     - Ridge Regression's moderate performance under both approaches suggests it is better suited for initial assessments where multicollinearity is a concern, but complex interactions are not central to the analysis. It is less effective for capturing broader, non-linear impacts, making it more suitable for simpler models or as part of an ensemble approach.

5. Lasso Regression

   - Performance with Project Details Only

     - Lasso Regression was performed with an MSE of 0.6141 and an R2 score of 0.9264 in the baseline approach. Known for its feature selection capability, Lasso can reduce the model to only the most significant variables by shrinking the coefficients of less important variables to zero. However,

this aggressive regularisation led to losing important variables and lower predictive accuracy.
  - The model struggled with the complex interactions inherent in the dataset, which limited its effectiveness in predicting direct social outcomes.

- Performance with SDG Integration
  - When integrated with SDG data, Lasso Regression's performance worsened, with an MSE of 1.4491 and an R2 score of 0.8102. The feature selection process significantly lost important variables related to broader socio-economic impacts, making it unsuitable for complex, multi-dimensional evaluations.
  - The limited performance with SDG integration indicates that while Lasso is useful for identifying core predictors, it is ineffective when deeper interactions between project attributes and SDG indicators are necessary for accurate predictions.

- Interpretation
  - Lasso Regression's ability to simplify models makes it useful in scenarios where dimensionality reduction is needed. However, its aggressive feature selection limits its applicability in broader socio-economic evaluations, where multi-dimensional impacts are crucial. It is more suitable for preliminary analyses or with other models that can capture complex relationships.

6. Support Vector Regression (SVR)
- Performance with Project Details Only
  - SVR had the weakest performance in the baseline approach, with an MSE of 0.8623 and an R2 score of 0.8967. The model struggled to handle the high-dimensional, non-linear relationships present in the dataset, indicating that it was not suitable for predicting direct social outcomes based solely on project attributes.
  - While useful in other contexts, the model's kernel-based approach was limited by the lack of broader contextual information, making it less effective in capturing the underlying complexity of project impacts.

- Performance with SDG Integration
  - SVR showed moderate improvement with SDG integration, achieving an MSE of 0.7871 and an R2 score of 0.8969. The broader data helped the model capture some indirect impacts, but its performance remained limited by the kernel's ability to generalise in a high-dimensional space.
  - The modest improvement suggests that SVR benefits from additional contextual data but lacks the flexibility to handle complex, multi-dimensional interactions.

- Interpretation

   - SVR's performance across both approaches indicates that it is not the best choice for predicting social outcomes in carbon credit projects, especially when broader socio-economic impacts are considered. It suits simpler tasks with fewer variables and more linear relationships.

7. AdaBoost

   - Performance with SDG Integration

     - AdaBoost was only tested with the SDG-integrated approach, where it delivered excellent performance, achieving near-zero MSE and an R2 score of 1.0. The model's ensemble nature allows it to combine weak learners into a strong predictive framework, making it highly adaptable to complex datasets.
     - The near-perfect accuracy indicates that AdaBoost can capture various interactions between project attributes and SDG indicators. However, the perfect R2 score also raises concerns about overfitting, suggesting that while the model maximises accuracy, it may not generalise well to new data.

   - Interpretation

     - AdaBoost's performance highlights its potential for use in detailed socio-economic evaluations. However, its tendency to overfit requires careful tuning and validation, making it more suitable for exploratory analyses than final predictions.

8. Stacking Model

   - Performance with SDG Integration

     - The Stacking Model, which combines predictions from multiple base models, performed strongly with an MSE of 0.0691 and an R2 score of 0.9850. It leverages the strengths of different models, allowing for a more balanced and accurate prediction of direct and indirect impacts.
     - The Stacking Model's ability to integrate the outputs of models like XGBoost, Gradient Boosting, and Random Forest makes it highly adaptable to diverse datasets, capturing complex interactions effectively. The ensemble approach allows it to mitigate the weaknesses of individual models, resulting in a more stable and reliable prediction.

   - Interpretation

     - The Stacking Model's strong performance suggests it is an effective tool for comprehensive socio-economic evaluations, particularly in contexts where both direct and indirect impacts are crucial. Its adaptability makes it valuable for projects with complex interactions and diverse data sources.

9. Overall Insights and Implications for Stakeholders

   - Project Developers

- Models like Random Forest, Gradient Boosting, and XGBoost offer robust tools for predicting social impacts based on core project attributes. These models are useful for early-stage evaluations, helping developers identify key project features that drive direct outcomes.
- For broader evaluations, models integrated with SDG data, especially Gradient Boosting and XGBoost, provide valuable insights into multi-dimensional impacts, aligning project design with global sustainability goals.

- Policymakers
  - The results emphasise the importance of integrating SDG indicators in policy evaluations of carbon credit projects. Using models that account for broader socio-economic outcomes, policymakers can better prioritise funding and support for projects that align with long-term sustainability goals.
- Investors
  - Models with integrated SDG data offer a clearer understanding of projects' broader impacts, making them more attractive for sustainable investments. Ensemble models, such as Stacking, provide a balanced evaluation that accounts for diverse impacts, supporting informed decision-making.

**Conclusion of Model Performance**

The two modelling approaches highlight how carbon credit projects contribute to social outcomes. Models trained only on project details provide useful insights into direct impacts, while those integrating SDG data offer a more holistic evaluation of socio-economic contributions. Models like **XGBoost** and **Gradient Boosting** excelled across both approaches, proving the most adaptable and accurate. Integrating SDG data significantly enhances predictive capacity, aligning evaluations with global sustainability goals and supporting strategic decision-making.

## *5.3 Implementation of Gen AI Auditor System*

This study's third objective is to design and implement a **Gen AI Auditor System** to verify whether carbon credit projects align with UN SDG claims. The AI system uses structured prompts to automate, scale, and improve the transparency of the SDG verification process. This section outlines the architecture, methodology, implementation, and outcomes of the Gen AI Auditor.

### 5.3.1 Overview of the Gen AI Auditor System

The **Gen AI Auditor** employs natural language processing (NLP) techniques to evaluate the SDG alignment of carbon credit projects. By using structured prompts, the system offers clear, consistent, and automated assessments, making it a reliable tool for stakeholders aiming to verify the social impacts of projects.

**Key Features of the System**

- **Automation**: The AI system reduces manual workload by rapidly processing multiple project descriptions simultaneously.
- **Scalability**: The system adapts to diverse project descriptions and geographies, enabling large-scale verification.
- **Transparency**: The use of structured prompts ensures clear and traceable verification outcomes, with confidence scores indicating the certainty of each decision.

### 5.3.2 System Architecture and Design

The **Gen AI Auditor System** comprises several key components that work together to verify SDG alignment:

1. **Project Submission Module**
    - This module is the entry point, where project owners submit descriptions, methodologies, and claimed SDGs.
    - It collects and standardises project data, preparing it for AI analysis.
2. **Data Pre-processing Unit**
    - To ensure data consistency, the pre-processing phase involves tokenisation, normalisation, and text cleaning.
    - This step reduces biases and enhances the accuracy of AI-based verification.
3. **AI Analysis Module with Structured Prompts**
    - The system's core uses structured prompts to guide the AI in evaluating SDG claims.
    - Prompts include initial validation, follow-up checks, chain-of-thought analysis, and consistency verification to ensure comprehensive assessments.
4. **Human Verifier Review Module**
    - Human verifiers review AI outputs to ensure automated results align with the expected verification standards.
    - This layer adds a quality control step, confirming that AI-based evaluations are accurate and credible.
5. **Final Decision Output**

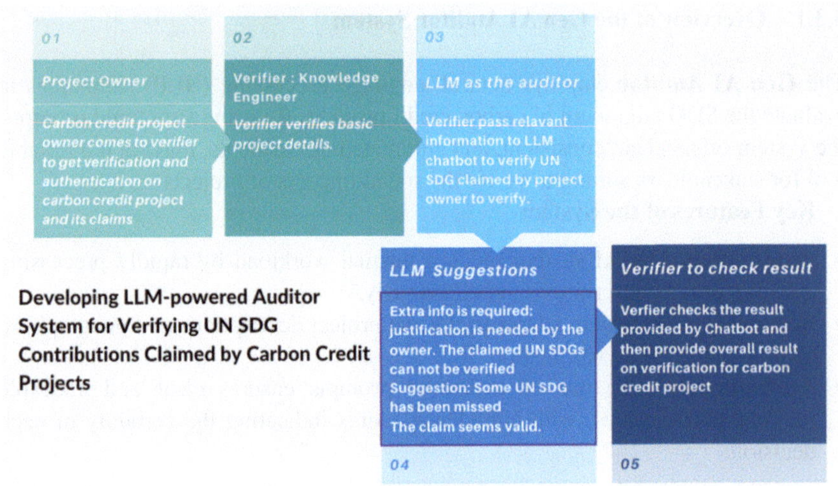


**Fig. 11** Architecture of gen AI audit (Author's work)

- The output includes binary results (yes/no) for SDG alignment and confidence scores that reflect the AI's certainty (Fig. 11).

*This diagram shows the components and data flow within the Gen AI Auditor system, illustrating how project submissions are processed, analysed, and verified.*

### 5.3.3 Structured Prompts for SDG Verification

The Gen AI Auditor uses structured prompts to guide the AI's evaluation process. These prompts are categorised into different types, each designed to ensure clear and consistent verification of project claims.

**Types of Structured Prompts**

1. **Rule-Based Simple Prompts**
   - These prompts provide a basic understanding of each SDG, helping the AI establish whether the project aligns with the claimed SDG.
   - *Example Prompt:*

     "SDG 3 aims to ensure health and well-being for all. Does the project contribute to this goal by improving healthcare access in targeted regions?" (Fig. 12)

2. **One-Shot and Few-Shot Prompts**
   - One-shot prompts use a single example to help the AI understand the context of SDG alignment.

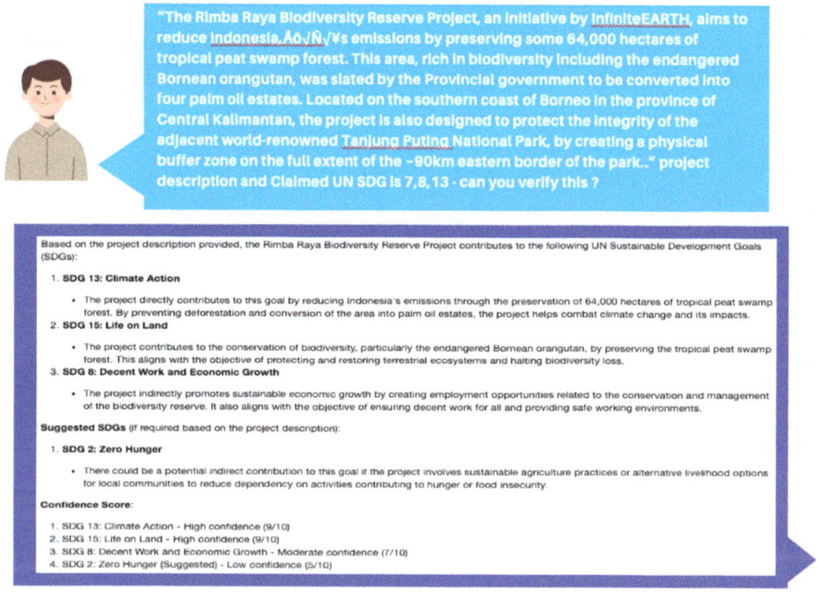


**Fig. 12** Rule-based simple prompts (Author's work)

- Few-shot prompts provide multiple instances to guide the AI's deeper analysis of project descriptions.
- *Example One-Shot Prompt:*(Fig. 13)
- *Example Few-Shot Prompt:* (Fig. 14)

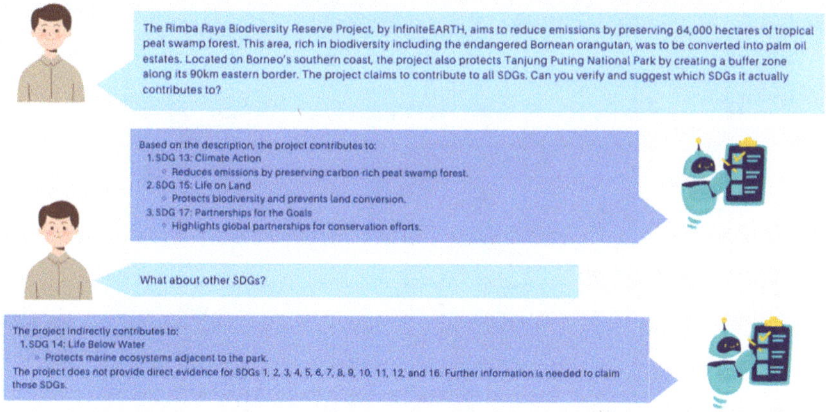


**Fig. 13** One shot prompts (Author's work)

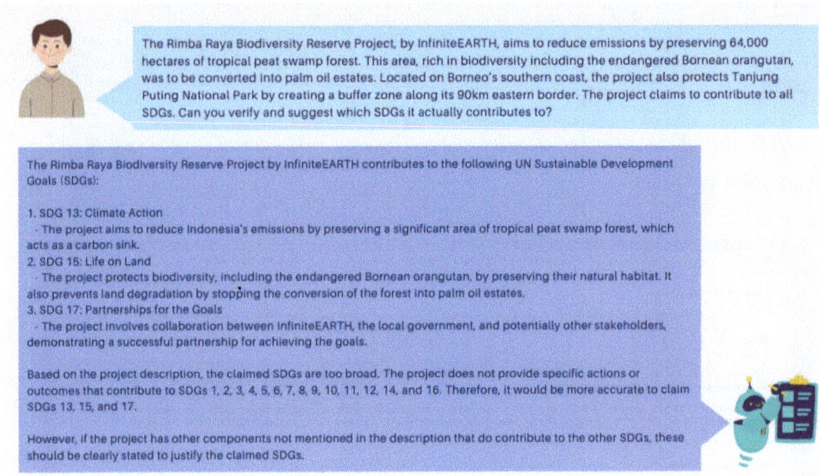


**Fig. 14** Few shot prompts (Author's work)

3. **Chain-of-Thought Prompts**
    - These prompts prompt the AI to articulate its reasoning step-by-step, ensuring transparent evaluations.
    - *Example Prompt:* (Fig. 15)
4. **Contrastive Chain-of-Thought Prompts**

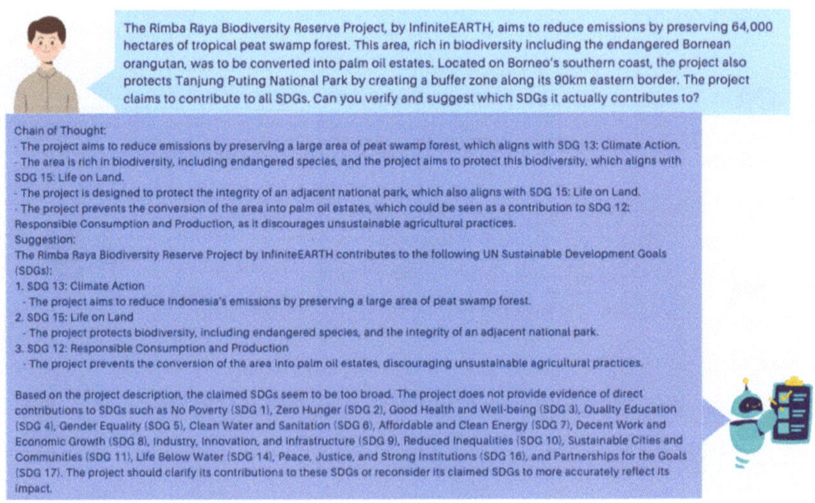


**Fig. 15** Chain of thought prompts (Author's work)

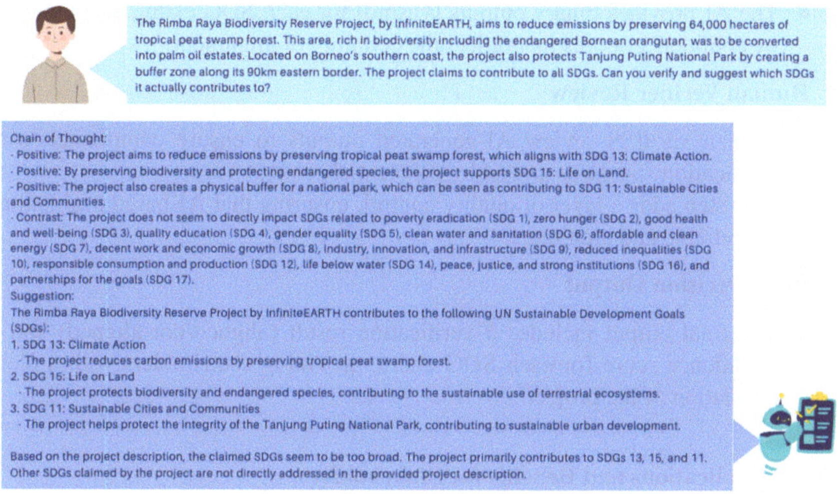


**Fig. 16** Contrastive chain of thought prompts (Author's work)

- Contrastive chain-of-thought prompts encourage the AI to explore both supporting and opposing viewpoints regarding a project's SDG alignment, ensuring a more comprehensive and stable evaluation. This contrastive chain-of-thought prompt helps the AI not only affirm the project's alignment but also identify potential weaknesses, leading to a more nuanced and reliable assessment.
- *Example Prompt:* (Fig. 16)

### 5.3.4 Implementation Process

The Gen AI Auditor follows a systematic process for verifying SDG alignment, ensuring accuracy and scalability in evaluations:

1. **Project Submission**
   - Project owners submit relevant documentation, including project descriptions, objectives, methodologies, and claimed SDGs.
   - The system collects and formats the data for consistent analysis.
2. **Data Preprocessing**
   - The submitted data undergoes tokenization, normalization, and text cleaning, ensuring compatibility with AI processing.
   - Preprocessing reduces noise and prepares the data for accurate AI evaluation.
3. **AI-Based Analysis Using Structured Prompts**
   - The AI auditor uses a sequence of structured prompts to verify SDG claims.

- The AI provides binary outputs (yes/no) for each SDG claim, accompanied by confidence scores that reflect the certainty of alignment.

4. **Human Verifier Review**
   - Human verifiers review AI-generated outputs to ensure compliance with verification standards and address potential errors.
   - This step adds a layer of quality control, ensuring that AI-based evaluations are reliable and credible.
5. **Final Decision Output**
   - The final output includes a verification result (aligned/not aligned) and a confidence score for each SDG claim, providing stakeholders with a clear evaluation of the project.

### 5.3.5 Implications and Benefits for Stakeholders

The Gen AI Auditor offers several benefits:

- **For Project Developers:**
  - Provides rapid and consistent feedback on SDG alignment, allowing for more targeted project improvements.
- **For Policymakers and Investors:**
  - Offers a scalable tool for assessing projects' broader socio-economic impacts, supporting informed decision-making.

## 6 Conclusion

This study has presented an innovative integration of statistical modelling and AI-based verification systems for assessing the socio-economic impacts of carbon credit projects, specifically focusing on their alignment with the UN Sustainable Development Goals (SDGs). By combining predictive models with automated verification, the proposed framework offers a comprehensive approach to evaluating the broader impacts of sustainability projects.

**Summary of Findings**

1. **Performance of Statistical Models**
   - The application of various statistical models demonstrated robust capabilities in predicting social outcomes based on project-specific characteristics. When trained solely on project attributes, models provided accurate estimates of direct social impacts.

- Including SDG-related data improved model accuracy significantly, allowing for a more holistic understanding of the projects' multi-dimensional impacts. Models such as **XGBoost** and **Gradient Boosting** achieved high levels of precision, making them suitable for comprehensive socio-economic evaluations.

2. **Development and Implementation of Gen AI Auditor**

- The **Gen AI Auditor system** successfully automated the verification process for SDG alignment, using structured prompts to evaluate project descriptions. The system achieved a verification accuracy rate exceeding **95%** and a consistency rate of **92%**.
- With an average confidence score of **0.87**, the Gen AI Auditor provided reliable assessments, demonstrating its potential as a scalable, transparent tool for evaluating carbon credit projects' compliance with global sustainability goals.

**Implications and Contributions**

The integration of statistical models and AI-based verification has considerable implications for the evaluation of carbon credit projects:

- **Accuracy and Scalability**: The use of AI and statistical modeling not only enhances the accuracy of social impact predictions but also ensures that SDG compliance is verified consistently and at scale. This is particularly valuable for project developers, investors, and policymakers seeking reliable insights into projects' socio-economic impacts.
- **Automated Compliance Verification**: The Gen AI Auditor's structured prompt methodology supports automated verification, reducing manual workload while maintaining transparency and accountability. The system's ability to deliver clear, structured outputs aligned with UN SDG requirements supports effective decision-making and resource allocation.

**Challenges and Recommendations**

While the results are promising, certain limitations were observed:

- **Data Quality**: The accuracy of both the statistical models and AI-based verification is highly dependent on the quality and completeness of project documentation. Incomplete or ambiguous descriptions can limit model performance and reduce the accuracy of AI-based evaluations.
- **Model Interpretability**: Despite achieving high accuracy, the models' interpretability remains a challenge, particularly when analyzing complex social outcomes and interactions. This could limit the system's transparency, which is critical for stakeholder trust.
- **Overfitting Risks**: The structured prompts used by the Gen AI Auditor could lead to overfitting, making the system overly reliant on specific phrasing and potentially affecting generalizability across diverse project datasets.

**Future Improvements**

- **Enhancing Explainability**: Implementing advanced explainable AI techniques to increase the interpretability of statistical models and AI verification outputs, making the decision-making process more transparent and comprehensible to stakeholders.
- **Broadening Dataset Diversity**: Expanding the scope of datasets used to train models and AI systems, ensuring that evaluations are generalisable across different project types and geographical contexts.
- **Incorporating Blockchain Technology**: Integrating blockchain technology could improve data integrity and provide an additional layer of transparency, further enhancing the credibility of AI-based verification systems.

**Final Remarks**

Integrating statistical models with AI-based verification systems significantly advances the carbon credit project evaluation field. This combined approach achieves high accuracy in social impact predictions and establishes an automated, scalable, and transparent framework for SDG verification. By supporting informed decision-making, this framework contributes to global sustainability efforts, offering a reliable tool for assessing the broader socio-economic impacts of carbon credit projects.

The findings underscore the transformative potential of AI and statistical modelling in driving accurate, scalable, and transparent evaluations, aligning with IEEE's standards for innovative research and technological solutions that address global challenges.

## References

1. Waldegren, L.T.: Carbon Credits: Origins, Effectiveness and Future (Doctoral dissertation, Lund University, Environmental and Energy Systems Studies). Lund University (2012)
2. Liu, Y., Zhang, Y., Wang, F., Gao, P.: Are carbon emissions trading and green financial instruments synergistic? PLOS ONE **17**(7) (2022)
3. Li, Y., Antwi-Afari, M.F., Anwer, S., et al.: Artificial intelligence in net-zero carbon emissions for sustainable building projects: a systematic literature review. Buildings **14**(9), 2752 (2024). https://doi.org/10.3390/buildings14092752
4. Fraccascia, L., Albino, V., Garavelli, A.C.: Integrating artificial intelligence in environmental sustainability assessments: toward comprehensive carbon credit evaluations. J. Clean. Prod. **393**, 136762 (2023). https://doi.org/10.1016/j.jclepro.2023.136762
5. Kitchenham, B., Charters, S.: Guidelines for performing systematic literature reviews in software engineering. EBSE Tech. Rep. (2007). https://doi.org/10.1145/1134285
6. Lee, D.-H., Kim, D.-H., Kim, S.-I.: Characteristics of forest carbon credit transactions in the voluntary carbon market. Clim. Policy **6**(3), 245–257 (2018)
7. Jennifer L.: Carbon credits and the sustainable development goals: aligning climate action with global priorities. Carbon Credits. https://www.carboncredits.com
8. Xu, Y., Zhang, L., Liu, W.: A comparative study of XGBoost and random forest in predicting non-linear social outcomes. Semantic Scholar (2023). https://pdfs.semanticscholar.org

9. Smith, B., Kistruck, G., Cannatelli, B.: The impact of project size, geographic location, and stakeholder engagement on social value measurement in social enterprises. J. Bus. Ethics **133**, 677–689 (2016). https://doi.org/10.1007/s10551-015-2890-0
10. Bachmann, N., Tripathi, S., Brunner, M., Jodlbauer, H.: The contribution of data-driven technologies in achieving the sustainable development goals. Sustainability **14**(5), 2497 (2022). https://doi.org/10.3390/su14052497
11. Peffers, K., Tuunanen, T., Rothenberger, M.A., Chatterjee, S.: A design science research methodology for information systems research. J. Manag. Inf. Syst.Manag. Inf. Syst. **24**(3), 45–77 (2007). https://doi.org/10.2753/MIS0742-1222240302
12. Doomah, Z., Seeboo, A., Fowdur, T.P.: Achieving SDG targets in the land transport sector using intelligent transportation systems. In Fowdur, T.P., Rosunee, S., Ah King, R.T.F., Jeetah, P., Gooroochurn, M. (eds.) Artificial Intelligence, Engineering Systems and Sustainable Development, pp. 81–93. Emerald Publishing Limited (2024). https://doi.org/10.1108/978-1-83753-540-820241007
13. Hsain, A., El Housni, H.: Large language model-powered chatbots for internationalizing student support in higher education (2024). https://doi.org/10.48550/arXiv.2403.14702
14. Dörgő, G., Sebestyén, V., Abonyi, J.: Evaluating the interconnectedness of the sustainable development goals based on causality analysis of sustainability indicators. Sustainability **10**(10), 3766 (2018). https://doi.org/10.3390/su10103766
15. Feor, L., Clarke, A., Dougherty, I.: Social impact measurement: a systematic literature review and future research directions. World **4**(4), 816–837 (2023). https://doi.org/10.3390/world4040051
16. Reich, Y., Konda, S., Monarch, I., Levy, S., Subrahmanian, E.: Evaluating large-language-model chatbots in community projects. AI EDAM **37**(2), 145–163 (2023). https://www.cambridge.org
17. Sebestyén, V., Dörgő, G., Abonyi, J.: Network model-based analysis of the goals, targets, and indicators of sustainable development for strategic environmental assessment. J. Environ. Manag. **238**, 126–135 (2019). https://doi.org/10.1016/j.jenvman.2019.03.054
18. Steingard, D.: Assessing SDG impact: elevating and measuring the societal impact of research. AACSB Insights (2023). https://www.aacsb.edu
19. Lou, J., Hultman, N., Patwardhan, A., Qiu, Y.L.: Integrating sustainability into climate finance by quantifying the co-benefits and market impact of carbon projects. Commun. Earth Environ. **3**(1), 137 (2022). https://doi.org/10.1038/s43247-022-00468-9
20. Center for Global Sustainability.: Integrating sustainability into climate finance by quantifying the co-benefits of carbon projects. University of Maryland School of Public Policy (2023). https://www.cgs.umd.edu
21. Von Avenarius, A., Devaraja, T.S., Kiesel, R.: An empirical comparison of carbon credit projects under the clean development mechanism and verified carbon standard. Climate **6**(2), 49 (2018). https://doi.org/10.3390/cli6020049
22. Liao, Z.: Assessing sustainable impacts of green energy projects for the development of renewable energy technologies: a triple bottom line approach. Processes **11**(8), 2228 (2023). https://doi.org/10.3390/pr11082228)
23. Limmeechokchai, B., Winyuchakrit, P., Pita, P.: The role of discount rate and social cost of carbon in long-term climate policy on renewable electricity in Thailand. IEEE Conf. (2022). https://doi.org/10.1109/ICUE53082.2022
24. Xu, R., Lin, X., Dong, Q., Chen, Y.: Constructing trustworthy and safe communities on a blockchain-enabled social credits system. In: Proceedings of MobiQuitous'18 (2018). https://doi.org/10.1145/3266157.3266174

# Looking Beyond the COVID-19 Pandemic: Navigating Strategic Shift for Achieving the New Normal in the Global Infrastructure Sector

**Naresh Gupta, Indra Gunawan, and Rajeev Kamineni**

**Abstract** The global infrastructure sector, crucial for economic development, has confronted unprecedented challenges, including supply chain disruptions and financial exigencies, amid the COVID-19 pandemic. This chapter investigates the response strategies employed by project managers, focusing on the adoption of digital and sustainable techniques to revive construction supply chains. Drawing on a systematic literature review of 69 articles, this research explores disruptions faced by construction enterprises, identifying crucial strategic shifts towards digital transformation and sustainable growth in the post-pandemic landscape. Despite context-specific countermeasures, findings indicate a lack of sustained resilience in infrastructure organisations, with short-term interventions proving insufficient for long-term success amid future disruptions. The chapter advocates for proactive strategic shifts influenced by evolving government policies, economic paradigms, environmental considerations, resource allocation, and geopolitical dynamics. The research highlights the need for transformative improvements through the adoption of digital technologies and sustainable practices, aligning with megatrends and forward-looking strategies. By thematically organising the literature, the chapter underscores the importance of long-term strategic pivots for achieving true resilience and sustained growth. The implications extend to construction enterprises, project managers, and industry practitioners, offering a roadmap for transitioning from crisis management paradigms to sustained progress and resilience in the global infrastructure sector. This chapter contributes novel insights to academic discourse, addressing a significant gap and adding intrinsic value to broader infrastructure studies by emphasising the critical role of digital and sustainable techniques in reviving construction supply chains.

N. Gupta (✉) · I. Gunawan · R. Kamineni
Adelaide Business School, University of Adelaide, Adelaide, SA, Australia
e-mail: naresh.gupta@adelaide.edu.au

I. Gunawan
e-mail: indra.gunawan@adelaide.edu.au

R. Kamineni
e-mail: rajeev.kamineni@adelaide.edu.au

 
N. D. K. Nguyen et al. (eds.), *Digital Transformation and Sustainability: Methods and Applications*, Information Systems Engineering and Management 47,
https://doi.org/10.1007/978-3-031-91278-8_4

**Keywords** Infrastructure sector · Post-pandemic · Strategic shifts · Digital techniques · Sustainability · Digital supply chain

## 1 Introduction

The infrastructure sector is a critical component of a country's economic landscape, playing a pivotal role in fostering economic growth and development [1]. It acts as a vital nexus between industries, serving as the backbone of a country's economy [2]. Its multifaceted impact on the economy and far-reaching implications on stimulating various economic activities are undeniable. Governments and policymakers prioritise infrastructure investment to drive economic growth and improve the quality of life for their citizens [1]. According to the World Bank, global infrastructure investment needs are projected to reach $94 trillion by 2040 to support population growth, urbanisation, and economic development [3]. This includes a significant demand in emerging economies, with Asia alone expected to require $50 trillion by 2030 to meet its urbanisation needs [4]. As technology and societies continue to evolve, the role of infrastructure in fostering economic prosperity remains indispensable, making it a top priority for nations looking to secure their economic future [5].

The COVID-19 pandemic severely disrupted the global infrastructure sector, reducing or altogether halting operations due to lockdowns and labour shortages, which led to project delays and cost overruns [6]. According to the Organisation for Economic Co-operation and Development (OECD), global infrastructure spending dropped by approximately 25% in 2020, with significant delays in project timelines and budget allocations [7]. The pandemic also caused supply chain disruptions, material shortages, and price escalations, complicating budget adherence for project managers [8]. Financial constraints forced governments to prioritise healthcare over infrastructure, resulting in reduced investment flows [9]. Private investors became cautious, further diminishing capital for infrastructure projects, with McKinsey estimating a $1 trillion shortfall in global infrastructure investment as a direct result of the pandemic [10]. Many small and mid-level enterprises, which normally operate on thin margins, closed, causing a loss of capacity and expertise, and hindering the sector's recovery and adaptation to the new economic landscape.

In the aftermath of the pandemic, the infrastructure sector encountered significant uncertainty, necessitating meticulous risk evaluation and management by project managers and stakeholders to address potential disruptions, financial constraints, and fluctuating demand [11]. As governments and investors renew their commitment to infrastructure development, the industry must adapt to new realities, including increased use of digital technologies for remote project management, enhanced risk mitigation strategies, and a stronger focus on sustainability and resilience in infrastructure planning [12]. Moving forward, the infrastructure sector must recover from the pandemic's impacts and strategically shift towards a new normal by embracing digital transformation to improve efficiency and productivity in project management and execution [13]. As the global economy rebounds post-pandemic, the infrastructure sector must navigate these challenges and embrace strategic shifts to emerge stronger and more resilient, meeting the evolving needs of societies and economies worldwide.

The Swiss Re Institute report [14] provides a comprehensive estimate of global infrastructure investment needs and gaps, especially in the post-pandemic recovery period from 2021 to 2040. The report emphasises that the global investment gap is as high as $15 trillion. Although the total global infrastructure investment is estimated to be $66 trillion, the funds needed to meet infrastructure needs are insufficient. Emerging markets need $43 trillion in investment, with a gap of $10 trillion, primarily driven by demand in Asia's rapidly growing economies [14]. Table 1 shows that developed markets are expected to invest $23 trillion but still face a $5 trillion gap. The report underscores the need to increase investment in smart and resilient infrastructure to address upcoming environmental and economic challenges and enhance global resilience.

This research investigates how resilient businesses in the infrastructure sector are adapting to the "new normal" post-COVID-19 by examining the strategies such as supply chain management, alternative sourcing, and digital technologies to ensure project sustainability. The chapter explores the role of AI, machine learning, remote collaboration, and innovative construction methods in facilitating these adaptations by delving into strategic shifts undertaken by infrastructure stakeholders and highlighting the embrace of digital transformation, data-driven decision-making, and technologies like BIM and digital twins. It evaluates the impact of these strategic shifts on the sector's resilience and long-term sustainability, providing insights into the sector's post-pandemic trajectory and contributing to discussions on building resilient enterprises to prepare for future crises beyond COVID-19.

Thus, the specific objectives of this research endeavour are:

**Table 1** Infrastructure investments and gap (post-pandemic period estimate: 2021–2040)

| Infrastructure sector | Infrastructure investments (In trillion USD) | Infrastructure investment gap (In trillion USD) |
|---|---|---|
| a. China | 22 | 1 |
| b. Other Asian Countries | 13 | 4 |
| = I. Emerging Asia (a + b) | 35 | 5 |
| II. Africa | 3 | 2.5 |
| III. Latin America | 3 | 2 |
| IV. Emerging Europe | 2 | 0.5 |
| = A. Emerging markets (I + II + III + IV) | 43 | 10 |
| B. Advanced markets | 23 | 5 |
| = Global sector (A + B) | 66 | 15 |

*Source* [14]

1. To analyse how the global infrastructure sector has adapted to the challenges posed by the COVID-19 pandemic through the implementation of digital techniques and sustainable practices.
2. To assess the impact of strategic shifts on the infrastructure sector's resilience, sustainability, and long-term prospects in achieving the "new normal".

## 2 Theoretical Review

### 2.1 *Immediate Focus: Resilience and Survival*

Resilience and survival strategies encompass strategic responses in financial, marketing, and safety domains [15]. The concept of resilience in the infrastructure sector involves its ability to withstand shocks and operate effectively in the face of expected or unexpected disruptions, such as the COVID-19 pandemic. This resilience enables the industry to adapt to unforeseen challenges by developing customised responses to threats that could jeopardise its survival. Furthermore, it allows for transformative actions to take advantage of such disruptive events [16]. While the global easing of COVID-19 containment measures has facilitated a return to near-normal levels of construction activity, it has also increased the risk of virus exposure within the industry. Therefore, it is crucial to take short-term measures to combat the outbreak at construction sites.

Multiple organisations, including the World Health Organisation (WHO) and the Occupational Safety and Health Administration (OSHA), have proposed various measures to address the pandemic's evolving nature. For instance, WHO has categorised safety measures into six groups: general guidelines, preventive measures (e.g., cleaning, hand washing, respiratory hygiene), pre-entry screening, transportation-related precautions, home-return precautions, and infection response plans with stress management techniques [17]. Agyekum et al., [18] found that construction companies enhanced education and personal protective equipment provision, along with regular inspections. Olanrewaju et al., [19] emphasised the importance of these measures for minimising virus spread and maintaining productivity at construction sites. Gamil and Alhagar [20] pointed out that construction companies have adopted strategies such as task reassignment, task automation, continuous communication with subcontractors, regular briefings, and stakeholder collaboration as immediate responses.

### 2.2 *Long-Term Focus: Recovery and Growth*

Safety guidelines implemented in response to the pandemic have significantly impacted the construction industry and its supply chains, highlighting the need for sustainable solutions. The Construction Leadership Council of the UK has proposed a three-stage recovery approach: restart, reset, and reinvention phases, focusing on embracing technology to drive sustainability, enhancing production, and transforming the industry through collaborative innovation. Ebekozein et al., [21] emphasised the importance of integrating digital technologies, supported by government initiatives, to achieve sustainability goals in the construction sector. They highlighted the need for skills development, budget allocations for sustainable infrastructure, and strategies to adopt advanced digital capabilities and omnichannel support mechanisms.

Proactive construction companies are developing innovative solutions to overcome pandemic challenges and advance sustainability, focusing on supply chain stabilisation, new technologies, and off-site construction methods [22]. The global construction industry is poised for growth, with a projected CAGR of 6% from 2020 to 2024, driven by automation, government initiatives, and increased awareness of sustainable practices [23]. Advancing sustainability and organisational resilience, as identified by Pascua and Chang-Richards [24], relies on strong leadership, a technologically skilled workforce, robust supply chains, and collaborative partnerships.

### *2.3 Identification of Research Gaps*

The COVID-19 pandemic has globally impacted various sectors, prompting entities to adopt specific strategies for addressing challenges and ensuring continuity [19]. Existing literature offers a broad view of business continuity with limited sector-specific analysis and lacks an understanding of how regulatory and economic factors influence strategic responses in the infrastructure sector [22, 23]. Despite recognising the importance of digital transformation, the barriers to adopting these technologies are underexplored, highlighting the need to digitalise the supply chain to enhance resilience and operational efficiency [25] [17, 26]. Additionally, while sustainability and resilience are deemed crucial, there is a lack of detailed exploration into specific sustainability initiatives and their impacts on infrastructure development [19]. Addressing these gaps provides an opportunity to probe into the interrelationship between digitalisation, sustainability, and resilience in infrastructure development, especially concerning the strategic shift towards the post-pandemic "new normal".

## 3 Research Methodology

A systematic literature review was conducted following Briner and Denyer's [27] five-stage methodology. Systematic reviews are designed to reduce bias and enhance the thoroughness and transparency of the review process, setting them apart from narrative and integrative reviews [28, 29]. This method permits the identification, analysis, and interpretation of all pertinent evidence in an impartial and somewhat replicable manner [28].

Several factors influence the level of repeatability, such as the transparency of search terms used across different database searches and the timing of these searches. Additionally, researchers must carefully balance innovation, novelty, and rigor when developing theories on central relationships [28, 30], ensuring the systematic approach remains intact and a comprehensive view of the evidence is maintained.

A systematic review is especially advantageous in the realm of the post-COVID-19 global infrastructure sector, where existing literature may be fragmented and lacking coherence [31]. As a result, narrative and integrative reviews are less likely to thoroughly identify, analyse, and interpret all available evidence pertinent to our specific research question in the manner a systematic review can achieve [28]. Thus, it offers a more organised and consolidated overview of the current understanding regarding the navigation of strategic shifts towards attaining the 'new normal' in the global infrastructure sector and its supply chains post-COVID-19. This process enables the establishment of an evidence base to enhance both research understanding and industry practices [29, 32]. The Research Methodology used in this research is presented in Fig. 1.

(i) **Identification**—The first step involved establishing the inclusion/exclusion criteria, identifying the relevant databases, academic journals, and other sources, and defining the key search terms and phrases to search for literature. The inclusion & exclusion criteria are set to include only those articles in English language that were published in scientific, peer-reviewed journals during or after 2020. Three widely used academic databases were identified: Google Scholar, Scopus, and Web of Science.

(ii) **Searching**—Systematic searches were conducted across identified academic databases to find all potentially relevant sources, using the set of search terms, such as "survival measures to respond to the COVID-19 pandemic", "recovery strategies facilitating the transition to the new normal", "resilient transformation in the infrastructure sector", and "strategic shifts in the infrastructure sector", existing in the title or abstract, either exactly or in part, to ensure comprehensive coverage. The search strategy and results were documented to allow transparency and reproducibility of the review. The search resulted in a total of 218 articles from all the databases.

(iii) **Screening**—Subsequently, the identified sources were screened to determine their relevance and reliability. The screening process involved two stages: inclusion/exclusion criteria screening and title/abstract screening. In the first stage, the identified articles were reviewed to exclude duplicate and low-rated articles. In the second stage, titles and abstracts of the identified articles were reviewed to exclude articles not consistent with the objectives of the current study. In the first stage, 94 duplicate and low-rated articles were removed and in the second stage 43 irrelevant articles were removed.

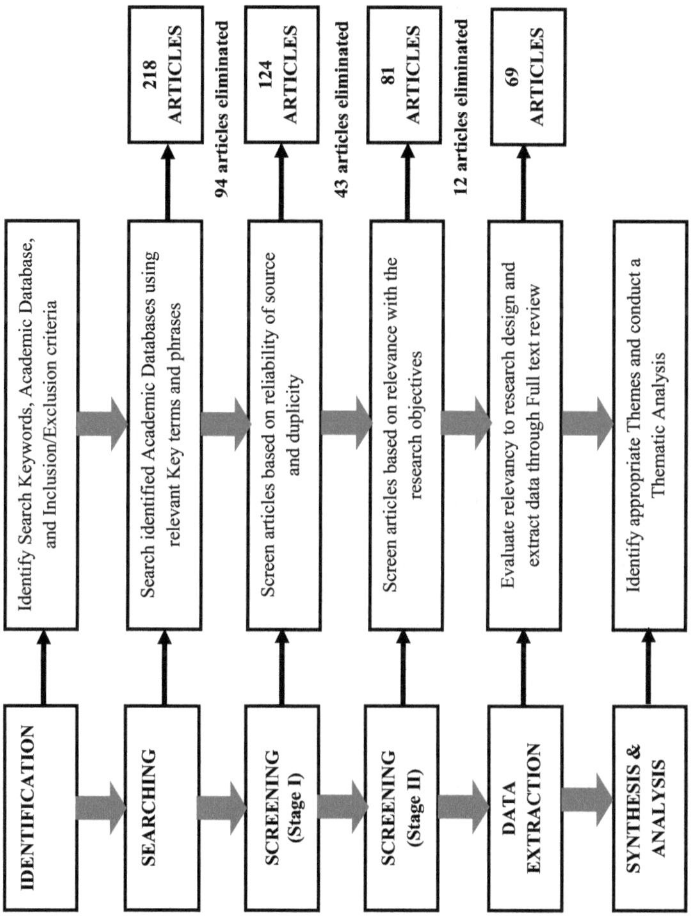


**Fig. 1** Research methodology. *Source* Author's Work

(iv) **Data extraction**—This stage involved a detailed assessment of the full text of the remaining 81 studies, including research methodology, sample characteristics, key findings, and other important information. At this stage, 12 articles, not relevant to the study design, were eliminated.
(v) **Synthesis and analysis**—Finally, the findings from the remaining 69 studies were synthesised and organised into a structured format, using the thematic analysis technique, to facilitate further analysis and comparison. The synthesis process involved identifying common themes, patterns, and discrepancies across the selected studies. This helped provide insights into the existing state of knowledge in the research area.

# 4 Analysis, Findings, and Discussion

## *4.1 Descriptive Analysis*

A descriptive analysis was performed to dissect the selected studies. The analysis is based on three aspects: year of publication, journal name, and continent of origin of the study. Table 2 summarises the dissection of all 69 studies, based on the year of publication.

The data in the table above shows a growing trend towards adopting better and innovative strategies in recent years. This shift aims to minimise vulnerabilities and mitigate negative impacts, enhancing preparedness for future disruptions.

Table 3 summarises the dissection of all 69 studies according to the name/title of the journals, listing the frequency of articles in different journals to identify those that published the most articles related to strategies to mitigate the impact of the pandemic.

Table 4 summarises the dissection of all 69 studies according to the continent of origin of the study, highlighting the frequency of articles from various continents. This comprehensive overview assesses global perspectives and concerns regarding the adoption of innovative strategies to mitigate the consequences of the pandemic, showcasing a diverse range of viewpoints from across the world.

**Table 2** Descriptive analysis of selected studies by year of publication

| Year | No. of studies | % of studies |
|---|---|---|
| 2020 | 2 | 2.9 |
| 2021 | 11 | 15.9 |
| 2022 | 24 | 34.8 |
| 2023 | 22 | 31.9 |
| 2024 | 10 | 14.5 |
| Total | 69 | 100 |

*Source* Studies Reviewed

**Table 3** Descriptive analysis of selected studies by name of journal

| Journal title | Frequency |
|---|---|
| Ain shams engineering journal | 2 |
| Buildings | 4 |
| Earth and environmental science | 2 |
| Environmental science and pollution research | 2 |
| Frontiers | 4 |
| Heliyon | 2 |
| International journal of construction management | 3 |
| Journal of business research | 2 |
| Journal of construction engineering and management | 6 |
| Journal of engineering, design, and technology | 2 |
| Journal of management in engineering | 3 |
| NICMAR journal of construction management | 2 |
| Smart infrastructure and construction | 2 |
| Sustainability | 10 |
| Single articles from different high-impact journals (such as social sciences; automation in construction; construction economics and building; Asian business and management; journal of transport and supply chain management; journal of global accounting; Malaysian construction research journal; etc.) | 23 |
| Total | 69 |

*Source* Studies Reviewed

**Table 4** Descriptive analysis of selected studies by continent of origin

| Continent | No. of studies | % of studies |
|---|---|---|
| North America | 9 | 13.1 |
| Europe | 13 | 18.8 |
| Middle East and Africa | 10 | 14.5 |
| Asia | 21 | 30.4 |
| Australia | 5 | 7.3 |
| Global | 11 | 15.9 |
| Total | 69 | 100 |

*Source* Studies Reviewed

The analysis presented in the above table indicates that developing economies are paying greater focus on research studies to adopt innovative methods and strategies to minimise the negative impacts of the pandemic.

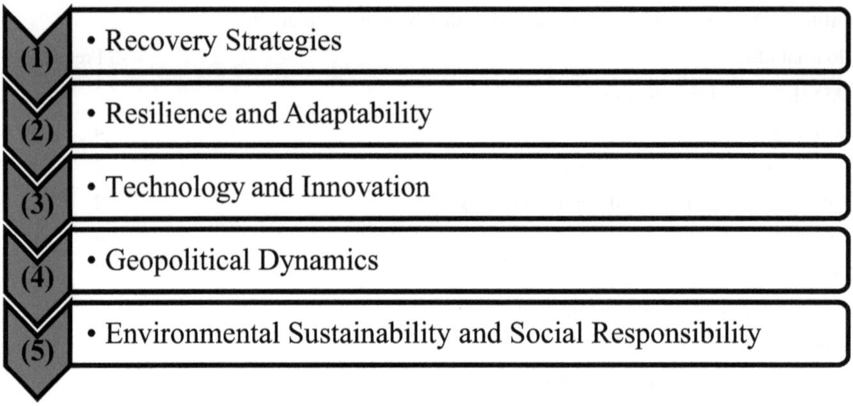


**Fig. 2** Identified themes. *Source* Author's Work

## 4.2 *Identified Themes*

The research employed thematic analysis to scrutinise identified articles and pinpoint recurrent themes and sub-themes, encompassing common topics, ideas, or patterns of meaning. The process involved five steps: becoming familiar with the studies, coding relevant information, analysing data to generate themes, reviewing the themes, and deciding and naming them. The researcher used NVivo qualitative data analysis tool for Windows (version 14) for keyword occurrence frequency analysis and draw relevant themes and sub-themes for this study.

The study emphasises the imperative need for the global infrastructure sector to transition towards a "new normal". This transition demands a shift in focus from traditional practices to more resilient, sustainable, and technologically advanced approaches. Embracing digitalisation, innovation, remote work capabilities, social responsibility, and advanced project management tools emerged as pivotal strategic components of this transition. Based on its systematic thematic analysis, the study identified five major themes, as presented in Fig. 2.

## 4.3 *Recovery Strategies*

Recovery strategies emerged as a set of plans, actions, and initiatives aimed at restoring and revitalising the infrastructure sector following the COVID-19 crisis. The research identified key strategies to help the sector bounce back, adapt, and thrive in the aftermath of significant challenges posed by the pandemic, as presented in Fig. 3.

The Table 5 presents a description of the relevant studies reviewed to identify and generate the key recovery strategies.

**Fig. 3** Identified recovery strategies. *Source* Author's Work

The COVID-19 pandemic significantly altered the global landscape, particularly impacting the infrastructure sector. This sector experienced disruptions in operations, finances, and development. To navigate these challenges, recovery strategies have become crucial to revitalise the infrastructure sector in the aftermath of the pandemic. These strategies, when properly executed and combined, play a pivotal role in fostering the sector's recuperation and fortifying its resilience in a post-pandemic world [23].

The recovery strategies are not solely aimed at addressing the immediate impacts of the pandemic but are forward-looking, seeking to position the infrastructure sector for sustainable growth, innovation, and adaptability [42]. By doing so, they aim to equip the sector to better face and overcome future challenges that might arise, whether they are similar health crises, climate-related disruptions, technological advancements, or other unforeseen circumstances.

The execution of these strategies not only aids in rebuilding what was affected but also aims to lay the groundwork for a more robust, adaptive, and forward-thinking infrastructure landscape. This resilience and preparedness are essential in ensuring that the infrastructure sector can continue to function effectively and efficiently

**Table 5** Recovery strategies identified in selected studies

| Relevant studies reviewed | (1) | (2) | (3) | (4) | (5) | (6) | (7) |
|---|---|---|---|---|---|---|---|
| Lagat et al. [23] | √ | √ | √ | √ | √ | | √ |
| Zhao et al. [33] | √ | √ | √ | √ | | | |
| Raj et al. [11] | | | | √ | | √ | |
| Nosike et al. [34] | | | √ | | | √ | √ |
| Salami et al. [35] | √ | √ | | | | | |
| Nebolsina et al. [36] | | | | √ | √ | | |
| Assaad et al. [37] | | | | √ | | √ | √ |
| Kam et al. [38] | √ | √ | √ | √ | | √ | |
| Nassereddine et al. [39] | | | | | | √ | √ |
| Gan et al. [40] | √ | | | | √ | | |
| Ibrahim et al. [41] | √ | √ | | √ | | √ | |
| Eissa and El-adaway et al. [42] | | √ | | √ | | √ | |
| Yin and Ran et al. [43] | | | √ | √ | | | |
| Hadjielias et al. [44] | √ | | | √ | | √ | |
| Oo et al. [45] | √ | √ | | | | | √ |

*Source* Studies Reviewed

despite unforeseen challenges, contributing to the overall economic recovery and societal well-being [46].

## *4.4 Resilience and Adaptability*

Resilience and adaptability are critical for construction companies in the post-pandemic landscape due to the industry's inherent susceptibility to economic fluctuations, supply chain disruptions, market uncertainties, and regulatory changes. Resilience emerged as a key theme in the discussions. Companies that swiftly adapted to the changing landscape showcased a higher degree of resilience [31]. The ability to pivot operations, adopt alternative and digital supply chains, and implement remote work strategies became defining factors for success. This research identified the major strategic qualities to enhance them, as per Fig. 4.

The Table 6 presents a description of the relevant studies reviewed to identify the resilience and adaptability strategies.

The contemporary business landscape, especially within sectors like construction, is facing a confluence of challenges including fluctuating demand, project delays, and financial uncertainties, often stemming from various external factors like the ongoing impact of the COVID-19 pandemic, geopolitical instabilities, and supply chain disruptions. In response, companies are compelled to adopt a more resilient and innovative approach in managing costs and resources effectively.

**Fig. 4** Identified resilience and adaptability strategies. *Source* Author's Work

Resilient and adaptable companies display the capacity to swiftly adapt to disruptions and vulnerabilities in their supply chains. They do this by swiftly identifying alternate suppliers or, in some cases, by developing in-house resources to ensure the continuity of their operations [31]. This adaptability involves being open to strategic shifts that increase a company's resilience and capacity to adapt. By doing so, companies can position themselves for sustainable growth and success, while navigating uncertainties and maintaining operational stability [64].

A core facet of resilience is the embrace of technological advancements. Companies, including those in the construction industry, are leveraging digital project management tools, remote collaboration platforms, and automation to enhance their efficiency and productivity [33]. These technological solutions enable streamlined operations, even in situations where physical presence or traditional methods might not be viable, as seen during the pandemic. They facilitate smoother project management, effective communication, and streamlined workflows [60].

**Table 6** Resilience and adaptability strategies identified in selected studies

| Relevant studies reviewed | (1) | (2) | (3) | (4) | (5) | (6) | (7) |
|---|---|---|---|---|---|---|---|
| Trabucco and Giovanni et al. [47] | | √ | √ | | | | |
| Iannone and Caruso et al. [48] | √ | √ | | | | √ | |
| Chih et al. [16] | √ | √ | | | √ | | √ |
| Zhao et al. [33] | | | √ | √ | | √ | √ |
| Nassereddine et al. [39] | | | | √ | √ | | √ |
| Zamani et al. [49] | √ | | | | | | √ |
| Mohite et al. [50] | √ | | √ | √ | | | |
| King et al. [51] | √ | √ | | | | | |
| King et al. [52] | √ | √ | √ | √ | √ | | |
| Lagat et al. [23] | | | | √ | √ | | √ |
| Zhu et al. [53] | | | | √ | | √ | |
| Basseal et al. [54] | | | | | | √ | √ |
| Nosike et al.[34] | √ | √ | √ | | √ | | √ |
| Sharma et al. [55] | √ | | √ | | √ | | |
| Liang et al. [56] | √ | | √ | | | | √ |
| Ernstsen et al. [57] | | √ | √ | | √ | √ | |
| Al-Mhdawi et al. [58] | | | | √ | | √ | |
| Xue et al. [59] | √ | √ | √ | √ | | | √ |
| Zhang et al. [60] | √ | | √ | | √ | | √ |
| Pretorius et al. [61] | | √ | √ | √ | | √ | |
| Chathuranga et al. [62] | √ | | √ | | √ | | |
| Maqbool et al. [63] | | | | | √ | | √ |
| Isang and Ebiloma et al. [64] | | √ | √ | | | √ | |

*Source* Studies Reviewed

## 4.5 Technology and Innovation

This research underscores the critical role of technology and innovation in shaping the future of the infrastructure sector. In the infrastructure sector, there have been notable strategic shifts towards embracing technology and innovation, especially in the post-pandemic scenario. These shifts not only address immediate challenges but also ensure long-term resilience, sustainability, and competitiveness in the rapidly evolving landscape [65]. The selected literature identifies the major strategic shifts as per Fig. 5, evident in the infrastructure sector in its attempts to embrace technological and innovative qualities.

The Table 7 presents a description of the relevant studies reviewed to identify the technology and innovation strategies.

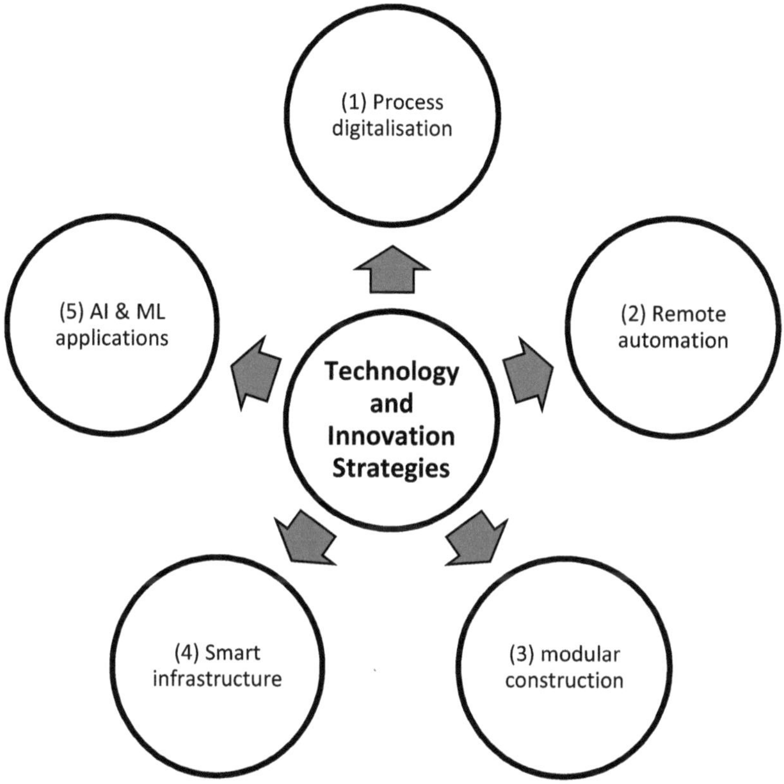


**Fig. 5** Identified technology and innovation strategies. *Source* Author's Work

Technology streamlines processes by reducing manual work and optimising resource allocation, thereby cutting costs, and improving overall efficiency. The adoption of advanced technologies such as Artificial Intelligence (AI), the Internet of Things (IoT), and data analytics isn't just about modernising; it significantly improves operational efficiency and sets the stage for advanced strategies like predictive maintenance and risk mitigation by offering a deeper understanding of patterns and potential challenges [68]. They help in making informed decisions, thereby reducing risks, and enhancing project outcomes.

The adoption of digital technologies such as Building Information Modelling (BIM) has revolutionised the way construction projects are planned and executed. BIM allows for a comprehensive digital representation of a project's characteristics and performance. It facilitates better coordination and collaboration among various stakeholders throughout the construction lifecycle, leading to smoother processes and reduced errors [67]. Remote monitoring, made possible by such digital tools, further streamlines construction processes, allowing for real-time oversight and control, even from a distance [75].

**Table 7** Technology and innovation strategies identified in selected studies

| Relevant studies reviewed | (1) | (2) | (3) | (4) | (5) |
|---|---|---|---|---|---|
| Ahmed et al. [66] | √ | | | √ | √ |
| Ebekozien et al. [21] | √ | | | √ | |
| Nassereddine et al. [39] | | √ | √ | | √ |
| Radzi et al. [67] | | √ | | | √ |
| Gartoumi et al. [68] | √ | √ | √ | √ | √ |
| Wang and Chen et al. [69] | √ | √ | | | √ |
| Li et al. [70] | | | √ | √ | √ |
| Rocha et al. [71] | √ | √ | √ | √ | √ |
| Subramanian et al. [72] | √ | | | √ | |
| Sharma et al. [55] | √ | √ | | | |
| Amankwah-Amoah et al. [73] | √ | √ | | | √ |
| Nguyen et al. [74] | | | √ | √ | |
| Zhu et al. [75] | √ | √ | | √ | √ |
| Brennan et al. [76] | √ | √ | | √ | √ |
| El Khatib et al. [77] | √ | √ | | | √ |
| Osorio-Gómez et al. [78] | | √ | √ | √ | √ |

Source: Studies Reviewed

Furthermore, the integration of renewable energy solutions and sustainable infrastructure practices has become a central focus in redefining the construction industry. This shift towards sustainability involves incorporating eco-friendly materials, energy-efficient systems, and environmentally conscious practices [68]. With a global emphasis on reducing carbon footprints and promoting sustainability, construction companies are increasingly integrating renewable energy solutions like solar panels or wind turbines into their projects. Sustainable practices not only benefit the environment but also align with changing regulations and market demands, ensuring that constructions are future-proof and more cost-effective in the long run [66].

## 4.6 *Geopolitical Dynamics*

This study recognised the impact of geopolitical dynamics and economic fluctuations on the infrastructure sector. As such, the review observed several major strategic shifts in the infrastructure sector, concerning geopolitical dynamics in the post-pandemic landscape. Undoubtedly, these key strategic shifts, as highlighted in Fig. 6, are crucial for the sector's development and sustainability.

The Table 8 presents a description of the relevant studies reviewed to identify the key geopolitical dynamics leading to the post-pandemic strategic navigation.

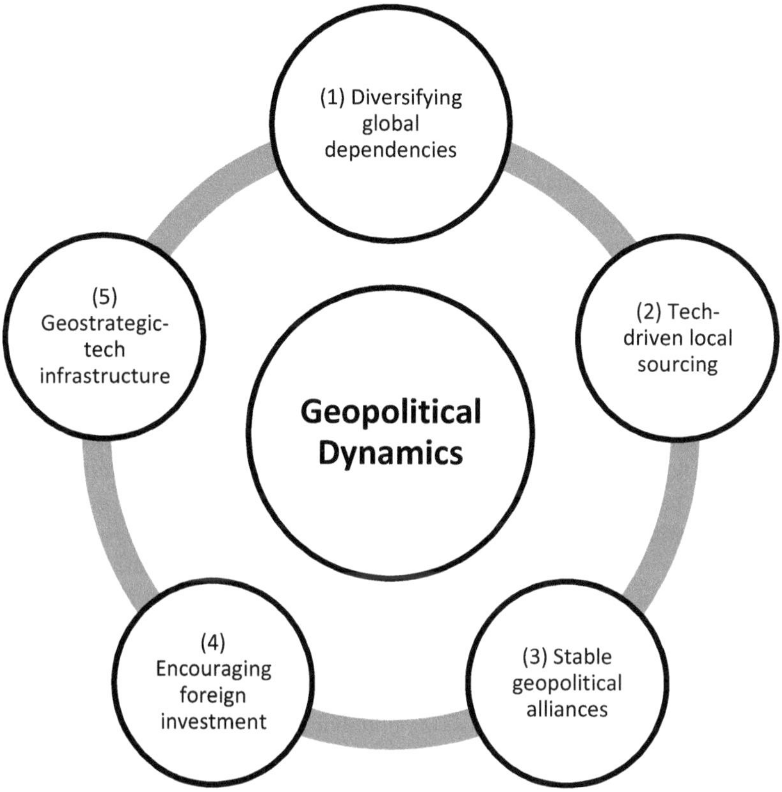


**Fig. 6** Identified geopolitical dynamics. *Source* Author's Work

**Table 8** Strategies concerning geopolitical dynamics identified in selected studies

| Relevant studies reviewed | (1) | (2) | (3) | (4) | (5) |
|---|---|---|---|---|---|
| Du et al. [1] | | | | √ | √ |
| Raj et al. [11] | √ | √ | | | |
| Radzi et al. [67] | √ | √ | | √ | |
| Zhu et al. [53] | √ | √ | | | |
| Chen et al. [79] | √ | √ | | | |
| Yu and Wang et al. [80] | | | √ | √ | √ |
| Seuring et al. [81] | | | √ | √ | √ |
| Adhi and Muslim et al. [82] | | | | √ | √ |
| Langston and Crowley et al. [83] | | | √ | | √ |
| O'Brien et al. [84] | √ | √ | √ | √ | |
| Topalidis et al. [85] | √ | √ | | | √ |

*Source* Studies Reviewed

In recent times, there has been a notable shift towards prioritising local sourcing and domestic production within various industries, including infrastructure development. This shift aims to reduce dependence on foreign suppliers, thereby enhancing economic independence and stability. The emphasis on local sourcing helps in mitigating risks associated with global supply chain disruptions, geopolitical tensions, and economic uncertainties [53].

To support infrastructure projects and bolster economic growth, it is crucial to encourage foreign investments and establish robust public–private partnerships. These partnerships bring together the strengths of both the sectors, leveraging private sector efficiency and innovation alongside public sector resources and regulatory frameworks (Pantic et al., 2021). Foreign investments inject capital and expertise into infrastructure development, helping to address funding gaps and facilitating the execution of large-scale projects.

Collaborative efforts, multilateral agreements, and policy interventions play a significant role in ensuring stability and fostering growth in the infrastructure domain. Cooperation between governments, international organisations, and private entities is vital in creating an environment conducive to investment and development. Multilateral agreements can provide a framework for trade, investment, and regulatory standards, encouraging smoother international collaboration and reducing barriers to entry for various stakeholders [81].

The significance of international cooperation is underscored by its role in mitigating risks and fostering a conducive environment for investment and development in the infrastructure sector. By working collaboratively across borders, nations can pool resources, share expertise, and jointly tackle challenges related to infrastructure development. Such cooperation not only mitigates geopolitical and economic risks but also stimulates economic growth and fosters stability in the sector [81].

These strategies are instrumental in making the infrastructure sector more resilient to uncertainties and potential disruptions. By diversifying and digitalising supply chains, fostering partnerships, and engaging in international cooperation, the sector becomes less vulnerable to shocks and more adaptable to changing global scenarios. It helps in creating a robust foundation that can withstand challenges and ensure continued progress and development in the infrastructure domain.

## *4.7 Environmental Sustainability and Social Responsibility*

The review observed several major strategic shifts concerning promotion of environmental sustainability and social responsibility in the post-pandemic landscape. These shifts are not only ethical imperatives but also crucial for the sector's long-term viability, success, societal impact, and environmental stewardship [86]. They address long-term challenges, contribute to societal well-being, and promote environmentally conscious practices crucial for a sustainable future. The key strategic shifts leading to environmental sustainability and social responsibility are highlighted in Fig. 7.

**Fig. 7** Identified environmental sustainability and social responsibility sub-themes. *Source* Author's Work

The Table 9 presents a description of the relevant studies reviewed to identify the key environmental sustainability and social responsibility dimensions leading to the post-pandemic strategic navigation.

In the realm of infrastructure development, there's a growing emphasis on integrating environmental sustainability and social responsibility into project planning and execution. This shift signifies recognition of the significant impact infrastructure projects have on the environment and local communities [92].

The research underscores that for future projects to be successful, it's imperative to prioritise stakeholder engagement, community involvement, and the adoption of eco-friendly practices [74]. Involving stakeholders and communities in the planning process is crucial as it builds trust, fosters collaboration, and ensures that infrastructure projects align with the needs and expectations of the people they will ultimately serve [82]. This participatory approach leads to more effective, sustainable, and community-friendly outcomes.

Additionally, there's an increasing emphasis on aligning infrastructure projects with Sustainable Development Goals (SDGs). These internationally recognised goals address global challenges such as poverty, inequality, climate change, environmental

**Table 9** Strategies concerning environmental sustainability and social responsibility

| Relevant studies reviewed | (1) | (2) | (3) | (4) | (5) |
|---|---|---|---|---|---|
| Oke and Aliu et al. [26] | √ | √ | √ | | √ |
| Ebekozien et al. [21] | √ | √ | | | √ |
| Xia et al. [86] | √ | √ | | √ | √ |
| Fallah Shayan et al. [87] | √ | | | √ | |
| Afieroho et al. [88] | | | √ | √ | |
| Nguyen et al. [74] | √ | √ | √ | | |
| Alraja et al. [89] | √ | | | √ | |
| Adhi and Muslim et al. [82] | √ | | √ | √ | |
| Sojobi and Liew et al. [65] | √ | √ | | √ | |
| Ayat et al. [90] | | √ | √ | √ | |
| Duflot et al. [91] | √ | √ | | √ | |
| Zhu et al. [75] | √ | | | √ | |
| Jalil et al.[92] | √ | | | √ | √ |
| Elavarasan et al. [93] | | | √ | √ | √ |
| Kumar et al. [94] | √ | √ | | √ | √ |
| Cortes et al. [95] | | √ | | √ | |
| Zhang et al. [96] | √ | √ | | √ | √ |

*Source* Studies Reviewed

degradation, peace, and justice [89]. Inclusivity and diversity within the infrastructure sector have gained prominence as integral factors. Ensuring that projects are not only environmentally sustainable but also promote inclusivity, diversity, and social equity is becoming a fundamental part of project design and execution.

Strategies that focus on sustainability and resilience are not only essential for the success of infrastructure projects but are also crucial in addressing post-pandemic challenges. These strategies aim to protect the environment, mitigate climate change, and create infrastructure that can withstand future environmental challenges [65]. The aftermath of the pandemic has highlighted the significance of addressing environmental concerns and creating infrastructure that not only meets immediate needs but also sustains and protects the environment for future generations.

By incorporating eco-friendly practices, focusing on community needs, aligning with global sustainability goals, and ensuring diversity and inclusivity, infrastructure development becomes more than just constructing physical structures. It becomes a catalyst for positive change, promoting a sustainable, inclusive, and resilient future. This approach ensures that infrastructure not only serves the present but also contributes positively to the well-being of future generations and the planet as a whole.

# 5 Conclusions, Implications, and Future Avenues for Research

## 5.1 Conclusions

A comprehensive review of the post-pandemic landscape of the global infrastructure industry reveals key insights across themes to help navigate the strategic transformation required to successfully transition to the new normal. The review identified seven strategic shifts incorporating digital risk assessment, tech-driven action plans, digital inventory and resource management, reshaping digital supply chain, cyber-resilient insurance, innovative digital project management, and remote work digitalisation, which emerged as critical components for the sector's post-pandemic recovery. These strategies, when combined and applied cohesively, are instrumental in ensuring the sector's resilience and sustainable growth.

The chapter highlights resilience and adaptability as fundamental qualities essential for companies in the infrastructure sector. Strategies to enhance resilience and adaptability, such as agile financial innovation, sustainable tech integration, digital tool adoption, tech-enhanced supply chain diversification, digital skills training, tech-driven risk and scenario planning, and digital collaboration culture, play a pivotal role in enabling companies to swiftly adapt to disruptions and uncertainties, ultimately ensuring their success and sustainability. The significance of technology and innovation in shaping the future of the infrastructure sector was a prevalent finding. The shift towards process digitalisation, remote automation, modular construction, smart infrastructure, and artificial-intelligence/machine-learning applications emerged as key drivers for efficiency, cost reduction, and enhanced project outcomes.

The impact of geopolitical dynamics and economic fluctuations on the infrastructure sector was recognised, necessitating strategic shifts to mitigate adverse consequences. Strategic shifts focused on diversifying global dependencies, tech-driven local sourcing, stable geopolitical alliances, encouraging foreign investment, and geostrategic-tech infrastructure, which are vital for the sector's stability and growth. The emphasis on environmental sustainability and social responsibility emerged as a crucial aspect to ensure the long-term viability and success of the sector. Strategies promoting green-tech infrastructure, resilient and adaptive design, digital community engagement, smart eco-technologies, and compliance with environmental standards were found fundamental for the sector's alignment with Sustainable Development Goals (SDGs) and ensuring inclusivity and diversity. To conclude, the global infrastructure sector faces a critical juncture in its evolution, requiring a paradigm shift to navigate the challenges posed by the COVID-19 pandemic and adapt to the "new normal". The identified strategic themes highlight the imperative need for a multifaceted approach in order to thrive and achieve sustained growth in a post-pandemic world. The integration of recovery strategies, resilience and adaptability, technological innovation, geopolitical acumen, and a strong commitment to environmental sustainability and social responsibility is essential for the sector's future success.

Collaborative efforts, leveraging technological advancements, including the development of a robust digital supply chain, and strategic planning will be crucial to drive the industry forward.

This chapter underscores the vital importance of a comprehensive and integrated approach that addresses not only immediate recovery needs but also long-term sustainability and resilience. By embracing change, fostering innovation, adopting digital tools, and prioritising sustainability, the global infrastructure sector can effectively position itself to overcome challenges, embrace opportunities, and contribute to a more resilient, sustainable, and technologically advanced future.

## *5.2 Research Implications*

These findings have significant implications across research, practical, and social domains. The chapter lays the groundwork for further in-depth research into specific areas that emerged as critical in the post-pandemic landscape of the global infrastructure sector. Research focusing on the effectiveness and implementation of recovery strategies, the impact of technology and innovation on infrastructure projects, the role of the digital supply chain, the influence of geopolitical dynamics on infrastructure development, and the long-term consequences of environmental sustainability in infrastructure can be avenues for future exploration. The continuous evaluation and adaptation of these strategies within the rapidly evolving landscape will be crucial for sustained growth and development.

From a practical perspective, the identified strategic shifts hold immense significance for companies and stakeholders in the infrastructure sector. Implementing comprehensive recovery strategies encompassing risk assessment, reshaping the digital supply chain, innovative methods, and remote work policies will be vital for post-pandemic recovery and future resilience. Companies need to prioritise resilience and adaptability by implementing strategies like agile financial planning, sustainable construction practices, and embracing digital tools to swiftly respond to uncertainties. The integration of technology and innovation, such as digitalisation and sustainable infrastructure practices, is essential for enhancing efficiency and project outcomes. Embracing these findings in practical applications will ensure the sector's robustness and sustainability in the face of future challenges.

The implications of the findings go beyond the infrastructure industry and extend to the social fabric. The emphasis on environmental sustainability, community engagement, and social responsibility demonstrates the sector's commitment to society and the environment. By promoting sustainable, green infrastructure, and adhering to stringent environmental regulations, the sector contributes to mitigating climate change and ensuring a healthier environment for communities. Additionally, the alignment of infrastructure projects with Sustainable Development Goals (SDGs) emphasises inclusivity, diversity, and societal well-being, ensuring that communities are active participants in infrastructure development. Furthermore, these initiatives create employment opportunities, foster community growth, and contribute to societal progress.

These implications suggest that a multifaceted approach, integrating research insights into practical applications and social responsibilities, is imperative for steering the global infrastructure sector toward a more resilient, sustainable, and socially responsible future in the post-pandemic era.

## *5.3 Future Avenues for Research*

Based on the insights gleaned from the findings of this chapter, several avenues for future research within the global infrastructure sector can be identified. Analysing how strategies such as reshaping digital supply chains, remote work policies, and technological innovations might contribute to sustained growth, resilience, and adaptability would provide valuable insights for future crisis preparedness. Future studies could also focus on assessing the effectiveness of strategies aimed at enhancing resilience and adaptability. This might include an in-depth exploration of agile financial strategies, digital supply chain diversification, and the impact of technological tools on the industry's resilience to future disruptions. Further, conducting comparative studies across different regions and infrastructure sub-sectors can provide a comprehensive understanding of the effectiveness of various strategies. Case analyses of successful and less successful implementations can offer practical insights for stakeholders.

Future research initiatives across these areas will contribute to a more comprehensive understanding of the challenges, opportunities, and transformations in the global infrastructure sector, thereby providing valuable guidance for industry stakeholders and policymakers.

## References

1. Du, X., Zhang, H., Han, Y.: How does new infrastructure investment affect economic growth quality? Empirical evidence from China. Sustainability **14**(6), 3511 (2022)
2. Pheng, L.S., Hou, L.S.: The economy and the construction industry. In: Construction quality and the economy: a study at the firm level. Springer Nature, pp. 21–54 (2019). PMC COVID-19 Collection: PMC7124044
3. World Bank.: Infrastructure investment needs to 2040. World Bank Publications (2020)
4. Chen, J., Silva, E.A.: Urbanisation and infrastructure: a comprehensive analysis of global trends. Habitat International (2020)
5. Allen, C., Metternicht, G., Wiedmann, T.: Prioritising SDG targets: assessing baselines, gaps, and interlinkages. Sustain. Sci. **14**(2), 421–438 (2018)

6. Osman, L.A., Mohamud, A.O.: The impact of time delay on construction projects cost overrun: a case study on some of construction projects in Mogadishu-Somalia. J. Posit. Sch. Psychol. **6**(4), 3884–3895 (2022)
7. OECD.: The impact of COVID-19 on global infrastructure investment. Organisation for Economic Co-operation and Development (2021)
8. Rani, H.A., Farouk, A.M., Anandh, K.S., Almutairi, S., Rahman, R.A.: Impact of COVID-19 on construction projects: the case of India. Buildings **12**(6), 762 (2022)
9. Cao, M., Duan, K., Cao, M., Ibrahim, H.: Exploring the interrelationships between public health, fiscal decentralisation, and local government debt in China. Healthcare **11**(14), 2103 (2023)
10. McKinsey Global Institute.: Global Infrastructure Investment Shortfall Due to COVID-19. McKinsey and Company (2021)
11. Raj, A., Mukherjee, A.A., de Sousa Jabbour, A.B.L., Srivastava, S.K.: Supply chain management during and post-COVID-19 pandemic: mitigation strategies and practical lessons learned. J. Bus. Res. **142**, 1125–1139 (2022)
12. Ferrari, M.: Reflexive governance for infrastructure resilience and sustainability. Sustainability **12**(23), 10224 (2020)
13. Smith and Jones.: Navigating the new normal. 10th Annual Healthcare Marketing Trends (2023). https://smithandjones.com/wp-content/uploads/2020/09/Smith-Jones-2023-Trends-Report.pdf
14. Swiss Re Institute Report.: Infrastructure investment to be a key driver of growth in emerging markets post COVID-19 crisis (2020). https://www.swissre.com/media/press-release/nr-20200617-sigma-3-2020.html.
15. Jahanshahi, A.A., Mendoza, M.I.R., Aguilar-Rodríguez, I.E.: Unveiling resilience: strategic responses, success factors and challenges faced by Latin American female entrepreneurs amidst the COVID-19 pandemic. IEEE Eng. Manage. Rev. (2023). https://doi.org/10.1109/EMR.2023.3327742
16. Chih, Y.Y., Hsiao, C.Y.L., Zolghadr, A., Naderpajouh, N.: Resilience of organisations in the construction industry in the face of COVID-19 disturbances: dynamic capabilities perspective. J. Manag. Eng.Manag. Eng. **38**(2), 1–16 (2022)
17. PAHO/WHO.: COVID-19: Prevention measures at construction sites. Pan American Health Organisation and World Health Organisation, pp. 1–17 (2020)
18. Agyekum, K., Kukah, A.S., Amudjie, J.: The impact of COVID-19 on the construction industry in Ghana: the case of some selected firms. J. Eng. Des. Technol. **20**(1), 222–244 (2022)
19. Olanrewaju, A.L., AbdulAziz, A.R., Preece, C.N., Shobowale, K.: Evaluation of measures to prevent the spread of COVID-19 on the construction sites. Clean. Eng. Technol. **5**, 100277 (2021)
20. Gamil, Y., Alhagar, A.: The impact of pandemic Crisis on the survival of construction industry: a case of COVID-19. Mediterr. J. Soc. Sci.. J. Soc. Sci. **11**, 122–128 (2020)
21. Ebekozien, A., Aigbavboa, C., Aigbedion, M.: Construction industry post-COVID-19 recovery: Stakeholders perspective on achieving sustainable development goals. Int. J. Constr. Manag.Manag. **23**(8), 1376–1386 (2021)
22. Freeman, M.A.: Back to (construction) work after COVID. Shutts and Bowen LLP (2021). https://www.shutts.com/flconstruction/back-to-construction-work-after-covid/
23. Lagat, D., Njuguna, M., Mutanu, M., Mwende, A., Magak, A., Nyakondo, S.: Strategies towards business continuity for construction industry in Kenya during and after COVID-19 pandemic. Curr. Urban Stud.. Urban Stud. **11**, 367–382 (2023)
24. Pascua, M.C., Chang-Richards, A.: Investigating the resilience of civil infrastructure firms in New Zealand. Procedia Eng. **212**, 286–293 (2018)
25. ILO.: COVID-19 Action Checklist for the Construction Industry, pp. 6–9. Available at: https://www.ilo.org/wcmsp5/groups/public/---ed_protect/---protrav/---safework/documents/instructionalmaterial/wcms_764847.pdf (2020)
26. Oke, A.E., Aliu, J.: Strategies for the implementation of environmental economic practices for sustainable construction in a developing economy. Int. J. Constr. Manag.Manag. (2024). https://doi.org/10.1080/15623599.2024.2338689

27. Briner, R.B., Denyer, D.: Systematic review and evidence synthesis as a practice and scholarship tool. In: Rousseau, D.M. (ed.) Handbook of Evidence-Based Management: Companies, Classrooms, and Research, pp. 112–129. New York University Press, New York (2012)
28. Fan, D., Breslin, D., Callahan, J.L., Iszatt-White, M.: Advancing literature review methodology through rigour, generativity, scope, and transparency. Int. J. Manag. Rev.Manag. Rev. **24**(2), 171–180 (2022)
29. Rojon, C., Okupe, A., McDowall, A.: Utilization and development of systematic reviews in management research: what do we know and where do we go from here? Int. J. Manag. Rev.Manag. Rev. **23**(2), 191–223 (2021)
30. Breslin, D., Gatrell, C., Bailey, K.: Developing insights through reviews: reflecting on the 20th anniversary of the International Journal of Management Reviews. Int. J. Manag. Rev.Manag. Rev. **22**(1), 3–9 (2020)
31. Katsaliaki, K., Galetsi, P., Kumar, S.: Supply chain disruptions and resilience: a major review and future research agenda. Ann. Oper. Res.Oper. Res. **319**, 965–1002 (2022)
32. Sombultawee, K., Lenuwat, P., Aleenajitpong, N., Boon-itt, S.: COVID-19 and supply chain management: a review with bibliometric. Sustainability **14**, 3538 (2022)
33. Zhao, N., Hong, J., Lau, K.H.: Impact of supply chain digitalisation on supply chain resilience and performance: a multi-mediation model. Int. J. Prod. Econ. **259**, 108817 (2023)
34. Nosike, R.J., Nosike, U.C., Ojobor, O.S.: Digital transformation and anal agility in post-COVID-19 pandemic era. J. Glob. Account. **9**(4), 410–426 (2024)
35. Salami, B.A., Ajayi, S.O., Oyegoke, A.S.: Coping with the Covid-19 pandemic: an exploration of the strategies adopted by construction firms. J. Eng. Des. Technol. **20**(1), 159–182 (2022)
36. Nebolsina, E.: The impact of the COVID-19 pandemic on the business interruption insurance demand in the United States. Heliyon **7**(11), e08357 (2021)
37. Assaad, R.H., El-adaway, I., Hastak, M., Needy, K.: The COVID-19 pandemic: a catalyst and accelerator for offsite construction technologies. J. Manag. Eng.Manag. Eng. **38**(6), 04022062 (2022)
38. Kam, K.J., Lim, T.S., Lee, L.H., Soon, L.T., Leong, B.T., Kamarazaly, M.A.: Survival strategies and challenges faced by subcontractors in the Malaysian construction industry. Malays. Constr. Res. J. **15**(1), 22–37 (2022)
39. Nassereddine, H., Seo, K.W., Rybkowski, Z.K., Schranz, C., Urban, H.: Propositions for a resilient, post-COVID-19 future for the AEC industry. Front. Built Environ. **7**, 687021 (2021)
40. Gan, V.J.L.: BIM-based building geometric modeling and automatic generative design for sustainable offsite construction. J. Constr. Eng. Manag. **148**(10) (2022). https://doi.org/10.1061/(ASCE)CO.1943-7862.0002369
41. Ibrahim, F.S., Esa, M., Kamal, E.M.: Strategies to minimise the impact of COVID-19 on the construction industry: a case study of construction site clusters in Malaysia. Constr. Econ. Build. **22**(3), 21–42 (2022)
42. Eissa, R., El-adaway, I.H.: Accelerating the circular economy transition: a construction value chain-structured portfolio of strategies and implementation insights. J. Constr. Eng. Manag. **150**(8) (2024). https://doi.org/10.1061/JCEMD4.COENG-1455.
43. Yin, W., Ran, W.: Supply chain diversification, digital transformation, and supply chain resilience: configuration analysis based on fsQCA. Sustainability **14**(13), 7690 (2022)
44. Hadjielias, E., Christofi, M., Tarba, S.: Contextualising small business resilience during the COVID-19 pandemic: evidence from small business owner-managers. Small Bus. Econ. **59**, 1351–1380 (2022)
45. Oo, B.L., Lim, B.T.H., Kim, S.: Working from home during COVID-19 and beyond: exploring the perceptions of consultants in construction. Buildings **13**(1), 166 (2023)
46. Raoufi, M., Fayek, A.R.: Identifying actions to control and mitigate the effects of the COVID-19 pandemic on construction organisations: preliminary findings. Public Works Manage. Policy **26**(1), 47–55 (2021)
47. Trabucco, M., De Giovanni, P.: Achieving resilience and business sustainability during COVID-19: the role of lean supply chain practices and digitalisation. Sustainability **13**(22), 12369 (2021)

48. Iannone, B., Caruso, G.: "Sustainab-lisation": Sustainability and digitalisation as a strategy for resilience in the coffee sector. Sustainability **15**(6), 4893 (2023)
49. Zamani, S.H., Rahman, R.A., Fauzi, M.A., Yusof, L.M.: Effect of COVID-19 on building construction projects: impact and response mechanisms. IOP Conf. Ser. Earth Environ. Sci. **682**, 012049 (2021)
50. Mohite, R.: Resilience and recovery: a case study on the post-COVID-19 impact on micro, small, and medium enterprises in Maharashtra. Shodha Prabha **48**(3), 166–175 (2023)
51. King, S.S., Rahman, R.A., Fauzi, M.A., Haron, A.T.: Mechanisms for addressing the impact of COVID-19 on infrastructure projects. IOP Conf. Ser. Earth Environ. Sci. **682**, 012047 (2021)
52. Harouache, A., Abkar, M.M.A., Gamil, Y., Gamil, Y., Al-Shameri, A.S.A., Gabir, A.A.M.: Influence of supply chain management on the sustainable construction industry in Algeria. Front. Built Environ. **10**, 1233266 (2024)
53. Zhu, G., Chou, M.C., Tsai, C.W.: Lessons learned from the COVID-19 pandemic exposing the shortcomings of current supply chain operations: a long-term prescriptive offering. Sustainability **12**(14), 5858 (2020)
54. Basseal, M., Bennett, C.M., Collignon, P., Currie, B.J., Durrheim, D.N., Leask, J., McBryde, E.S., McIntyre, P., Russell, F.M., Smith, D.W., Sorrell, T.C., Marais, B.J.: Key lessons from the COVID-19 public health response in Australia. Viewpoint **30**, 100616 (2023)
55. Sharma, M., Luthra, S., Joshi, S., Joshi, H.: Challenges to agile project management during COVID-19 pandemic: an emerging economy perspective. Oper. Manag. Res.. Manag. Res. **15**, 461–474 (2022)
56. Liang, C-J., Wang, X., Kamat, V.R., Menassa, C.C.: Human–robot collaboration in construction: classification and research trends. J. Constr. Eng. Manag. 147(10) (2021). https://doi.org/10.1061/(ASCE)CO.1943-7862.0002154
57. Ernstsen, S.N., Whyte, J., Thuesen, C., Maier, A.: How innovation champions frame the future: three visions for digital transformation of construction. J. Constr. Eng. Manag.Manag. **147**(1), 05020022 (2020)
58. Al-Mhdawi, M.K.S., Brito, M., Onggo, B.S., Qazi, A., O'Connor, A.: COVID-19 emerging risk assessment for the construction industry of developing countries: evidence from Iraq. Int. J. Constr. Manag.Manag. (2023). https://doi.org/10.1080/15623599.2023.2169301
59. Xue, Y., Le, Y., Jiang, K., Gan, X.: Collaborative management framework for interorganizational risk in megaprojects based on meta network. J. Manag. Eng. **40**(5) (2024). https://doi.org/10.1061/JMENEA.MEENG-6073
60. Zhang, J., Zhao, W., Cheng, B., Li, A., Wang, Y., Yang, N., Tian, Y.: The impact of digital economy on the economic growth and the development strategies in the post-COVID-19 era: Evidence from countries along the Belt and Road. Front. Public Health **10**, 856142 (2022)
61. Pretorius, O.R., Drewes, J.E., Engelbrecht, W.H., Malan, G.C.: Developing resilient supply chains in the Southern African development community: lessons from the impact of COVID-19. J. Transp. Supply Chain. Manag. **16**, a737 (2022)
62. Chathuranga, S., Jayasinghe, S., Antucheviciene, J., Wickramarachchi, R., Udayanga, N., Weerakkody, W.A.S.: Practices driving the adoption of agile project management methodologies in the design stage of building construction projects. Buildings **13**(4), 1079 (2023)
63. Maqbool, R., Rashid, Y., Altuwaim, A., Shafiq, M.T., Oldfield, L.: Coping with skill shortage within the UK construction industry: scaling up training and development systems. Ain Shams Eng. J. (2023). https://doi.org/10.1016/j.asej.2023.102396
64. Isang, I.W., Ebiloma, D.O.: Challenges and strategies for sustainable project performance in the post-COVID era in Nigeria: a thematic analysis approach. Front. Eng. Built Environ. **3**(1), 32–47 (2023)
65. Sojobi, A.O., Liew, K.M.: Multi-objective optimisation of high-performance bio-inspired prefabricated composites for sustainable and resilient construction. Compos. Struct.Struct. **279**, 114732 (2022)
66. Ahmed, T., Karmaker, C.L., Nasir, S.B., Moktadir, M.A., Paul, S.K.: Modelling the artificial intelligence-based imperatives of industry 5.0 towards resilient supply chains: a post-COVID-19 pandemic perspective. Comput. Ind. Eng. 177, 109055 (2023)

67. Radzi, A.R., Rahman, R.A., Almutairi, S.: Modelling COVID-19 impacts and response strategies in the construction industry: PLS-SEM approach. Int. J. Environ. Res. Public Health **19**(9), 5326 (2022)
68. Gartoumi, K.I., Aboussaleh, M., Zaki, S.: Building information modelling a key for construction industry recovery post-Covid-19. Proc. Inst. Civ. Eng.-Eng. Sustain. **176**(2), 82–93 (2023)
69. Wang, T., Chen, H.M.: Integration of building information modelling and project management in construction project life cycle. Autom. Constr.. Constr. **150**, 104832 (2023)
70. Li, L., Luan, H., Yin, X., Dou, Y., Yuan, M., Li, Z.: Understanding sustainability in off-site construction management: state of the art and future directions. J. Constr. Eng. Manag. **148**(11) (2022). https://doi.org/10.1061/(ASCE)CO.1943-7862.0002396
71. Rocha, P.F., Ferreira, N.O., Pimenta, F., Pereira, N.B.: Impacts of prefabrication in the building construction industry. Encyclopedia **3**(1), 28–45 (2023)
72. Subramanian, S.L., Jadhav, N., Raajkumar, L., Goyal, S.: Application of IoT at different stages of construction and adaptability for the Indian market. NICMAR J. Constr. Manag. **37**(4), 116–125 (2022)
73. Amankwah-Amoah, J., Khan, Z., Wood, G., Knight, G.: COVID-19 and digitalisation: the great acceleration. J. Bus. Res. **136**, 602–611 (2021)
74. Nguyen, Q.C., Perera, S., Ginigaddara, B., Nguyen, D.T.M., Rahmawati, R., Operio, J.H., Nguyen, D.H.T.: An evaluation of offsite construction recoveries after the pandemic: the case of the southeast Asian region. Buildings **13**, 50 (2023)
75. Zhu, H., Hwang, B-G., Ngo, J., Tan, J.P.S.: Applications of smart technologies in construction project management. J. Constr. Eng. Manag. **148**(4) (2022). https://doi.org/10.1061/(ASCE)CO.1943-7862.0002260.
76. Brennan, T.M., Jesson, J.E., Deese, A.S., Rodriguez, E., Bechtel, A.J.: Internet of things application for smart construction assets. Smart Infrastruct. Constr. **174**(3), 79–87 (2021)
77. El Khatib, M., Al Hashimi, S., Al Ketbi, M.: Impact of remote work on project management cases from UAE construction sector. J. Serv. Sci. Manag.Manag. **16**(1), 30–43 (2023)
78. Osorio-Gómez, C.C., Herrera, R.F., Prieto-Osorio, J.M., Pellicer, E.: Conceptual model for implementation of digital transformation and organizational structure in the construction sector. Ain Shams Eng. J. **15**(7), 102749 (2024)
79. Chen, C.: Advantages and barriers of modular construction method in constructing buildings. Smart Infrastruct. Constr. **176**(2), 75–84 (2023)
80. Yu, M., Wang, N.: The influence of geopolitical risk on international direct investment and its countermeasures. Sustainability **15**(3), 2522 (2023)
81. Seuring, S., Brandenburg, M., Sauer, P.C., Schünemann, D.S., Warasthe, R., Aman, S., Qian, C., Petjak, K., Neutzling, D.M., Land, A., Khalid, R.U.: Comparing regions globally: impacts of COVID-19 on supply chains—a Delphi study. Int. J. Oper. Prod. Manag.Oper. Prod. Manag. **42**(8), 1077–1108 (2022)
82. Adhi, A.B., Muslim, F.: Development of stakeholder engagement strategies to improve sustainable construction implementation based on lean construction principles in Indonesia. Sustainability **15**(7), 6053 (2023)
83. Langston, C., Crowley, C.: Fiscal success: creating quality infrastructure in a Post-COVID world. Sustainability **14**(3), 1642 (2022)
84. O'Brien, T., Zargar, K., Clancy, N., McLaren, B.: Navigating supply chain challenges in Saudi Arabia's construction industry. Glob. Arbitr. Rev. **2024**, 1–18 (2024)
85. Topalidis, G.T., Kartalis, N.N., Velentzas, J.R., Sidiropoulou, C.G.: New developments in geopolitics: a reassessment of theories after 2023. Social Sciences **13**, 109 (2024)
86. Xia, X., Wu, X., BalaMurugan, S., Karuppiah, M.: Effect of environmental and social responsibility in energy-efficient management models for smart cities infrastructure. Sustain. Energy Technol. Assess. **47**, 101525 (2021)
87. Fallah Shayan, N., Mohabbati-Kalejahi, N., Alavi, S., Zahed, M.A.: Sustainable development goals (SDGs) as a framework for corporate social responsibility (CSR). Sustainability **14**(3), 1222 (2022)

88. Afieroho, U.E., Li, Y., Han, Y., Soomro, M.A., Radujkovic, M.: Transformational community engagement in urban infrastructure public-private partnerships: a governmentality approach to create social value. Buildings **13**(5), 1225 (2023)
89. Alraja, M.N., Imran, R., Khashab, B.M., Shah, M.: Technological innovation, sustainable green practices, and SMEs sustainable performance in times of crisis (COVID-19 pandemic). Inf. Syst. Front. **24**, 1081–1105 (2022)
90. Ayat, M., Qureshi, S.M., Kang, C.: An enhanced framework for reliably managing private participation in infrastructure ICT projects in developing countries. J. Eng. Des. Technol. **22**(2), 480–500 (2024)
91. Duflot, R., Baumeister, S., Burgas, D., Eyvindson, K., Trivino, M., Blattert, C., Kuparinen, A., Potterf, M.: Building up an ecologically sustainable and socially desirable post-COVID-19 future. Sustain. Sci. **16**, 1397–1403 (2021)
92. Jalil, F., Yang, J., Rehman, S.U., Khan, M.M.: Post-COVID-19's impact on green supply chain management and sustainable e-commerce performance: the moderating role of big data analytics. Environ. Sci. Pollut. Res.Pollut. Res. (2023). https://doi.org/10.1007/s11356-023-30581-x
93. Elavarasan, R.M., Pugazhendhi, R., Shafiullah, G.M., Kumar, N.M., Arif, M.T., Jamal, T., Chopra, S.S., Dyduch, J.: Impacts of COVID-19 on sustainable development goals and effective approaches to manoeuvre them in the post-pandemic environment. Environ. Sci. Pollut. Res.Pollut. Res. **29**, 33957–33987 (2022)
94. Kumar, S., Shukla, A., Garg, V., Borhade, R., Vankhande, V., Mathur, A.: ITC green centre: a case study of a green Building in India. NICMAR J. Constr. Manag. **37**(4), 270–278 (2022)
95. Cortes, D., Traxler, A.A., Greiling, D.: Sustainability reporting in the construction industry—status quo and directions of future research. Heliyon **9**(11), e21682 (2023)
96. Zhang, R., Tang, Y., Liu, G., Wang, Z., Zhang, Y.: Unveiling the impact of innovation on competitiveness among construction projects: moderating and mediating role of environmental regulation and sustainability. Ain Shams Eng. J. **15**(3), 102558 (2024)

# Fostering Supply Chain Resilience in VUCA World with Blockchain Technology: A Bibliometric and Systematic Review

**Thi Huong Tran, Thanh Tuyen Nguyen, Sebastian Kummer, and Chi Cuong Hoang**

**Abstract** Supply chains today are facing many challenges from a volatile, uncertain, complex, and ambiguous business environment (VUCA world). As a result, enhancing supply chain resilience has become a prevalent topic in both industry and academia. Practitioners and researchers have proposed promising solutions to achieve resilience capability, including agility, visibility, flexibility, collaboration, information sharing, and many emerging technologies. Blockchain technology is considered a silver bullet for building a transparent, traceable, and resilient supply chain. This study aims to synthesize overall aspects of blockchain applications in different fields of supply chain management to enhance resilience capability. To this end, we use a bibliometric and systematic review to investigate 117 high-quality research articles collected from the Scopus database. The bibliometric analysis explores the global research landscape with trends of research papers over time, the most influential authors, and interactions between various research streams. The systematic review provides a comprehensive overview of developed frameworks and roadmaps for a blockchain-enabled supply chain in different sectors, as well as critical factors, challenges, and strategies to take advantage of blockchain in proactively and reactively facing disruptions. In addition, this study explores research gaps and provides an outlook for future research in the field of fostering supply chain resilience with blockchain technology.

**Keywords** Blockchain technology · Supply chain resilience · VUCA world · Bibliometric analysis · Systematic review

T. H. Tran
Hanoi University of Science and Technology, Hanoi, Vietnam

T. T. Nguyen · S. Kummer
Vienna University of Economics and Business, Vienna, Austria

C. C. Hoang (✉)
Van Lang University, Ho Chi Minh City, Vietnam
e-mail: cuong.hc@vlu.edu.vn

 
N. D. K. Nguyen et al. (eds.), *Digital Transformation and Sustainability: Methods and Applications*, Information Systems Engineering and Management 47,
https://doi.org/10.1007/978-3-031-91278-8_5

# 1 Introduction

Supply chain disruptions stemming from unexpected events such as natural disasters, cyber-attacks, and pandemics have highlighted the crucial need for resilience [3, 37]. Industry 4.0 technologies, including blockchain, artificial intelligence, cloud computing, and big data analytics, have emerged as vital tools for enhancing supply chain resilience [76, 96]. However, the adoption of technologies has also introduced new cyber risks, with increased opportunities for hackers, as seen in high-profile cases like the Toll Group ransomware attack.

Blockchain technology (BCT) addresses key cybersecurity concerns by providing decentralized, tamperproof records of transactions, thereby reducing vulnerability to cyber-attacks [3, 74]. Its cryptographic, transparent, and immutable nature ensures enhanced data security, transparency, and traceability across supply chains. Blockchain's decentralized structure eliminates single points of failure and offers a secure means of tracking transactions. BCT also supports private and restricted memberships, further improving supply chain security [57, 65].

A growing body of research underscores BCT's role in bolstering supply chain resilience (SCR) [17]. However, existing studies often focus narrowly on specific applications of blockchain [66], limiting a comprehensive understanding of its overall impact on SCR [74]. Currently, no literature review has thoroughly examined the multifaceted applications of blockchain in fostering SCR [36]. Given the recognized importance of blockchain in contemporary supply chains [37], a systematic literature review (SLR) is necessary to synthesize and analyse the relevant literature cohesively.

Therefore, our study seeks to undertake an SLR to synthesize the existing knowledge of BCT and SCR and offers a comprehensive direction for future research using a new framework based on SCR stages and bibliometric network analysis. We articulate the following research questions:

i. What major themes emerge in the application of blockchain for SCR?
ii. What knowledge gaps exist in the current literature, and what future research opportunities can enhance blockchain understanding?

The rest of the study is structured as follows: Section 2 overviews the related works. Section 3 outlines the methodology employed for the literature review. Section 4 provides findings from bibliometric analysis and network analysis of the reviewed papers. Section 5 presents the results of systematic content analysis, and Sect. 6 concludes the chapter by summarizing our research contributions.

# 2 Related Works

The prevalence of SCR and BCT has led to a considerable amount of empirical research, followed by a not small number of literature review papers on this topic. This section summarizes the recent literature review articles applying BCT to foster

SCR in a risky business environment. Choi et al. [12] examined previous research on the benefits of BCT to supply chain risk analysis in the air logistics sector. Collart and Canales [16] reviewed the application of blockchain and how it can enhance the SCR in the US fresh produce industry. Bibliometric analysis has been frequently used to survey literature, such as the research of Kumar et al. [42] and Etemadi et al. [23]. Beck et al. [7] and Singh et al. [78] carried out systematic reviews to explore major themes and how blockchain technology can deal with challenges in SCR. The predominant literature reviews focused on agriculture or food supply chain and leaned forward the bibliometric analysis. There is a lack of systematic literature review investigating various types of blockchain-enabled supply chains and utilizing both bibliometric and systematic analysis. The study aims to fill the research gap by employing descriptive and network analysis along with systematic content analysis, revealing publication trends, top contributing journals, and theoretical perspectives while highlighting knowledge themes such as implementation framework, the impact mechanism of BCT on SCR, critical factors and barriers in adopting BCT, and existing research gaps for future investigation.

# 3 Methodology

This study employed a combined methodology of SLR and bibliometric network analysis. SLR is well-established for providing a comprehensive and unbiased understanding of a given research field, offering deep insights into past and current research efforts [20, 51]. This enables researchers to identify promising future research directions and develop conceptual frameworks based on previous findings. Meanwhile, the bibliometric approach serves as a valuable complement, providing information on a wide range of topics defined by enormous amounts of bibliometric and bibliographic data [50]. A structured procedure for retrieving and selecting reviewed articles was followed, adhering to the guidelines outlined by Tranfield et al. [81], which was summary in Fig. 1.

## *3.1 Database Selection*

The SCOPUS database was selected to identify and shortlist relevant research articles due to its extensive coverage, encompassing articles from a wide range of reputable publishers, including IEEE, Elsevier, Emerald Insights, Wiley, Springer, Taylor & Francis, and many others [77]. In addition, while Web of Science covers 22,209 journals, books, and conference proceedings [14], SCOPUS boasts a broader scope with 27,950 active peer-reviewed journals [22].

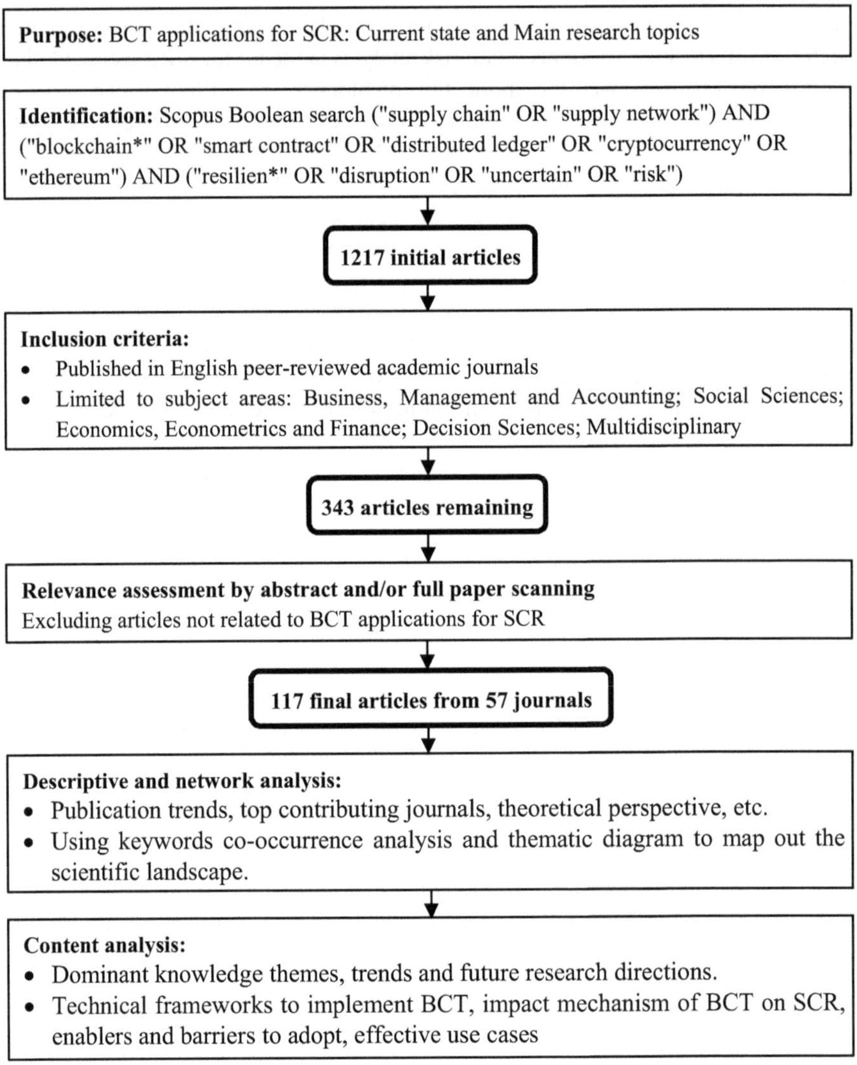


**Fig. 1** Systematic literature review and bibliometric network analysis methodology. *Source* Authors' own work

## 3.2 Keywords Selection

This study employed a three-level keyword strategy to refine search strings and select articles from Boolean search engines. An 'AND' connector was used within each level, while an 'OR' connector separated between levels. The first level included the terms "supply chain" and "supply network" to focus on articles within the supply chain management domain. At the second level, "blockchain*" and related terms, like "smart contract", "distributed ledger", "cryptocurrency", "Ethereum", "blockchain", "block chain" was used to identify articles discussing blockchain applications. The third level included "resilien*" relevant terms, such as "disruption" and "uncertain" and "risk". This search string was used to locate articles that matched the criteria within title, abstract, and keyword list.

### 3.3 Article Include and Exclude Criteria

The dataset development process involved several inclusion criteria. First, only English language papers published up to May 2024 were considered. Second, the scope was limited to peer-reviewed journal articles, excluding book chapters, conference proceedings, and working papers. Third, given the vast amount of literature on the topic, only research papers published in relevant subject areas, such as Business, Management and Accounting; Social Sciences; Economics, Econometrics and Finance; Decision Sciences; and Multidisciplinary, were included. Authors evaluated each potential data sample. During this process, nonrelevant papers were excluded based on subjective criteria selection: the abstract and/or full text of each paper was examined to assess its explicit focus on blockchain applications and resilience capability in the context of supply chain management. The final dataset used for further analysis includes 117 research papers from 57 journals.

### 3.4 Bibliometric Network Analysis

This study employs thematic diagram to map out the scientific landscape. Thematic diagram visually depicts studied topics, categorizing them based on their centrality (relevance degree) and density (development degree) [68]. Centrality indicates a cluster's connection to others, while density reflects internal term connections [15]. A central cluster is often a research focal point, signifying relevance and impact. Cluster positioning indicates strategic significance, reflecting both internal coherence and overall field relevance [44].

### 3.5 Systematic Content Synthesis

In addition to bibliometric analysis, the authors conducted systematic content analysis based on steps for SLR proposed by Tranfield et al. [81]. We coded the collected research papers by the research objective (main focus), technical frameworks to implement BCT, impact mechanism of BCT on SCR, enablers/critical factors and barriers/challenges/risks to adopting BCT in the supply chain, effective use cases where BCT comes into play to enhance SCR.

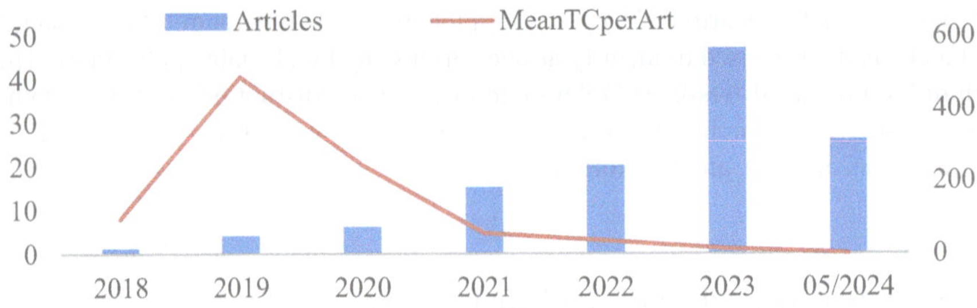


**Fig. 2** Distribution of reviewed articles by year (2018–05/2024). *Source* Authors' own work

# 4 Bibliometric Analysis

## *4.1 Publication Trends*

Figure 2 illustrates the number of publications on BCT and SCR. In particular, all publications in this domain were published between 2018 and May 2024, with the majority appearing in 2021 and 2023. This reflects its recent emergence as a popular research area. Notably, research output has accelerated significantly since 2021, increasing by 150% to reach a peak of 47 articles in 2023. The COVID-19 pandemic may have contributed to this surge by accelerating the advancement of Industry 4.0 and increasing the demand for resilience capacity [70]. In terms of mean citations per article, publications in 2019 had the highest number at 488, followed by 2020 at 240.

## *4.2 Top Contributing Journals*

A total of 117 articles were published across 57 journals. Figure 3 highlights the leading journals in the field of BCT and SCR. Notably, 33% of the identified journals published more than one paper in our sample. Among these, Annals of Operations Research and Sustainability (Switzerland) published the most articles from the dataset, with 13 papers each. Transportation Research Part E: Logistics and Transportation Review and International Journal of Production Research followed with six papers each. Operations Management Research and International Journal of Production Economics shared third place as the most prominent journals in the sample. These top 10 journals collectively contributed to 52% of the total articles considered in this study.

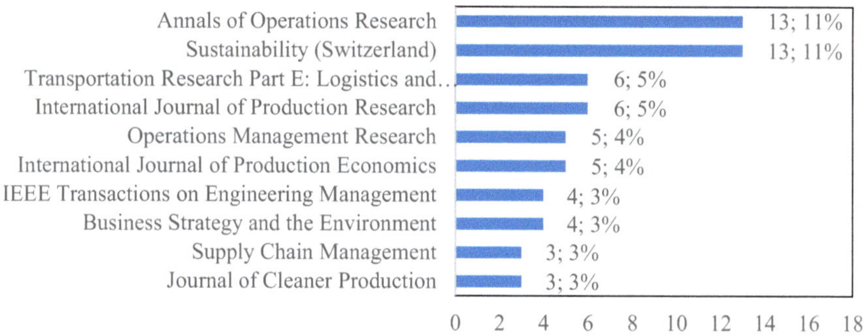


**Fig. 3** Distribution of reviewed articles by top journal. *Source* Authors' own work

## 4.3 Most Influential Authors

A total of 416 authors have contributed to the research field of BCT applications on SCR. Among these, the top 10 most impactful authors are presented in Fig. 4 with their number of articles and citations based on year. Kumar Anil leads with six publications, three in 2021 and three in 2023. While Kumar Anil and Dora Manoj have 6 and 2 publications, respectively, the remaining eight authors have three each. However, Queiroz Maciel stands out with an impressive 941 citations, accounting for nearly a third of the total citations among the top 10 authors. This is evident in Fig. 4, which shows a high number of citations for Queiroz Maciel's publications in 2019, 2021, and 2022, marked with bright blue nodes. His significant contributions to BCT applications in SCR, particularly his 2019 paper on blockchain adoption challenges and drivers in India and the USA supply chain [66], have garnered substantial attention. Together, these 10 authors account for 26% of all articles and 36% of all citations in the fields of BCT and SCR.

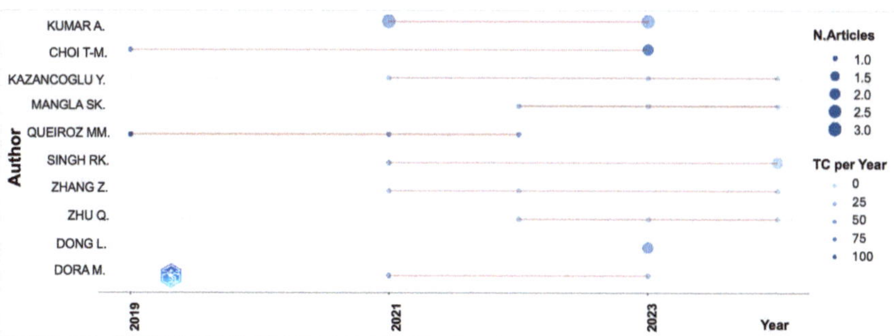


**Fig. 4** Annual publications of the top productive authors over time. *Source* Biblioshiny

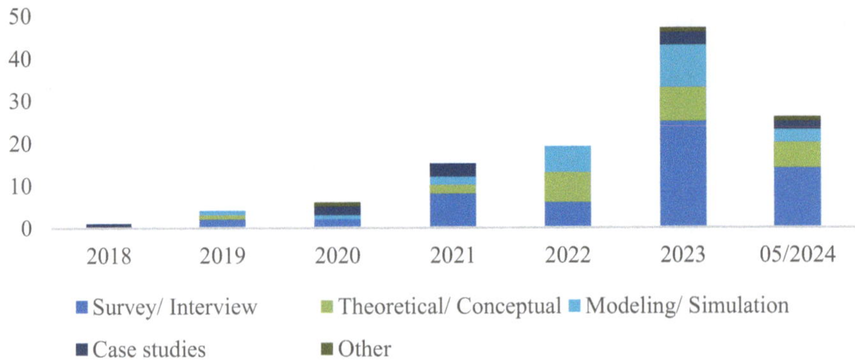


**Fig. 5** Distribution of reviewed articles over time by methodologies. *Source* Authors' own work

## 4.4 Research Methodologies

Seuring and Müller [72] identified survey/interview, theoretical and conceptual, modeling/simulation, and case studies as the predominant research approaches in operations and supply chain management literature. Figure 5 categorizes the 117 reviewed BCT and SCR articles, revealing a dominance of survey/interview methods (48%), followed by theoretical and conceptual (20%), modeling/simulation (20%), and case studies (9%). Since 2021, survey/interview methods have consistently been the most popular. Additionally, there has been a growing trend in the use of theoretical/conceptual and modeling/simulation approaches since 2019. In contrast, the number of case study papers has fluctuated between one and three during the past seven years.

## 4.5 Research Theories

Following [43], a comprehensive review of the theories used to explore BCT applications in SCR reveals a wide range of technology adoption and organizational capacity theories. The list of theories appearing at least twice in the analyzed data is shown in Fig. 6. Among these, Dynamic Capability (DC) and Resource-based view (RBV) emerged as the most frequently employed. DC, emphasizing dynamic organization's capacities to adapt to disruptions, aligns well with SCR [4, 58]. Therefore, DC is considered the most suitable theory for explaining the relationship between BCT applications and resilience development [27, 56]. RBV suggests that organizations can enhance resilience by increasing process transparency, information symmetry, risk management, collaboration, flexibility, and financial and information reliability, all of which can be facilitated by Blockchain [29, 45, 58]. Other theories, such as

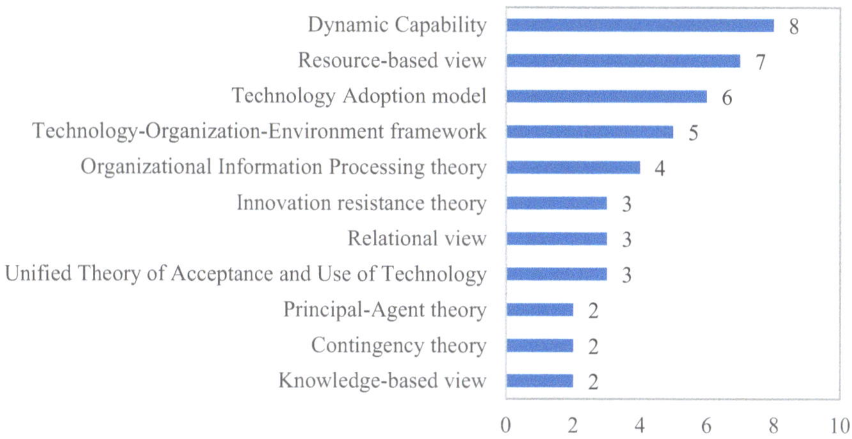


**Fig. 6** Distribution of reviewed articles by top theories. *Source* Authors' own work

Technology Adoption Model (TAM), Technology-Organization-Environment framework (TOE), and Organizational Information Processing theory (OIPT), offer valuable insights into other dimensions of BCT. TAM highlights perceived utility and ease of use as key factors influencing blockchain adoption [41, 66, 86]. TOE provides a comprehensive perspective on adoption factors, processes, implementation, and post-adoption of BCT in SCR [8, 30]. OIPT explains how information is effectively used for competitive advantage, particularly in uncertain environments. Therefore, BCT can transform the supply chain by enhancing transparency, leadership, and trust [19, 47]. The remaining theories, such as Knowledge-based View, Contingency theory, and Principal-Agent theory… have not yet been extensively explored in the context of BCT and SCR, may offer opportunities for future research.

## *4.6 Network Analysis*

Figure 7 presents a comprehensive analysis of research topics within the domain of BCT application in SCR through the strategic diagram, which illustrates the significance of each topic based on its density and centrality metrics. Six major study topics were defined and scattered throughout the four quadrants, including 'blockchain', 'sustainability', 'risk assessment', 'supply chain finance', 'digital supply chain' and 'distributed ledger'.

'*Blockchain*' emerged as the top theme in terms of centrality. This basic theme includes keywords such as supply chains, resilience, risk management, covid19, disruption, barriers, and so on, reflecting blockchain's role in developing and enhancing supply chain risk management and resilient [27, 53, 55]. Manupati et al. [53] developed a genetic algorithm model to predict and manage disruptions in smart contract environments, focusing on reactive measures for post-disruption scenarios.

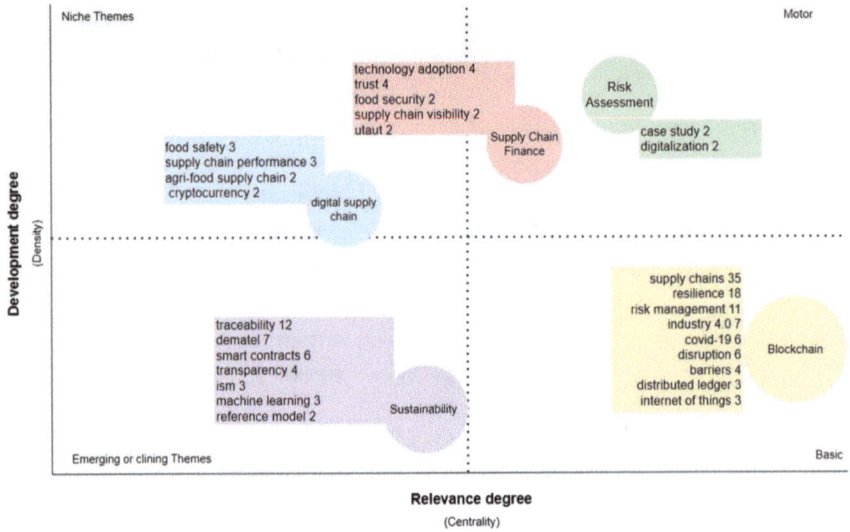


**Fig. 7** Thematic diagram. *Source* Biblioshiny

Yadav and Singh [93] proposed a mixed-integer nonlinear programming model integrated with machine learning to optimize digital procurement costs for supplier selection while ensuring transparency, traceability, security, and complete information. Lohmer et al. [48] explored potential risk-related blockchain application scenarios and their impact on existing supply chain resilience strategies using an agent-based simulation model. Other empirical studies investigate relationships between BCT, SCR and others factors like Artificial intelligence under the influence of environmental dynamism Gupta et al. [27], operational supply chain transparency and swift-trust [19], supply chain leadership [47], business model design [45], and so on.

'*Supply chain finance*', which includes keywords such as technology adoption, trust, smart contract, food security, supply chain visibility, UTAUT, is a motor theme. Blockchain technology, through ancillary smart contracts and decentralized applications, offers trust, efficiency, transparency, and secure financing services to buyers, suppliers, and financial institutions within the supply chain finance ecosystem [69, 82]. Blockchain can also mitigate financial institutions' credit risk and address financing challenges by reducing operational risk and improving business income, creating a strict regulatory environment that discourages default behavior among enterprises [80]. However, blockchain-enabled delegate financing increases risk-mitigation investments and benefits all three tiers of the supply chain only when the tier-2 supplier faces severe capital constraints with working capital below a threshold [18].

'*Risk assessment*' is another motor phrase, encompassing terms such as case study and digitalization. Research in this theme focuses on assessing the risks of blockchain adoption [12, 21, 39], including underdevelopment and unreadiness in terms of digitalization and standardization, as well as friction between industry's trust-based business models and blockchain's decentralized and transparent economic model [61]. Ekinci et al. [21] found that enterprises with well-established, longterm partnerships are more likely to adopt and leverage blockchain technologies. Klöckner et al. [39] highlighted the importance of data integrity, noting that data fed into the blockchain must be valid and synchronized with the actual physical state. Additionally, technological approaches linking products to their blockchain entries need further development and adoption.

'*Digital supply chain*', a niche theme, encompasses subthemes such as food safety, supply chain performance, agri-food supply chain, and cryptocurrency. This theme focuses on integrating blockchain with other digital technologies in agri-food supply chains to enhance supply chain performance [32, 73, 85]. Industry 4.0 technologies like blockchain and IIoT strengthen linkages among supply chain partners, leading to reduced risks, in-creased benefits, and improved supply chain performance [97]. Iftekhar and Cui [31] propose a tamper-proof tracing strategy using IoT and blockchain to ensure reliable traceability, promote authenticity, establish a new industry trust mechanism, improve corporate responsibility, and eliminate tampering practices throughout the supply chain, thereby protecting consumer safety and health. In addition, integrating ERP and blockchain technologies can benefit companies by enhancing sustainable corporate performance and facilitating successful supply chain integration [79].

The emerging theme such as '*sustainability*' encompass subthemes such as traceability, DEMATEL, transparency, ISM, machine learning, and reference model. Transparency and traceability are crucial for creating sustainable and resilient supply chains and effective corporate governance [2, 54]. Kamble et al. [35] found traceability to be the most significant reason for BCT implementation, followed by audibility, immutability, and provenance. McGrath et al. [54] demonstrate the effective application of control and relational orientations toward BCT use by multinational corporations (MNCs) in creating sustainability transparency and providing managerial guidance on design choices. Moreover, multi-criteria decision-making methodologies like DEMATEL and ISM are preferred for identifying top barriers to BCT adoption in sustainable supply chains and their relationships. Singh et al. [77] highlighted market-based risks, high sustainability costs, and underground economy use as the most influential barriers, while inadequate knowledge/employee training, contractual risk, and scalability issues are also pivotal. Gupta et al. [26] identified organizational barriers as the most crucial category for BCT adoption in secured and carbon–neutral logistics operations, followed by environmental barriers. Nisar et al. [62] revealed that Regulatory compliance, implementation costs, and complex supply chain networks were identified as primary influencing factors demanding attention for effective BCT integration in the fish supply chain.

# 5 Systematic Content Synthesis of Research on BCT Adoption Towards SCR

Figure 8 presents codes and findings from systematic content analysis of reviewed papers with main focuses (research objectives), technical frameworks to implement BCT in terms of independently or integrated with other technologies or management practices, advantages of BCT deployment, impact mechanism of BCT on SCR, enablers/critical factors and barriers/challenges/risks to adopting BCT in the supply chain, suitable use cases where BCT should be adopted to come into play in enhancing SCR.

## *5.1 Major Research Objectives*

There are six main focuses of studies in BCT-SCR literature.

- Group 1: Building technical framework/architecture to implement BCT towards SCR. Min [55] designed a detailed architecture to revolutionize the supply chain to enhance visibility and security with smart contract, asset tracking, error-free order fulfillment, and cyber security. Tsai and Shen [82] implemented a blockchain prototype using Hyperledger Fabric and Composer to tackle security risk. The technical frameworks and architectures have designed and tried in various context, such as floral supply chain [82], cold chain [31], agricultural supply chain [57], textile [28] and so on.

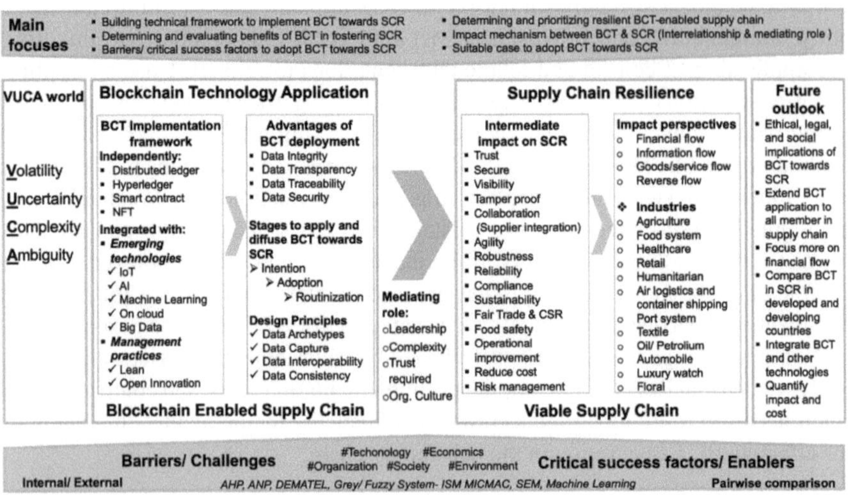


**Fig. 8** Systematic content synthesis of research on BCT adoption towards SCR. *Source* Authors' own work

- Group 2: Research puts effort into determining and evaluating the benefit of BCT in fostering SCR. Both qualitative and quantitative approaches are implemented to analyze the benefits of BCT. Wang et al. [87] interviewed 14 supply chain experts to investigate how BCT can transform the supply chain and build resilience capability. Cao et al. [9] and Zhu et al. [98] built mathematical models and simulations to quantify the benefit of BCT in reducing risk and enhancing SCR. Sathiya et al. [71] used DEMATEL technique to analyze benefits of BCT in the healthcare supply chain.
- Group 3: Many studies defined and analyzed barriers/critical success factors to adopt BCT towards SCR. Barriers and risks to adoption of BCT in supply chain have received much attention from researchers. Section 5.4 will analyze this topic in more detail.
- Group 4: There are several papers in the literature determined and prioritized resilient BCT-enabled supply chains in general [48] and food supply chains [74].
- Group 5: Exploration of impact mechanism between BCT and SCR. The interrelationship and mediating variables between BCT application and SCR with simulation and Structural Equation Model will be analyzed in Sect. 5.3.
- Group 6: Suitable application cases to adopt BCT toward SCR. Lohmer et al. [48] and Wolf et al. [88] stated that a long and complex supply chain with a high value of trust and automation required would be the cases BCT can come into play.

### *5.2 Technical Architecture and Framework to Implement BCT Towards SCR*

In line with the VUCA world and the emergence of blockchain, practitioners and researchers have proposed and prototyped frameworks and architectures to deploy BCT towards SCR in an independent manner with Distributed Ledger, Hyperledger, Smart contract, NFT as well as integration with other emerging technologies and management practice, for example, IoT [33, 82], AI [92], Big Data [28, 59], Machine Learning and Cloud Computing [89], Lean manufacturing [11], and Open Innovation [49].

The integration with these advanced digital supply chain technologies further strengthens the benefits of blockchain. ‘Machine learning’ enables blockchain to sense the surrounding environment, leading to actionable insights and interoperable solutions [27, 93]. A combination of blockchain and IoT technologies can act as a subsystem supporting realtime data sharing among supply chain partners, enhancing their ability to integrate and coordinate business processes to meet customer needs [92, 97]. Moreover, big data helps supply chains overcome scalability challenges often faced by blockchain, enabling realtime, automatic, scalable traceability and information sharing, and facilitating quick and reliable responses [34].

Outstanding advantages of BCT, such as data integrity, data transparency, data traceability, data security are demonstrated to facilitate resilient supply chains in almost reviewed papers. Wamba and Queiroz [86] suggested the stages to apply and diffuse BCT towards SCR, including three phases: intention, adoption, and routinization. Data archetypes, data capture, data interoperability, and data consistency are the principles that should be followed to design a resilient BCT-enabled supply chain [84].

## 5.3 Impact Mechanism of BCT on SCR

The adoption of BCT can foster SCR through intermediate impacts such as increasing trust among supply chain members, ensuring security and visibility, and tamper-proof in supply chain operations. As a result, collaboration and supplier integration are also easier to achieve, and supply chains have become more agile, robust, reliable, and sustainable.

Specifically, blockchain facilitates the identification and authentication of suppliers and distributors, fostering 'transparency' by securely sharing information across the supply chain [5, 69]. Additionally, blockchain-enabled internal integration streamlines operations, improving 'sustainability' through enhanced resource tracking, waste reduction, and optimized utilization. By monitoring inventory levels and demand trends, organizations can minimize overstocking and stockouts, leading to better resource allocation and reduced environmental impact [21, 63, 79]. 'Smart contracts' a key feature of blockchain, facilitate the efficient, safe, and transparent management of information, materials, and money, eliminating intermediaries [19, 25, 87], creating an immutable, transparent, and traceable ecosystem [27].

Other demonstrated benefits to SCR are compliance, fair trade, corporate social responsibility, food safety, operational improvement, reduced cost, and risk management. Blockchain technology enhances supply chain resilience by addressing risks, improving performance, and managing disruptions. The core functionalities of 'blockchain', including heightened transparency, traceability, and immutability, offer significant advantages [29]. By leveraging these capabilities, supply chains can address a wide range of critical 'risk management' challenges, such as cybersecurity, prevent fraud, detect counterfeit products, ensure tamper-proof data records, foster trust, enable advanced environmental monitoring, and facilitate dispute resolution [6, 61, 67]. Beyond risk management, blockchain also facilitates trust, collaboration, cost reduction, and increased transparency and traceability among supply chain partners, contributing to improved 'supply chain performance' [25, 64, 97].

In recent research, the role of BCT towards SCR has been investigated in almost all supply chains, including agriculture, fisheries, food systems, floral, healthcare, retail, air logistics and container shipping, port systems, textile, oil/petroleum, automobile, luxury watch, humanitarian supply chain. BCT not only impacts on goods and service flow but also financial flow, information flow and reverse flow.

The effect of BCT in SCR can be mediated by leadership [47], type of disruption, type of supply chain, complexity level, and trust requirements [48, 88].

## 5.4 Barriers and Enablers to Adopt BCT Towards SCR

Many researchers have investigated critical factors, enablers, and barriers to adopting BCT in supply chain management. The barriers can come from internal or external sites and can be classified into five groups as follows:

- Technology (immature, early stage of development, scalability issues, lack of standardization and interoperability, criminal activity, malicious attacks, poor user experience)
- Economics (high investment cost)
- Organization (lack of readiness, lack of exemplary use case, lack of knowledge/employee training, hard to bringing all party/supply chain members together, culture of an organization)
- Society (lack of trust in new technology, privacy and information disclosure issues, poor clarity regulatory provisions)
- Environment (Wasted resources or high energy consumption).

Together with expert interviews and qualitative analysis, semi-quantitative multi-criteria decision-making techniques like AHP, ANP, DEMATEL, Grey/Fuzzy System-ISM MICMAC, as well as Structural Equation Model, and Machine Learning were used to establish hierarchical levels and relationships among blockchain adoption enablers and barriers [35]. Previous research has identified traceability as the most significant reason for BCT implementation, followed by audibility, immutability, and provenance [35]. Four key enabler groups for successful blockchain-enabled tracing systems and new business model implementation include well-established technology standards, ease of adoption, risk management facilitation, and dependability and trust assurance [40]. However, market-based risks, high sustainability costs, and underground economy usage are major barriers to BCT adoption. Additionally, inadequate knowledge/employee training, contractual risk, and scalability issues play pivotal roles [77].

## 5.5 Research Gaps

While blockchain technology has been explored in supply network applications for some time, its potential to enhance supply chain resilience remains a promising area for future research. Among the 117 high-ranking articles included in this systematic review, many, such as Gupta et al. [27], Pattanayak et al. [64], Queiroz and Wamba [66]... called for longitudinal studies and broader applications across diverse contexts and industries, highlighting the need for more mature research in this field. Moreover, although this systematic review paper has identified various previous research themes on BCT's application to SCR, including implementation frameworks, advantages, strategic and design principles for deployment, and intermedia impact on SCR, the field is still nascent (Fig. 8). There are few established starting points for new research can be followed:

- The dark side of blockchain applications, such as legal, ethical, security, and social implications, particularly regarding supply chain governance, need to be explored more [1, 77, 87]. Legal ambiguities, ethical dilemmas, and security vulnerabilities pose significant challenges that must be addressed. As this technology continues to evolve, a comprehensive understanding of these issues is crucial. By proactively identifying and mitigating risks, future research can help harness the power of blockchain.
- Extending blockchain applications to all members in the supply chain and cross countries instead of a specific company or specific nation is strongly recommended [40, 46, 90]. While current applications have focused on specific companies or nations, the true power of blockchain lies in its ability to connect and secure the entire supply chain, irrespective of geographic boundaries. By focusing on researching the extension of blockchain's reach to all participants, from raw material suppliers to end consumers, future studies can contribute to the development of a blockchain-based global supply network.
- The contextual factors, including competitive pressures, environmental uncertainty, government intervention [4, 10, 19, 55], organizational culture [60, 94], can significantly impact the relationship between BCT adoption and SCR outcomes. Future research should delve deeper into the interplay between these contextual factors, BCT implementation, and SCR. Possible research questions can be: How do peer pressures within an industry influence the adoption of BCT for SCR? What role does government intervention play in shaping the adoption and success of BCT applications for SCR?
- Many previous researches had called for future avenues on the integration of BCT with other advanced digital technologies, such as business analytics [24, 55], IoT, AI and cloud computing [2, 27, 47, 59, 65], and machine vision, sensors, robotics [25, 52, 83]. By leveraging the synergy of these technologies, organizations can achieve greater transparency, traceability, and automation in their supply chains. This, in turn, can lead to improved risk management, faster response times to disruptions, and enhanced overall resilience.
- The new constructs and variables from emerging theories, like network theory, contingency theory, institution theory, management theories, relational view, and information governance theoretical relationships, also offer a variety of promising research directions [13, 64, 73, 98]. And, to gain deeper insights into the complex interplay between enablers, barriers, and strategies in BCT adoption, the use of emerging multi-criteria decision-making (MCDM) methodologies, such as VIKOR, TOPSIS, ANP [38, 52, 57, 73], PROMETHEE, ELECTRE [75, 95] are recommended.
- Future research should consider the unique needs of small and medium-sized eterprises (SMEs) [32, 59, 91]. As the backbone of many economies, SMEs often face resource constraints and vulnerability to disruptions. Future studies should explore how blockchain can be tailored to SMEs' specific challenges, such as

developing cost-effective implementation strategies, mitigating security risks, and fostering collaboration with larger partners.

Based on our findings, numerous avenues remain for researchers to explore. This includes technology-focused research on scalability and interoperability, as well as sociotechnical research on the development and societal implications of blockchain for enhancing individual, organizational, and societal resilience. While much of the current research concentrates on technical advancements, future research should also delve into models, architectures, societal impacts, and potential disruptions to existing business models arising from proposed blockchain use cases.

## 6 Conclusion

Our study systematically reviewed and synthesized research focused on fostering SCR with BCT in the VUCA world by examining the evolution of this research between 2018 and 2024. Through bibliometric analysis of 117 papers, distribution of research by year, top journals, top theories, top contributing authors, and trendy themes are identified and discussed. The systematic content analysis reveals the technical framework/architecture to implement BCT in the supply chain and the benefit and impact mechanism of BCT in enhancing SCR in different sectors and contexts. Although BCT has many benefits, it is considered a complex and costly technology. The semi-quantitative multi-criteria decision-making like AHP, ANP, DEMATEL, Grey/Fuzzy System-ISM MICMAC, as well as Structural Equation Model, and Machine Learning have been taken advantage of in investigating enablers, critical factors, challenges, and barriers. Our research has a limitation in terms of qualitative content analysis approach for content analysis. Future research can implement a quantitative approach, text mining, or other modern technologies to mine the literature better.

**Acknowledgements** This research is funded by the Vietnam Ministry of Education and Training with code B2024.BKA.20 "Research on solutions to enhance supply chain resilience capability for Vietnamese businesses in VUCA world"

## References

1. Ada, N., Ethirajan, M., Kumar, A., KEk, V, Nadeem SP, Kazancoglu, Y., Kandasamy, J.: Blockchain technology for enhancing traceability and efficiency in automobile supply chain—a case study. Sustainability **13**(24), 13667 (2021). https://doi.org/10.3390/su132413667
2. Alabaddi, Z., Obidat, A., Alziyadat, Z.: Exploring the effect of blockchain technology on supply chain resilience and transparency: evidence from the healthcare industry. Uncertain Supply Chain Manage. **11**(2), 787–798 (2023). https://doi.org/10.5267/j.uscm.2023.1.001

3. Alkhudary, M.M.Q.R., Féniès, P.: Mitigating the risk of specific supply chain disruptions through blockchain technology. Supply Chain Forum: Int. J. **25**(1), 1–11 (2024). https://doi.org/10.1080/16258312.2022.2090273
4. Al-Swidi, A.K., Al-Hakimi, M.A., Halbusi, H.A, Harbi, J.A.A, Al-Hattami, H.M.: Does blockchain technology matter for supply chain resilience in dynamic environments? the role of supply chain integration. PLOS ONE **19**(1), e0295452 (2024). https://doi.org/10.1371/journal.pone.0295452
5. Babu, E.S., Kavati, I., Nayak, S.R., Ghosh, U., Numay, W.A.: Secure and transparent pharmaceutical supply chain using permissioned blockchain network. Int. J. Log. Res. Appl. 1–28 (2022). https://doi.org/10.1080/13675567.2022.2045578
6. Bai, C., Zhu, Q., Sarkis, J.: Do blockchain capabilities help overcome supply and operational risks: insights from firm market returns during COVID-19. Omega **126**, 103049 (2024). https://doi.org/10.1016/j.omega.2024.103049
7. Beck, J., Birkel, H., Spieske, A., Gebhardt, M.: Will the blockchain solve the supply chain resilience challenges? insights from a systematic literature review. Comput. Ind. Eng. **185** (2023). https://doi.org/10.1016/j.cie.2023.109623
8. Boakye, E.A., Zhao, H., Ahia, B.N.K.: Blockchain technology prospects in transforming Ghana's economy: a phenomenon-based approach. Inf. Technol. Dev. **29**(2–3), 348–377 (2023). https://doi.org/10.1080/02681102.2022.2073580
9. Cao, Y., Yi, C., Wan, G., Hu, H., Li, Q., Wang, S.: An analysis on the role of blockchain-based platforms in agricultural supply chains. Transp. Res. Part E: Logist. Transp. Rev. **163** (2022). https://doi.org/10.1016/j.tre.2022.102731
10. Chaouni, B.A., Zekhnini, K., Cherrafi, A., Garza-Reyes, J.A., Kumar, A., El Baz, J.: Blockchain technology for viable circular digital supplychains: an integrated approach for evaluating the implementation barriers. Benchmark. Int. J. **30**(10), 4397–4424 (2023). https://doi.org/10.1108/BIJ-04-2022-0240
11. Chen, P.K.: Efficiency of lean practices and blockchain combinations for green supplier integration improvements in sustainable development. Sustain. Dev. **31**(1) (2023). https://doi.org/10.1002/sd.2409
12. Choi, T.M., Wen, X., Sun, X., Chung, S.H.: The mean-variance approach for global supply chain risk analysis with air logistics in the blockchain technology era. Transp. Res. Part E: Logist. Transp. Rev. **127** (2019). https://doi.org/10.1016/j.tre.2019.05.007
13. Chowdhury, S., Rodriguez-Espindola, O., Dey, P., Budhwar, P.: Blockchain technology adoption for managing risks in operations and supply chain management: evidence from the UK. Ann. Oper. Res. **327**(1), 539–574 (2023). https://doi.org/10.1007/s10479-021-04487-1
14. Clarivate: Web of Science Coverage Details (2024). https://clarivate.libguides.com/librarianresources/coverage
15. Cobo, M.J., López-Herrera, A.G., Herrera-Viedma, E., Herrera, F.: Science mapping software tools: review, analysis, and cooperative study among tools. J. Am. Soc. Inform. Sci. Technol. **62**(7), 1382–1402 (2011). https://doi.org/10.1002/asi.21525
16. Collart, A.J., Canales, E.: How might broad adoption of blockchain-based traceability impact the U.S. fresh produce supply chain? Appl. Econ. Perspect. Policy **44**(1) (2022). https://doi.org/10.1002/aepp.13134
17. Datta, S., Jauhar, S.K., Paul, S.K.: Leveraging blockchain to improve nutraceutical supply chain resilience under post-pandemic disruptions. Comput. Ind. Eng. **183** (2023). https://doi.org/10.1016/j.cie.2023.109475
18. Dong, L., Qiu, Y., Xu, F.: Blockchain-enabled deep-tier supply chain finance. Manuf. Serv. Oper. Manage. (2022). https://doi.org/10.1287/msom.2022.1123
19. Dubey, R., Gunasekaran, A., Bryde, D.J., Dwivedi, Y.K., Papadopoulos, T.: Blockchain technology for enhancing swift-trust, collaboration and resilience within a humanitarian supply chain setting. Int. J. Prod. Res. **58**(11), 3381–3398 (2020). https://doi.org/10.1080/00207543.2020.1722860
20. Duong, L.N.K., Chong, J.: Supply chain collaboration in the presence of disruptions: a literature review. Int. J. Prod. Res. **58**(11), 3488–3507 (2020). https://doi.org/10.1080/00207543.2020.1712491

21. Ekinci, E., Sezer, M.D., Mangla, S.K., Kazancoglu, Y.: Building sustainable resilient supply chain in retail sector under disruption. J. Clean. Prod. **434**, 139980 (2024). https://doi.org/10.1016/j.jclepro.2023.139980
22. Elsevier.: Scopus now includes 90 million + content records! (2024) https://blog.scopus.com/posts/scopus-now-includes-90-million-content-records
23. Etemadi, N., Borbon-Galvez, Y., Strozzi, F., Etemadi, T.: Supply chain disruption risk management with blockchain: a dynamic literature review. Information **12**(2) (2021). https://doi.org/10.3390/info12020070
24. Gausdal, A.H., Czachorowski, K.V., Solesvik, M.Z.: Applying blockchain technology: evidence from Norwegian companies. Sustainability **10**(6), 1985 (2018). https://doi.org/10.3390/su10061985
25. Granillo, M.R., Hernández, I.J.G., Olivares-Benítez, E.: Logistics 4.0 in the agri-food supply chain with blockchain: a case study. Int. J. Logist. Res. Appl. **27**(10), 1766–1786 (2025). https://doi.org/10.1080/13675567.2023.2184467
26. Gupta, A., Singh, R.K., Kamal, M.M.: Blockchain technology adoption for secured and carbon neutral logistics operations: barrier intensity index framework. Ann. Oper. Res. (2024). https://doi.org/10.1007/s10479-024-05824-w
27. Gupta, S., Modgil, S., Choi, T.-M., Kumar, A., Antony, J.: Influences of artificial intelligence and blockchain technology on financial resilience of supply chains. Int. J. Prod. Econ. **261**, 108868 (2023). https://doi.org/10.1016/j.ijpe.2023.108868
28. Hader, M., Tchoffa, D., Mhamedi, A.E, Ghodous, P., Dolgui, A., Abouabdellah, A.: Applying integrated blockchain and big data technologies to improve supply chain traceability and information sharing in the textile sector. J. Ind. Inf. Integr. **28**, 100345 (2022). https://doi.org/10.1016/j.jii.2022.100345
29. Hong, L., Hales, D.N.: How blockchain manages supply chain risks: evidence from Indian manufacturing companies. Int. J. Logist. Manage. **35**(5), 1604–1627 (2024). https://doi.org/10.1108/IJLM-05-2023-0178
30. Hubenova, T., Lindeque, J.P., Peter, M.K.: Explaining the non-adoption of blockchain technology in global value chains: a micro-foundational perspective. J. Ind. Bus. Econ. **51**(2), 397–429 (2024). https://doi.org/10.1007/s40812-023-00296-8
31. Iftekhar, A., Cui, X.: Blockchain-based traceability system that ensures food safety measures to protect consumer safety and COVID-19 free supply chains. Foods **10**(6) (2021). https://doi.org/10.3390/foods10061289
32. İndap, Ş, Tanyaş, M.: Blockchain applications for traceability and food safety in agri-food supply chain: cherry product application. J. Enterp. Inf. Manage. (2023). https://doi.org/10.1108/JEIM-03-2023-0165
33. Irannezhad, E., Faroqi, H.: Addressing some of bill of lading issues using the Internet of Things and blockchain technologies: a digitalized conceptual framework. Marit. Policy Manage. **50**(4) (2023). https://doi.org/10.1080/03088839.2021.1930223
34. Kalbouneh, N.Y., Bataineh, K.A., Al-Hamad, A.A.-S.A., Dwakat, M.K.Al, Abualoush, S., Almasarweh, M.S., Al-Smadi, R.W.: The effects of the blockchain technology and big data analytics on supply chain performance: the mediating effect supply chain risk management. Uncertain Supply Chain Manage. **11**(3), 903–914 (2023). https://doi.org/10.5267/j.uscm.2023.5.008
35. Kamble, S.S., Gunasekaran, A., Sharma, R.: Modeling the blockchain enabled traceability in agriculture supply chain. Int. J. Inf. Manage. **52**, 101967 (2020). https://doi.org/10.1016/j.ijinfomgt.2019.05.023
36. Kashem, M.A., Shamsuddoha, M., Nasir, T., Chowdhury, A.A.: Supply chain disruption versus optimization: a review on artificial intelligence and blockchain. Knowledge **3**(1) (2023). https://doi.org/10.3390/knowledge3010007
37. Katsaliaki, K., Galetsi, P., Kumar, S.: Supply chain disruptions and resilience: a major review and future research agenda. Ann. Oper. Res. **319**(1), 965–1002 (2022). https://doi.org/10.1007/s10479-020-03912-1

38. Kazancoglu, Y., Ozbiltekin-Pala, M., Sezer, M.D., Luthra, S., Kumar, A.: Resilient reverse logistics with blockchain technology in sustainable food supply chain management during COVID-19. Bus. Strateg. Environ. **32**(4), 2327–2340 (2023). https://doi.org/10.1002/bse.3251
39. Klöckner, M., Schmidt, C.G., Fink, A., Flückiger, L., Wagner, S.M.: Exploring the physical–digital interface in blockchain applications: insights from the luxury watch industry. Transp. Res. Part E: Logist. Transp. Rev. **179**, 103300 (2023). https://doi.org/10.1016/j.tre.2023.103300
40. Kuei, S.C., Chen, M.C.: Enablers of blockchain adoption on supply chain with dynamic capability perspectives with ISM-MICMAC analysis. Ann. Oper. Res. (2023). https://doi.org/10.1007/s10479-023-05684-w
41. Kumar, N., Upreti, K., Upreti, S., Alam, M.S., Agrawal, M.: Blockchain integrated flexible vaccine supply chain architecture: excavate the determinants of adoption. Hum. Behav. Emer. Technol. **3**(5), 1106–1117 (2021). https://doi.org/10.1002/hbe2.302
42. Kumar, S., Lim, W.M., Sivarajah, U., Kaur, J.: Artificial intelligence and blockchain integration in business: trends from a bibliometric-content analysis. Inf. Syst. Front. (2022). https://doi.org/10.1007/s10796-022-10279-0
43. Kumar, S.R., Mishra, R., Gupta, S., Mukherjee, A.A.: Blockchain applications for secured and resilient supply chains: a systematic literature review and future research agenda. Comput. Ind. Eng. **175**, 108854 (2023). https://doi.org/10.1016/j.cie.2022.108854
44. Lascialfari, M., Magrini, M.-B., Cabanac, G.: Unpacking research lock-in through a diachronic analysis of topic cluster trajectories in scholarly publications. Scientometrics **127**(11), 6165–6189 (2022). https://doi.org/10.1007/s11192-022-04514-3
45. Li, G., Xue, J., Li, N., Ivanov, D.: Blockchain-supported business model design, supply chain resilience, and firm performance. Transp. Res. Part E: Logist. Transp. Rev. **163**, 102773 (2022). https://doi.org/10.1016/j.tre.2022.102773
46. Liu, S., Hua, G., Kang, Y., Cheng, T.C.E., Xu, Y.: What value does blockchain bring to the imported fresh food supply chain? Transp. Res. Part E: Logist. Transp. Rev. **165** (2022). https://doi.org/10.1016/j.tre.2022.102859
47. Liu, Y., Fang, W., Feng, T., Xi, M.: Blockchain technology adoption and supply chain resilience: exploring the role of transformational supply chain leadership. Supply Chain Manage. Int. J. (2024). https://doi.org/10.1108/SCM-08-2023-0390
48. Lohmer, J., Bugert, N., Lasch, R.: Analysis of resilience strategies and ripple effect in blockchain-coordinated supply chains: an agent-based simulation study. Int. J. Prod. Econ. **228** (2020). https://doi.org/10.1016/j.ijpe.2020.107882
49. Lotfi, R., Hazrati, R., Aghakhani, S., Afshar, M., Amra, M., Ali, S.S.: A data-driven robust optimization in viable supply chain network design by considering open innovation and blockchain technology. J. Clean. Prod. **436** (2024). https://doi.org/10.1016/j.jclepro.2023.140369
50. Majumdar, A., Agrawal, R., Raut, R.D., Narkhede, B.E.: Two years of COVID-19 pandemic: Understanding the role of knowledge-based supply chains towards resilience through bibliometric and network analyses. Oper. Manage. Res. **16**(3), 1105–1121 (2023). https://doi.org/10.1007/s12063-022-00328-x
51. Malacina, I., Teplov, R.: Supply chain innovation research: a bibliometric network analysis and literature review. Int. J. Prod. Econ. **251**, 108540 (2022). https://doi.org/10.1016/j.ijpe.2022.108540
52. Mangla, S.K., Kazançoğlu, Y., Yıldızbaşı, A., Öztürk, C., Çalık, A.: A conceptual framework for blockchain-based sustainable supply chain and evaluating implementation barriers: a case of the tea supply chain. Bus. Strateg. Environ. **31**(8), 3693–3716 (2022). https://doi.org/10.1002/bse.3027
53. Manupati, V.K., Schoenherr, T., Ramkumar, M., Panigrahi, S., Sharma, Y., Mishra, P.: Recovery strategies for a disrupted supply chain network: Leveraging blockchain technology in pre- and post-disruption scenarios. Int. J. Prod. Econ. **245**, 108389 (2022). https://doi.org/10.1016/j.ijpe.2021.108389
54. McGrath, P., McCarthy, L., Marshall, D., Rehme, J.: Tools and technologies of transparency in sustainable global supply chains. Calif. Manage. Rev. **64**(1), 67–89 (2021). https://doi.org/10.1177/00081256211045993

55. Min, H.: Blockchain technology for enhancing supply chain resilience. Bus. Horiz. **62**(1), 35–45 (2019). https://doi.org/10.1016/j.bushor.2018.08.012
56. Mohamed, S.K., Haddad, S., Barakat, M., Rosi, B.: Blockchain technology adoption for improved environmental supply chain performance: the mediation effect of supply chain resilience, customer integration, and green customer information sharing. Sustainability **15**(10), 7909 (2023). https://doi.org/10.3390/su15107909
57. Mukherjee, A.A., Singh, R.K., Mishra, R., Bag, S.: Application of blockchain technology for sustainability development in agricultural supply chain: justification framework. Oper. Manage. Res. **15**(1), 46–61 (2022). https://doi.org/10.1007/s12063-021-00180-5
58. Nagariya, R., Mukherjee, S., Baral, M.M., Chittipaka, V.: Analyzing blockchain-based supply chain resilience strategies: resource-based perspective. Int. J. Product. Perform. Manage. **73**(4), 1088–1116 (2024). https://doi.org/10.1108/IJPPM-07-2022-0330
59. Narwane, V.S., Raut, R.D., Mangla, S.K., Dora, M., Narkhede, B.E.: Risks to big data analytics and blockchain technology adoption in supply chains. Ann. Oper. Res. **327**(1), 339–374 (2023). https://doi.org/10.1007/s10479-021-04396-3
60. Nayal, K., Raut, R.D., Narkhede, B.E., Priyadarshinee, P., Panchal, G.B., Gedam, V.V.: Antecedents for blockchain technology-enabled sustainable agriculture supply chain. Ann. Oper. Res. **327**(1), 293–337 (2023). https://doi.org/10.1007/s10479-021-04423-3
61. Nguyen, S., Shu-Ling Chen, P., Du, Y.: Risk assessment of maritime container shipping blockchain-integrated systems: an analysis of multi-event scenarios. Transp. Res. Part E: Logist. Transp. Rev. **163**, 102764 (2022). https://doi.org/10.1016/j.tre.2022.102764
62. Nisar, U., Zhang, Z., Wood, B.P., Ahmad, S., Ellahi, E., Haq, S.I.U., Alnafissa, M., Abd-Allah, E.F.: Unlocking the potential of blockchain technology in enhancing the fisheries supply chain: an exploration of critical adoption barriers in China. Sci. Rep. **14**(1), 10167 (2024). https://doi.org/10.1038/s41598-024-59167-4
63. Pandey, A.K., Daultani, Y., Pratap, S.: Blockchain technology enabled critical success factors for supply chain resilience and sustainability. Bus. Strateg. Environ. **33**(2), 1533–1554 (2024). https://doi.org/10.1002/bse.3548
64. Pattanayak, S., Ramkumar, M., Goswami, M., Rana, N.P.: Blockchain technology and supply chain performance: the role of trust and relational capabilities. Int. J. Prod. Econ. **271**, 109198 (2024). https://doi.org/10.1016/j.ijpe.2024.109198
65. Petratos, P.N., Faccia, A.: Fake news, misinformation, disinformation and supply chain risks and disruptions: risk management and resilience using blockchain. Ann. Oper. Res. **327**(2) (2023). https://doi.org/10.1007/s10479-023-05242-4
66. Queiroz, M.M., Wamba, S.F.: Blockchain adoption challenges in supply chain: an empirical investigation of the main drivers in India and the USA. Int. J. Inf. Manage. **46**, 70–82 (2019). https://doi.org/10.1016/j.ijinfomgt.2018.11.021
67. Rauniyar, K., Wu, X., Gupta, S., Modgil, S., de Sousa Jabbour, A.B.L.: Risk management of supply chains in the digital transformation era: contribution and challenges of blockchain technology. Ind. Manage. Data Syst. **123**(1), 253–277 (2023). https://doi.org/10.1108/IMDS-04-2021-0235
68. Rejeb, A., Rejeb, K., Zrelli, I., Süle, E., Iranmanesh, M.: Blockchain technology in the renewable energy sector: a co-word analysis of academic discourse. Heliyon **10**(8), e29600 (2024). https://doi.org/10.1016/j.heliyon.2024.e29600
69. Rogerson, M., Parry, G.C.: Blockchain: case studies in food supply chain visibility. Supply Chain Manage. Int. J. **25**(5), 601–614 (2020). https://doi.org/10.1108/SCM-08-2019-0300
70. Sarfraz, Z., Sarfraz, A., Iftikar, H.M., Akhund, R.: Is COVID-19 pushing us to the fifth industrial revolution (society 5.0)? Pak. J. Med. Sci. **37**(2) (2021). https://doi.org/10.12669/pjms.37.2.3387
71. Sathiya, V., Nagalakshmi, K., Jeevamalar, J., Babu, R.A., Karthi, R., Acevedo-Duque, Á., Lavanya, R., Ramabalan, S.: Reshaping healthcare supply chain using chain-of-things technology and key lessons experienced from COVID-19 pandemic. Socioecon. Plann. Sci. **85**, 101510 (2023). https://doi.org/10.1016/j.seps.2023.101510

72. Seuring, S., Müller, M.: From a literature review to a conceptual framework for sustainable supply chain management. J. Clean. Prod. **16**(15), 1699–1710 (2008). https://doi.org/10.1016/j.jclepro.2008.04.020
73. Sharma, G.D., Tiwari, A.K., Chopra, R., Dev, D.: Past, present, and future of block-chain in finance. J. Bus. Res. **177**, 114640 (2024). https://doi.org/10.1016/j.jbusres.2024.114640
74. Sharma, M., Joshi, S., Luthra, S., Kumar, A.: Managing disruptions and risks amidst COVID-19 outbreaks: role of blockchain technology in developing resilient food supply chains. Oper. Manage. Res. **15**(1–2) (2022). https://doi.org/10.1007/s12063-021-00198-9
75. Sharma, M., Patidar, A., Anchliya, N., Prabhu, N., Asok, A., Jhajhriya, A.: Blockchain adoption in food supply chain for new business opportunities: an integrated approach. Oper. Manage. Res. **16**(4), 1949–1967 (2023). https://doi.org/10.1007/s12063-023-00416-6
76. Shoaib, M., Zhang, S., Ali, H.: A bibliometric study on blockchain-based supply chain: a theme analysis, adopted methodologies, and future research agenda. Environ. Sci. Pollut. Res. **30**(6), 14029–14049 (2022). https://doi.org/10.1007/s11356-022-24844-2
77. Singh, R.K., Mishra, R., Gupta, S., Mukherjee, A.A.: Blockchain applications for secured and resilient supply chains: a systematic literature review and future research agenda. Comput. Ind. Eng. **175**, 108854 (2023). https://doi.org/10.1016/j.cie.2022.108854
78. Singh, V.K., Singh, P., Karmakar, M., Leta, J., Mayr, P.: The journal coverage of web of science, Scopus and dimensions: a comparative analysis. Scientometrics **126**(6), 5113–5142 (2021). https://doi.org/10.1007/s11192-021-03948-5
79. Sislian, L., Jaegler, A.: Linkage of blockchain to enterprise resource planning systems for improving sustainable performance. Bus. Strateg. Environ. **31**(3), 737–750 (2022). https://doi.org/10.1002/bse.2914
80. Sun, R., He, D., Su, H.: Evolutionary game analysis of blockchain technology preventing supply chain financial risks. J. Theor. Appl. Electron. Commer. Res. **16**(7), 2824–2842 (2021). https://doi.org/10.3390/jtaer16070155
81. Tranfield, D., Denyer, D., Smart, P.: Towards a methodology for developing evidence-informed management knowledge by means of systematic review. Br. J. Manage. **14**(3), 207–222 (2003). https://doi.org/10.1111/1467-8551.00375
82. Tsai, W.C., Shen, C.W.: Using a smart contract for the floral supply chain. Asia Pac. Manage. Rev. (2024). https://doi.org/10.1016/j.apmrv.2023.12.004
83. Tsang, Y.P., Fan, Y., Lee, C.K.M., Lau, H.C.W.: Blockchain sharding for e-commerce supply chain performance analytics towards Industry 5.0. Enterp. Inf. Syst. **18**(4) (2024). https://doi.org/10.1080/17517575.2024.2311807
84. Tsolakis, N., Niedenzu, D., Simonetto, M., Dora, M., Kumar, M.: Supply network design to address United Nations sustainable development goals: a case study of blockchain implementation in Thai fish industry. J. Bus. Res. **131** (2021). https://doi.org/10.1016/j.jbusres.2020.08.003
85. Vivaldini, M.: Blockchain in operations for food service distribution: steps before implementation. Int. J. Logist. Manage. **32**(3), 995–1029 (2021). https://doi.org/10.1108/IJLM-07-2020-0299
86. Wamba, S.F., Queiroz, M.M.: Industry 4.0 and the supply chain digitalisation: a blockchain diffusion perspective. Prod. Plan. Control **33**(2–3), 193–210 (2022). https://doi.org/10.1080/09537287.2020.1810756
87. Wang, Y., Singgih, M., Wang, J., Rit, M.: Making sense of blockchain technology: how will it transform supply chains? Int. J. Prod. Econ. **211**, 221–236 (2019). https://doi.org/10.1016/j.ijpe.2019.02.002
88. Wolf, D.E., Louw, L., Palm, D.: An analysis of blockchain versus relational databases for digitalising information flows in global supply chains using the analytic network process. Int. J. Prod. Res. **62**(14) (2024). https://doi.org/10.1080/00207543.2023.2282751
89. Wong, S., Yeung, J.-K.-W., Lau, Y.-Y., So, J.: Technical sustainability of cloud-based blockchain integrated with machine learning for supply chain management. Sustainability **13**(15) (2021). https://doi.org/10.3390/su13158270

90. Wünsche, J.F., Fernqvist, F.: The potential of blockchain technology in the transition towards sustainable food systems. Sustainability (Switzerland) **14**(13). (2022). https://doi.org/10.3390/su14137739
91. Xiong, Y., Lam, H.K.S., Kumar, A., Ngai, E.W.T., Xiu, C., Wang, X.: The mitigating role of blockchain-enabled supply chains during the COVID-19 pandemic. Int. J. Oper. Prod. Manage. **41**(9), 1495–1521 (2021). https://doi.org/10.1108/IJOPM-12-2020-0901
92. Xu, J., Lou, J., Lu, W., Wu, L., Chen, C.: Ensuring construction material provenance using Internet of Things and blockchain: learning from the food industry. J. Ind. Inf. Integr. **33**, 100455 (2023). https://doi.org/10.1016/j.jii.2023.100455
93. Yadav, S., Singh, S.P.: Machine learning-based mathematical model for drugs and equipment resilient supply chain using blockchain. Ann. Oper. Res. (2024). https://doi.org/10.1007/s10479-023-05761-0
94. Yang, W., Xie, C., Ma, L.: Dose blockchain-based agri-food supply chain guarantee the initial information authenticity? an evolutionary game perspective. PLOS ONE **18**(6), e0286886 (2023). https://doi.org/10.1371/journal.pone.0286886
95. Yontar, E.: The role of blockchain technology in the sustainability of supply chain management: grey based dematel implementation. Clean. Logist. Supply Chain **8**, 100113 (2023). https://doi.org/10.1016/j.clscn.2023.100113
96. Zamani, E.D., Smyth, C., Gupta, S., Dennehy, D.: Artificial intelligence and big data analytics for supply chain resilience: a systematic literature review. Ann. Oper. Res. **327**(2) (2023). https://doi.org/10.1007/s10479-022-04983-y
97. Zelbst, P.J., Yang, L., Green, K.W., Sower, V.E.: Linkages between technologies and supply chain performance: benefits, power and risk reduction. Supply Chain Manage. Int. J. **29**(1), 207–218 (2024). https://doi.org/10.1108/SCM-03-2023-0131
98. Zhu, Q., Kouhizadeh, M., Sarkis, J.: Formalising product deletion across the supply chain: blockchain technology as a relational governance mechanism. Int. J. Prod. Res. **60**(1) (2022). https://doi.org/10.1080/00207543.2021.1987552

60. Wünsche, J.F., Fernqvist, F.: The potential of blockchain technology in the transition towards sustainable food systems. Sustainability (Switzerland) 14(13), (2022). https://doi.org/10.3390/su14137739

# Blockchain and Cryptocurrencies: Revolutionizing Finance in the Digital Age

**M. Rajarajeswari, C. Gayathri, Divya Shree, and M. S. Nishaanth**

**Abstract** The financial landscape of the digital age is undergoing a revolutionary change, propelled by blockchain technology and cryptocurrencies. Decentralized ledger systems, made possible by the introduction of blockchain technology, allow for safe, transparent, and unchangeable transactions. Cryptocurrency has transformed the financial markets by providing decentralized digital currencies that allow for international trade and financial inclusion. Shifting the focus to the transformative influence of blockchain and cryptocurrencies on finance, this chapter delves into the core principles of blockchain technology and provides an in-depth analysis of the economics surrounding digital currencies. The findings of the study have been arrived at by utilizing both primary and secondary data. The primary data has been collected from 384 respondents and found that there is a significant difference between demographic variables and impact. Further, the study reveals that transaction efficiency and stability, adoption barriers, security and resource management, knowledge and compliance gaps, and regulatory challenges are the challenges faced by investors. The study extends by exploring the moderate impact of the requirement demanded by investors for digital currencies and Bitcoin. The study concludes by providing suggestions to the stakeholders with respect to an awareness program, incentive option, and regulatory framework.

**Keywords** Blockchain · Cryptocurrencies · Digital finance · Bitcoin · Financial innovation · Online payments · Economic implications

---

M. Rajarajeswari · C. Gayathri (✉) · D. Shree
Department of Commerce, PSGR Krishnammal College for Women, Coimbatore, Tamil Nadu, India
e-mail: 22phdce003@psgrkcw.ac.in

M. S. Nishaanth
KGISL Institute of Technology, Coimbatore, Tamil Nadu, India



N. D. K. Nguyen et al. (eds.), *Digital Transformation and Sustainability: Methods and Applications*, Information Systems Engineering and Management 47,
https://doi.org/10.1007/978-3-031-91278-8_6

# 1 Introduction

The global financial ecosystem has seen a significant transformation in recent years due to the emergence of cryptocurrencies and blockchain technology (IBM), which have put traditional financial institutions under pressure and redefined the nature of value exchange and transactions. Since its launch as the first decentralized digital currency in 2009 under the alias Satoshi Nakamoto by a computer programmer, or group of programmers [48], Bitcoin has grown quickly to become a complex and diverse business with dozens of other cryptocurrencies and a wide range of blockchain applications. These developments constitute a fundamental change in the way one sees and handles financial assets, going beyond simple technological improvements [17].

Blockchain presents unparalleled prospects for developing systems that function autonomously of centralized authorities because of its decentralized [4], transparent, and unchangeable ledger structure [22, 26]. This has opened the door for the emergence of Decentralized finance (DeFi), which is the reimagining of conventional financial services [29] like borrowing, lending, and trading via decentralized protocols [10]. Conversely, cryptocurrencies have brought forth new kinds of digital assets that serve as repositories of wealth as well as means of exchange. As these technologies gain more and more popularity, they are having an impact on the financial markets and igniting debates about regulatory frameworks, the implications for the economy, and the future of money. The distinction between traditional and digital banking is becoming increasingly hazy as a result of the integration of blockchain technology into traditional finance and the emergence of innovations like Central Bank Digital Currencies (CBDCs) [30], and the tokenization of physical assets [58].

This study explores the origins, technological foundations, and ways that cryptocurrencies and blockchain are transforming financial practices, shedding light on the significant influence these technologies have had on the financial age. This study seeks to offer a thorough grasp of how these technologies are upending preexisting financial paradigms, what this means for the future of global finance and provide insightful insights into the opportunities and difficulties surrounding blockchain technology and cryptocurrencies in the emerging financial era.

## *1.1 Statement of the Problem*

Traditional financial systems have been significantly disrupted by the quick rise and spread of blockchain technology and cryptocurrencies, but it is still unclear what the full repercussions of these developments will be. Even though these technologies have the potential to be revolutionary, there is a significant lack of information in the literature about how they will ultimately affect global financial markets, legal frameworks, and the overall economy. With regard to stability, security, and accessibility in this new financial paradigm, the volatility of cryptocurrencies and the early stages of Decentralized Finance (DeFi) present serious concerns. In addition, the uneven adoption of these technologies across different countries and the absence of thorough regulatory monitoring all contribute to an unclear environment where the risks and benefits of these technologies are not fully understood or controlled. This research tries to methodically examine how blockchain technology and cryptocurrencies are changing the financial landscape, how they affect established banking systems, and pinpoint the critical elements that will shape their future trajectory, this research aims to address these issues. Thus, this study tries to enhance comprehension of how new technologies might be included in the international financial system in a manner that optimizes advantages while minimizing possible hazards.

### *1.2 Objectives*

1. To analyze the impact on traditional financial systems caused by diversification of the investments in Bitcoin.
2. To identify the challenges associated with digital currencies and blockchain networks.
3. To analyze the requirements demanded by digital currencies and Bitcoin investors for increasing the usage of Bitcoin.

### *1.3 Research Methodology*

This study has utilized quantitative research designs and both primary and secondary sources which are obtained from a variety of secondary sources, including academic papers published, research articles, databases, annual reports, blogs, and books on several websites. Further, the primary source data was collected from respondents in Tamil Nadu. Since the study focuses on an unknown population Cochran's sample size formula ($n = 1.96^2 (0.5) (0.5)/0.05^2$) has been utilized to collect data from 384 respondents. The study has used both descriptive and inferential statistics to analyze the primary data.

## 2 Digital Transformation of Finance

> The Internet is becoming the town square for the global village of tomorrow.—Bill Gates

In the ever-evolving digital age, the fusion of technology and finance has given birth to an extraordinary phenomenon: blockchain and cryptocurrencies. The rise of blockchain and cryptocurrencies is more than just a financial evolution; it is a revolution [36] that challenges conventional financial systems and, in the process, offers opportunities for greater financial inclusion [44], increased security [27], and the democratization of innovation and money [15].

## 2.1 The Early Days of Digital Transactions

The way we make payments has evolved significantly over the past few decades [3, 31], (Wikipedia).

(a) **Barter System (Pre-money Era)**: In ancient times, people engaged in barter, exchanging goods and services directly. This system had limitations, such as the "double coincidence of wants" problem, where both parties had to want what the other had to offer.
(b) **Commodity Money**: To overcome the limitations of barter, societies began to use commodity money, which had intrinsic value. Items like cowry shells, salt, and even grains served as mediums of exchange.
(c) **Metal Coins**: Around 600 BC, metal coins, typically made of gold, silver, or copper, became popular as they were durable, divisible, and had intrinsic value. Governments and rulers started minting standardized coins to facilitate trade and collect taxes.
(d) **Paper Money and Banking**: The transition from metal coins to paper money began in China during the Tang Dynasty (seventh century). Paper money was more convenient and easier to carry. As banking institutions emerged, they provided services like storage of valuable items and lending.
(e) **Fiat Currency and Central Banking**: Fiat currency, which has no intrinsic value and is backed by the government's trust and authority, became widespread. The first recorded use of fiat money was in Sweden in the seventeenth century. Central banks, like the Bank of England (founded in 1694), were established to issue and regulate national currencies.
(f) **Electronic Banking and Digital Transactions**: The twentieth century saw the rise of electronic banking, credit cards, and digital payment systems. These innovations made transactions faster and more efficient. The establishment of the SWIFT network in 1973 facilitated international financial transactions.
(g) **Internet and E-commerce**: The advent of the internet in the late twentieth century paved the way for online commerce and digital financial services. Companies like PayPal and Amazon revolutionized online payments.
(h) **Bitcoin, Cryptocurrencies, and Blockchain**: In 2009, an anonymous individual or group using the pseudonym Satoshi Nakamoto introduced Bitcoin, a decentralized digital currency based on blockchain technology. Blockchain, the underlying technology, is a distributed ledger that records transactions transparently and securely without the need for intermediaries.

## 2.2 *Introducing Cryptocurrency: A New Paradigm in Finance*

Cryptocurrencies represent a groundbreaking innovation in the world of finance. Unlike traditional currencies issued and regulated by governments, cryptocurrencies are digital or virtual currencies that rely on cryptography for security. They are decentralized, meaning they are not controlled by a central authority like a government or central bank [35].

The concept of cryptocurrency was first introduced in a paper titled "Bitcoin: A Peer-to-Peer Electronic Cash System", published in 2008 by an entity known as Satoshi Nakamoto, which is known as "Satoshi's Whitepaper" in the crypto world [37]. This whitepfaper proposed a novel way to create a digital currency that could be transferred directly between users in a secure and trustless manner, without the need for intermediaries. The currency itself was named Bitcoin.

### 2.2.1 Key Characteristics of Cryptocurrencies

The following are the main characteristics of cryptocurrencies [21, 38]:

1. **Digital Nature**: Cryptocurrencies exist only in digital form and have no physical counterpart, like coins or paper money.
2. **Decentralization**: They are typically maintained by a decentralized network of computers (nodes) rather than a central entity.
3. **Security**: Cryptocurrencies use cryptographic techniques to secure transactions and control the creation of new units.
4. **Anonymity**: Transactions are often pseudonymous, meaning they use public keys (addresses) rather than personal information.
5. **Global Accessibility**: Cryptocurrencies can be accessed and used by anyone with an internet connection, regardless of their location.
6. **Transparency**: Blockchain technology ensures that transaction history is recorded and can be viewed publicly.

### 2.2.2 Advantages of Cryptocurrencies

The advantages of cryptocurrencies are as follows [11, 50]:

- **Speed and Efficiency**: Cryptocurrency transactions, especially Bitcoin, are renowned for their speed and efficiency. Traditional cross-border transactions often involve intermediaries, leading to delays and fees. In contrast, Bitcoin enables swift and direct peer-to-peer transactions, reducing settlement times.
- **Lower Transaction Costs**: Cryptocurrency transactions can substantially lower costs compared to traditional banking and payment systems. Cross-border transfers often incur hefty fees and unfavorable exchange rates. Bitcoin eliminates or minimizes these costs, making it an attractive option for international transactions.

- **Enhanced Security**: Blockchain technology underpins cryptocurrencies, providing a high level of security. Transactions are recorded on a decentralized ledger, making them immutable and resistant to fraud. This inherent security feature bolsters trust in online payments.
- **Financial Inclusion**: Cryptocurrencies open doors to financial services for the unbanked and underbanked populations worldwide. People without access to traditional banking can participate in the digital economy through cryptocurrencies, fostering financial inclusion.
- **Anonymity**: With the ability to conduct transactions without disclosing personal information due to the fact that users are identified by cryptographic addresses rather than by their true identities, cryptocurrencies improve privacy. Particularly in areas with restricted financial freedom or where privacy is an issue, this anonymity offers a safe means to transact while shielding people from surveillance and lowering the chance of identity theft.
- **Transparency**: Blockchain technology, which underpins cryptocurrency operations, records every transaction on an open, unchangeable public ledger that is available to all. By making transactions transparent and traceable and preventing any one party from manipulating the system, this transparency promotes accountability. It establishes a trustless environment so users can interact with confidence and without the need for middlemen.

### 2.2.3 Limitations and Challenges

Despite their advantages, cryptocurrencies also face limitations and challenges in the online payment landscape [11, 43]:

- **Volatility**: Cryptocurrencies like Bitcoin are known for their price volatility. This can pose challenges for merchants and consumers who prefer stable prices for goods and services.
- **Regulatory Uncertainty**: The regulatory landscape for cryptocurrencies varies widely across countries. Some nations have embraced digital currencies, while others have imposed strict regulations or outright bans. This regulatory patchwork creates uncertainty for users.
- **Scalability Issues**: As cryptocurrencies gain popularity, scalability issues emerge. Bitcoin, for instance, faces challenges related to transaction processing speed and network congestion during periods of high demand.
- **User Education**: Widespread adoption of cryptocurrencies requires user education. Understanding the nuances of digital wallets, private keys, and transaction confirmations can be daunting for newcomers.

### 2.2.4 Exploring Diverse Cryptocurrency Types

The various types of cryptocurrencies are as follows (Source: Hayes [23], Maheshwari [41], Pele et al. [47], CFI [70], Byju's [71], Tschorsch & Scheuermann [64]).

1. **Bitcoin (BTC)**: Bitcoin is the pioneering cryptocurrency that introduced the concept of blockchain technology in 2009. It serves as a decentralized digital currency and a store of value, often compared to gold. Bitcoin's primary use case is as a medium of exchange and operates on a proof-of-work consensus mechanism. The drawback of Bitcoin is that it has limited scalability, slower confirmation times, and high transaction fees.
2. **Ethereum (ETH)**: Ethereum is a blockchain platform known for its smart contract functionality. It allows developers to create decentralized applications (DApps) on its blockchain, enabling a wide range of use cases, from decentralized finance (DeFi) to non-fungible tokens (NFTs). Ethereum's native cryptocurrency is called Ether.
3. **Ripple (XRP)**: Ripple is a blockchain platform designed for fast and low-cost cross-border payments. It aims to streamline international money transfers for banks and financial institutions. XRP is the native cryptocurrency used within the Ripple network. However, Ripple is less decentralized than Bitcoin or Ethereum.
4. **Litecoin (LTC)**: Often referred to as "silver" to Bitcoin's "gold", Litecoin is a peer-to-peer cryptocurrency created as a lighter and faster alternative. It is similar to Bitcoin but with shorter block generation times and a different hashing algorithm.
5. **Cardano (ADA)**: Cardano is a blockchain platform known for its focus on sustainability, scalability, and interoperability. It uses a proof-of-stake consensus mechanism and is designed to support smart contracts and DApps.
6. **Binance Coin (BNB)**: Binance Coin is the native cryptocurrency of the Binance exchange. It has a variety of use cases within the Binance ecosystem, including reduced trading fees and participation in token sales on the Binance Launchpad. It is vulnerable to regulatory changes.
7. **Polkadot (DOT)**: Polkadot is a multi-chain network that connects multiple blockchains, enabling them to interoperate and share information. It aims to create a more interconnected and efficient blockchain ecosystem that will provide cross-chain compatibility.
8. **Chainlink (LINK)**: Chainlink is a decentralized oracle network that connects smart contracts with real-world data. It plays a crucial role in enabling smart contracts to interact with external sources of information. The major use of Chainlink is in DeFi, insurance, and supply chain applications.
9. **Tezos (XTZ)**: Tezos is a blockchain platform known for its on-chain governance and self-amending capabilities. It allows stakeholders to vote on proposed protocol upgrades, making it adaptable and scalable.

10. **Solana (SOL)**: Solana is a high-performance blockchain platform designed for decentralized applications and cryptocurrencies. It employs a unique consensus mechanism called Proof of History to achieve high scalability and low transaction costs. It is mainly used for DeFi applications, providing speed and efficiency for activities like trading and lending. These are also advantageous for NFT platforms, enabling quick and cost-effective NFT creation and trading
11. **Avalanche (AVAX)**: Avalanche is a blockchain platform that prioritizes flexibility, speed, and scalability. It uses a consensus mechanism called Avalanche consensus and supports custom blockchain. These are mainly used for transparent and secure supply chain management, tracking products, verifying authenticity, and preventing fraud.
12. **NFTs (Non-Fungible Tokens)**: NFTs are unique digital assets that represent ownership of digital or physical items. They have gained popularity in the worlds of art, gaming, and entertainment, allowing artists and creators to tokenize and sell their work.

# 3 Bitcoin: Economics, Technology, and Governance

## *3.1 Bitcoin: The Pioneer Cryptocurrency*

Bitcoin, often referred to as digital gold or gold 2.0, was the first cryptocurrency and remains the most well-known and widely used [8]. It was introduced by an anonymous individual or group of individuals using the pseudonym Nakamoto [46]. The release of the Bitcoin whitepaper in 2008 laid the foundation for the cryptocurrency's development, and the first block, known as the "genesis block", was mined in January 2009 [56].

The key features of Bitcoin are its decentralized nature, built-in scarcity, and transparency [25]. Unlike fiat currencies, which can be printed by central authorities at will, Bitcoin has a fixed supply capped at 21 million coins. This scarcity is achieved through a process called mining, which involves solving complex mathematical puzzles [49].

## *3.2 Investigating Bitcoin's Economic Implications*

Bitcoin, as a decentralized digital currency, has garnered substantial attention for its economic implications [24].

### 3.2.1 Bitcoin in the Financial Systems

Bitcoin has introduced an alternative to traditional financial systems. It operates on a decentralized ledger, the blockchain, which enables peer-to-peer transactions without intermediaries [5, 20]. This disintermediation disrupts traditional banking and payment systems, providing users with more direct control over their finances [34, 51].

- **Financial Inclusion**: Bitcoin's borderless nature facilitates financial inclusion by allowing individuals without access to traditional banking services to participate in the digital economy. This has profound implications, particularly in regions with limited banking infrastructure.
- **Cross-Border Transactions**: Bitcoin excels in cross-border transactions, bypassing the complexities and delays associated with traditional international transfers. This is particularly advantageous for remittances and global commerce.
- **Reduced Transaction Costs**: Bitcoin transactions typically have lower fees compared to traditional banking and payment methods, making it attractive for microtransactions and international transfers.

### 3.2.2 Bitcoin as an Investment Asset

Bitcoin's value has experienced significant volatility [9], attracting investors seeking opportunities for capital appreciation. Its unique characteristics make it an appealing investment asset [65]:

- **Store of Value**: Some view Bitcoin as a digital store of value akin to gold. Its capped supply of 21 million coins creates scarcity, potentially hedging against inflation.
- **Speculative Asset**: Bitcoin's price volatility has drawn both institutional and retail investors, with many perceiving it as a speculative asset offering potential high returns.
- **Portfolio Diversification**: Institutional investors increasingly consider Bitcoin as a diversification tool within investment portfolios, given its low correlation with traditional assets.

## *3.3 Delving into Bitcoin's Technological Underpinnings*

Blockchain platforms are digital infrastructures that use distributed ledger technology to record and verify transactions across a network of computers. They enable the creation of decentralized applications (DApps), smart contracts, and various digital assets [45]. At the heart of Bitcoin is blockchain technology—a decentralized and immutable ledger that records all transactions [52]. The main features of blockchain include [25, 32]:

- **Decentralization**: The blockchain operates on a decentralized network of nodes, ensuring resilience against single points of failure and censorship.
- **Transparency**: All Bitcoin transactions are publicly recorded on the blockchain, enabling participants to verify the accuracy of transactions.
- **No Interface**: The absence of a centralized interface or intermediary for transactions. This allows users to interact directly with blockchain, enhancing decentralization and reducing the need for trusted third parties.
- **Immutability**: Once recorded on the blockchain, transactions cannot be altered, enhancing security and trust.

## *3.4 Governance and Regulation of Bitcoin*

The governance and regulation of Bitcoin present unique challenges and opportunities. Unlike traditional financial systems that are heavily regulated by governments and institutions, Bitcoin operates on a decentralized and pseudonymous network (Wikipedia). This decentralization allows users to transact without intermediaries [1]. However, it also presents challenges related to governance and regulation.

Changes to the Bitcoin protocol require broad agreement among miners, developers, node operators, and users [57]. While this decentralization ensures that no single entity can control or manipulate Bitcoin, it can also lead to debates and disagreements on proposed changes.

### 3.4.1 Bitcoin's Governance Structures and Challenges

Governance is a fundamental aspect of the Bitcoin network, influencing its evolution and decision-making processes. The challenges regarding this include [28, 61].

- **Coordination Issues**: Reaching consensus across these diverse, decentralized stakeholders can be slow and difficult, as each group has different incentives.
- **Power Concentration**: Despite decentralization, there are concerns about the influence of certain entities, such as large mining pools or major developers, that can sway decision-making.
- **Limited Formalized Governance**: Unlike other blockchains that have on-chain voting systems (e.g., Tezos, Polkadot), Bitcoin relies more on social consensus and informal governance mechanisms.
- **Complexity of Consensus**: Achieving consensus for upgrades can be slow due to the decentralized nature of the system. This was seen in the SegWit debate, which took years to resolve.
- **Regulatory Pressure**: As governments seek to regulate or control Bitcoin, developers, and the community face legal and operational challenges that could limit its growth or functionality.

- **Conflicting Interests**: Different stakeholder groups (developers, miners, businesses, governments) may have conflicting interests, complicating governance and decision-making.

# 4 Blockchain Smart Contracts Applications

Blockchain is the Tech. Bitcoin is Merely the First Mainstream Manifestation of Its Potential.—Marc Kenigsberg

## *4.1 The Concept of Smart Contracts*

Smart contracts, a revolutionary application of blockchain technology, have the potential to automate and secure a wide range of financial transactions. With the ability to execute agreements without intermediaries, these are poised to revolutionize traditional contract law and reshape industries [13].

Smart contracts are self-executing contracts with the terms and conditions directly written into code. These contracts automatically execute and enforce the agreed-upon terms when specific conditions are met, without the need for intermediaries (Utimaco). These enable automation, transparency, and trust in a wide range of applications. They are particularly well-suited for use cases where parties need to interact and transact with minimal reliance on traditional legal frameworks or intermediaries [63].

## *4.2 Real-World Applications of Smart Contracts*

In the financial sector, smart contracts can be used for automated lending, insurance, and asset management [32]. For supply chain management, it provides transparency and traceability, reducing fraud and errors, and ensuring quality [2]. In the legal industry, smart contracts can automate and enforce agreements, streamlining legal processes (Consensys).

- **Supply Chain Management**: Smart contracts enhance transparency and traceability in supply chains. They allow stakeholders to record and verify every step of a product's journey, from manufacturer to consumer. This is particularly valuable in industries like food and pharmaceuticals, where product authenticity and safety are critical.
- **Legal Agreements**: Smart contracts can streamline legal processes by automating contract execution and enforcement. They are utilized in areas such as wills,

property transfers, and intellectual property rights, reducing the need for costly intermediaries and minimizing the risk of disputes.
- **Decentralized Finance (DeFi)**: DeFi platforms leverage smart contracts to provide a wide range of financial services, including lending, borrowing, trading, and yield farming. These platforms operate without traditional intermediaries, offering users greater financial autonomy.

## *4.3 The Ethereum Ecosystem and the Rise of DeFi*

Ethereum, launched by Vitalik Buterin in 2015, has become a dominant player in the world of blockchain and smart contracts. It introduced a more flexible and versatile approach to blockchain by enabling developers to create custom applications and smart contracts. Ethereum's native cryptocurrency, Ether (ETH), serves as both a transactional currency and a digital asset within the platform [12].

The Ethereum ecosystem has witnessed exponential growth, primarily driven by the emergence of decentralized applications (DApps), smart contracts, and decentralized finance (DeFi) [42]. DApps are applications built on the Ethereum blockchain, offering services that range from decentralized exchanges to non-fungible tokens (NFTs) [7, 60]. DeFi, on the other hand, aims to recreate traditional financial services like lending, borrowing, and trading using decentralized platforms [6].

## *4.4 Exploring the Potential of Smart Contracts*

Smart contracts are self-executing digital contracts with predefined rules and conditions. They automatically execute and enforce agreements when predetermined conditions are met, eliminating the need for intermediaries [63], IBM and once deployed, they cannot be altered, providing a high level of security and trust [13]. The benefits and challenges of smart contracts are as follows [67].

### 4.4.1 Benefits of Smart Contracts

- **Improved Efficiency**: Smart contracts automate processes, reducing the time and resources required for contract execution and enforcement. This efficiency benefits industries ranging from logistics to legal services.
- **Reduced Costs**: By eliminating intermediaries, smart contracts reduce transaction costs. In financial services, for example, this can lead to lower fees for users.
- **Security**: Smart contracts' immutability and cryptographic security make them resistant to fraud and tampering. Users can trust that the terms of the contract will be executed precisely as programmed.

#### 4.4.2 Risks and Challenges

- **Technical Complexity**: Developing smart contracts requires a deep understanding of blockchain technology, which can be a barrier to entry for some organizations.
- **Security Vulnerabilities**: While smart contracts are secure when well-designed, vulnerabilities can arise from coding errors. High-profile incidents, such as the DAO hack, have underscored the importance of rigorous code audits and testing.
- **Legal Recognition**: The legal status of smart contracts varies across jurisdictions. Achieving legal recognition and enforcement can be a challenge in some regions.
- **Cannot be modified**: Once deployed, smart contracts cannot be altered, which can be problematic if there are coding errors or unforeseen loopholes. This makes it challenging to address bugs or security vulnerabilities after the contract is live.

#### 4.4.3 Ethical Considerations [55]

- **Automation and Job Displacement**: The automation of contract execution may displace jobs in industries reliant on manual contract processing, raising ethical questions about the impact on employment.
- **Transparency and Privacy**: Smart contracts on public blockchains are transparent and accessible to all participants. While this enhances trust, it may also compromise the privacy of contract parties, necessitating careful consideration in sensitive contexts.
- **Fairness and Accessibility**: Ensuring that smart contracts are accessible and fair to all parties is an ethical imperative. Addressing issues of accessibility and avoiding discriminatory algorithms are important ethical considerations.

## 5 Future of Digital Currencies

The landscape of digital currencies is continually evolving, and their role in finance is set to expand. As cryptocurrencies and blockchain technology continue to mature, they are poised to play a significant role in reshaping the financial industry in the digital age.

### *5.1 Stablecoins*

Stablecoins are a class of digital currencies designed to maintain a stable value, often pegged to fiat currencies like the US Dollar or assets like gold. They offer price stability compared to volatile cryptocurrencies like Bitcoin [18, 19].

*Advantages*:

- Reduced Volatility
- Fast Cross-Border Transactions
- Financial Inclusion for the unbanked and underbanked.

*Risks and Challenges*:

- Regulatory Scrutiny
- Concerns about transparency and control
- Intense Competition.

## 5.2 Central Bank Digital Currencies (CBDCs)

CBDCs are digital representations of a country's fiat currency issued and regulated by a central authority, such as a central bank or government [40, 62]. Thus, the Central Bank's Digital Currencies facilitate the adoption of fiscal and monetary policies and encourage financial inclusion [54].

*Advantages*:

- Enhanced trust in payments
- Financial Inclusion by promoting digital payments for underserved populations
- Reduced Counterfeiting.

*Risks and Challenges*:

- Privacy Concerns
- Impact on Commercial Banks
- Challenges related to global standards and regulations.

## 5.3 DeFi Tokens

DeFi tokens are native tokens of decentralized finance platforms, offering various financial services without intermediaries [14, 39].

*Advantages*:

- Decentralization
- Liquidity
- Fosters financial innovation.

*Risks and Challenges*:

- Smart contract vulnerabilities can lead to hacks and financial losses, emphasizing the importance of rigorous code audits
- DeFi platforms often operate in a regulatory gray area, facing uncertainty and security concerns
- Market Volatility.

## 6 Real-World Examples

### *6.1 Microlending with Cryptocurrencies—Kiva and Sierra Leone*

In Sierra Leone, a country with limited access to traditional banking, Kiva, a microlending platform, leveraged blockchain and cryptocurrencies to enable small-scale loans to entrepreneurs and farmers. Through this platform, individuals from around the world could lend cryptocurrency, providing much-needed capital for local businesses. This initiative empowered individuals in Sierra Leone to access funds and build businesses, ultimately contributing to economic development and poverty reduction [16].

### *6.2 Identity Verification—SelfKey*

SelfKey is a blockchain-based identity verification platform that allows users to securely manage their personal information and documents. Users can verify their identities once and then share this information securely with various service providers, such as banks and government agencies. SelfKey simplifies identity verification processes, reducing paperwork and the risk of identity theft. It also enhances user privacy and control over personal data (WBSFT [77]).

### *6.3 Supply Chain Transparency—IBM Food Trust and Walmart*

IBM Food Trust, powered by blockchain technology, allows retailers like Walmart to track the journey of food products from farm to shelf. This transparency helps identify and address issues such as contamination quickly, ensuring the safety of food products. Consumers benefit from safer food products, and producers can pinpoint and resolve supply chain issues more efficiently, reducing waste and increasing food quality [59].

### *6.4 Remittances with Bitcoin—Migrant Workers*

Many migrant workers face high remittance fees when sending money to their families in their home countries. Bitcoin and cryptocurrency-based remittance services, such as BitPesa (now known as Sendwave), have provided an alternative with lower fees. Migrant workers can send more of their earnings back home to support their families, reducing the financial burden of remittance fees [53].

### 6.5 Decentralized Finance (DeFi)—Uniswap and Yield Farmers

Uniswap is a decentralized exchange (DEX) that allows users to trade cryptocurrencies directly from their wallets. Yield farmers, individuals who provide liquidity to DeFi platforms, earn rewards in the form of interest or additional tokens. DeFi platforms like Uniswap have democratized finance, providing opportunities for individuals to earn passive income through decentralized and automated financial services [33].

Blockchain technology and cryptocurrencies are fundamentally transforming the financial in the digital age by offering decentralization, security, and efficiency, enabling peer-to-peer transactions, reducing intermediaries, and expanding financial inclusion. While they hold immense promise, they also pose challenges such as regulatory concerns and environmental impact, making it crucial to strike a balance between innovation and responsibility as they continue to revolutionize finance in the digital era [66].

## 7 Analysis Results

Table 1 reveals the trends across the demographic variables. 58.1% of respondents are male. The predominant age group is 26–35 years, accounting for 39.8%, which reflects a youthful workforce potentially influencing the investment market. Regarding education, 36.5% of respondents hold postgraduate degrees, suggesting a trend towards higher education. Occupation-wise, 57.3% of respondents highlight a preference for employment work in the private sector. The finance and banking sector emerges as the most significant, with 46.9% of participants engaged in it, underscoring its importance in the job market. Annual income analysis shows that 44.8% earn above Rs. 10 lakhs (Table 1).

Table 2 shows that the 150 respondents who are earning more than Rs. 10,00,000 are willing to invest the maximum amounts in Bitcoin. Further, the table expresses that 70 respondents earning above 10 lakhs are interested in investing below 10% of the total investment amount in Bitcoin (Table 2).

Table 3 shows that most of the respondents those who are working in "Finance and banking" sectors are highly interested in gathering knowledge from various sources. Further, these reasons influence most of the respondents to diversify their portfolio of invest in Bitcoin (Table 3).

**Table 1** Demographic variables

| Variables | Group | Frequency | Percent |
|---|---|---|---|
| Gender | Male | 223 | 58.1 |
| | Female | 161 | 41.9 |
| | Total | 384 | 100.0 |
| Age | Below 25 years | 45 | 11.7 |
| | 26–35 years | 153 | 39.8 |
| | 36–45 years | 87 | 22.7 |
| | 46–55 years | 72 | 18.8 |
| | Above 56 | 27 | 7.0 |
| | Total | 384 | 100.0 |
| Education | UG | 125 | 32.6 |
| | PG | 140 | 36.5 |
| | Doctorate | 67 | 17.4 |
| | Others | 52 | 13.5 |
| | Total | 384 | 100.0 |
| Occupation | Private sector | 220 | 57.3 |
| | Public sector | 103 | 26.8 |
| | Self-employed | 61 | 15.9 |
| | Total | 384 | 100.0 |
| Nature of sector employed | Education | 59 | 15.4 |
| | Healthcare | 37 | 9.6 |
| | Technology | 24 | 6.3 |
| | Finance and banking | 180 | 46.9 |
| | Others | 84 | 21.9 |
| | Total | 384 | 100.0 |
| Annual income | Below Rs. 5 lakhs | 28 | 7.3 |
| | 5–7 lakhs | 101 | 26.3 |
| | 7–10 lakhs | 83 | 21.6 |
| | Above 10 lakhs | 172 | 44.8 |
| | Total | 384 | 100.0 |

*Source* Primary Data

$\mathbf{H_0}$: There is no significant association between annual income and the nature of the sector employed.

Table 4 shows that 172 respondents are earning more than 10 lakhs per annum. Further, the table reveals that 99 respondents who are working in the "Finance and banking" sector are the high-income group respondents. The chi-square Table 5 reveals a value of 72.066 and a p-value of 0.000. Since this is less than 0.01, the null hypothesis is rejected at a 1% level of significance. Hence, there is a significant

**Table 2** Cross tabulation between annual income and percent of investment

| Annual income | Percentage of investment | | | | | Total |
|---|---|---|---|---|---|---|
| | Below 10% of the total investment amount | 10–20% of the total investment amount | 21–30% of the total investment amount | 31–40% of the total investment amount | Above 50% of the total investment amount | |
| Below Rs. 5 lakhs | 4 | 9 | 6 | 3 | 0 | 22 |
| 5–7 lakhs | 59 | 26 | 4 | 0 | 9 | 98 |
| 7–10 lakhs | 28 | 15 | 16 | 3 | 6 | 68 |
| Above 10 lakhs | 70 | 40 | 19 | 9 | 12 | 150 |
| Total | 161 | 90 | 45 | 15 | 27 | 338 |

*Source* Computed Data

association between annual income and the nature of the sector employed (Tables 4 and 5).

$\mathbf{H_0}$: There is no significant difference between demographic variables and impact on the traditional impact caused in traditional financial systems due to the increasing investment in bitcoin.

Table 6 reflects that there is no significant difference between gender and impact on the traditional impact caused in traditional financial systems due to the increasing investment in bitcoin. Further, the table reflects the significant difference between age, education, occupation, nature of the sector employed, annual income, and impact on the traditional impact caused in traditional financial systems due to the increasing investment in bitcoin, as the p-value of all these variables is less than 0.01. Hence, the null hypothesis is rejected for these variables at a 1% level of significance (Table 6).

Principal Component Analysis is performed for the challenges of verifying significant variables. The KMO Value surpasses the threshold of 0.7 (Table 7). The Eigenvalue associated with each linear factor is given in Table 8. Initially before extraction, there are 31 linear components and the first factor explains 26.160 of the variance whereas subsequent factors explain the small amount of variance. It is clear that the five factors can be extracted (Table 8).

Table 9 expresses that 21 variables are grouped based on their interrelationship among themselves. Factor 1 contains the maximum number of variables and explains most of the variance. Thus, it reveals that "**Transaction Efficiency and Stability**" is the major factor affecting the respondent's investment decision towards bitcoin. Factor 2, factor 3, factor 4, and factor 5 contain **Adoption Barriers, Security and Resource Management, Knowledge and Compliance Gaps, and Regulatory Challenges** (Table 9).

$\mathbf{H_0}$: There is no significant impact between requirements for upscaling of investment in digital currencies, bitcoin, and challenges faced in investing in digital currencies and bitcoin (Table 10).

**Table 3** Cross tabulation with multiple responses

| Sources for gathering knowledge | Nature of sector employed | Have you ever invested | | Total |
|---|---|---|---|---|
| | | Yes | No | |
| Financial news and websites | Education | 31 | 15 | 46 |
| | Healthcare | 19 | 6 | 25 |
| | Technology | 18 | 3 | 21 |
| | Finance and banking | 154 | 7 | 161 |
| | Others | 65 | 9 | 74 |
| | Total | 287 | 40 | 327 |
| Social media | Education | 7 | 9 | 16 |
| | Healthcare | 6 | 6 | 12 |
| | Technology | 3 | 3 | 6 |
| | Finance and banking | 62 | 3 | 65 |
| | Others | 10 | 0 | 10 |
| | Total | 88 | 21 | 109 |
| Bitcoin investment communities | Education | 7 | 0 | 7 |
| | Healthcare | 6 | 3 | 9 |
| | Technology | 3 | 3 | 6 |
| | Finance and banking | 69 | 7 | 76 |
| | Others | 22 | 3 | 25 |
| | Total | 107 | 16 | 123 |
| Professional advice from a financial advisor | Education | 11 | 6 | 17 |
| | Healthcare | 10 | 6 | 16 |
| | Technology | 12 | 3 | 15 |
| | Finance and banking | 68 | 3 | 71 |
| | Others | 13 | 0 | 13 |
| | Total | 114 | 18 | 132 |
| Online course or education platform | Education | 12 | 3 | 15 |
| | Healthcare | 10 | 0 | 10 |
| | Technology | 9 | 3 | 12 |
| | Finance and banking | 64 | 3 | 67 |
| | Others | 10 | 0 | 10 |
| | Total | 105 | 9 | 114 |
| Research papers and journals | Education | 3 | 0 | 3 |
| | Healthcare | 12 | 3 | 15 |
| | Technology | 3 | 0 | 3 |
| | Finance and banking | 48 | 7 | 55 |

(continued)

**Table 3** (continued)

| Sources for gathering knowledge | Nature of sector employed | Have you ever invested | | Total |
|---|---|---|---|---|
| | | Yes | No | |
| | Others | 21 | 3 | 24 |
| | Total | 87 | 13 | 100 |

*Source* Computed Data

**Table 4** Cross tabulation between annual income and nature of sector employed

| Annual income | Nature of sector employed | | | | | Total |
|---|---|---|---|---|---|---|
| | Education | Healthcare | Technology | Finance and banking | Others | |
| Below Rs. 5 lakhs | 3 | 0 | 9 | 9 | 7 | 28 |
| 5–7 lakhs | 18 | 10 | 3 | 52 | 18 | 101 |
| 7–10 lakhs | 22 | 15 | 3 | 20 | 23 | 83 |
| Above 10 lakhs | 16 | 12 | 9 | 99 | 36 | 172 |
| Total | 59 | 37 | 24 | 180 | 84 | 384 |

*Source* Computed Data

**Table 5** Chi-square tests

| | Value | df | Asymp. Sig. (2-sided) |
|---|---|---|---|
| Pearson chi-square | 72.066[a] | 12 | 0.000 |
| Likelihood ratio | 60.497 | 12 | 0.000 |
| Linear-by-linear association | 2.720 | 1 | 0.099 |
| N of valid cases | 384 | | |

*Source* Computed Data

Table 10 shows the mean value of both independent and dependent factors. The independent variable includes variables that are classified under the different heads of Transaction Efficiency and Stability, Adoption Barriers, Security and Resource Management, Knowledge and Compliance Gaps, and Regulatory Challenges. The table extends by showcasing the dependent variables listed under different heads of **User Experience and Accessibility, Education and Community Engagement, Regulation and Governance, Security and Integrity, Transparency and Fairness, and Infrastructure and Technology**).

Table 11 shows that there is a significant relationship between requirements and challenges faced by the respondents while making decisions about investments in digital currencies and bitcoin. The $r = 0.523$ which implies a moderate impact between requirements and challenges (Table 11).

Table 12 shows that there is a significant impact between requirements and challenges faced by the respondents while making decisions about investments in digital

**Table 6** ANOVA and t-test between impact and demographic variable

| Variables | Group | N | Mean | Std. deviation | T/F value | Sig | S/NS |
|---|---|---|---|---|---|---|---|
| Gender | Male | 223 | 4.00 | 0.445 | 0.843 | 0.400 | NS |
| | Female | 161 | 4.04 | 0.401 | | | |
| Age | Below 25 years | 45 | 3.80 | 0.405 | 8.119 | 0.000 | S** |
| | 26–35 years | 153 | 3.95 | 0.403 | | | |
| | 36–45 years | 87 | 4.06 | 0.466 | | | |
| | 46–55 years | 72 | 4.14 | 0.348 | | | |
| | Above 56 | 27 | 4.26 | 0.447 | | | |
| Education | UG | 125 | 4.01 | 0.449 | 5.813 | 0.001 | S** |
| | PG | 140 | 3.92 | 0.450 | | | |
| | Doctorate | 67 | 4.10 | 0.308 | | | |
| | Others | 52 | 4.17 | 0.382 | | | |
| Occupation | Private sector | 220 | 4.10 | 0.381 | 18.711 | 0.000 | S** |
| | Public sector | 103 | 3.81 | 0.525 | | | |
| | Self-employed | 61 | 4.07 | 0.250 | | | |
| Nature of sector employed | Education | 59 | 3.83 | 0.497 | 5.858 | 0.000 | S** |
| | Healthcare | 37 | 4.08 | 0.277 | | | |
| | Technology | 24 | 4.13 | 0.338 | | | |
| | Finance and banking | 180 | 4.09 | 0.440 | | | |
| | Others | 84 | 3.93 | 0.373 | | | |
| Annual income | Below Rs. 5 lakhs | 28 | 3.79 | 0.418 | 5.865 | 0.001 | S** |
| | 5–7 lakhs | 101 | 3.93 | 0.453 | | | |
| | 7–10 lakhs | 83 | 4.07 | 0.376 | | | |
| | Above 10 lakhs | 172 | 4.08 | 0.419 | | | |

*NS* Not Significant
S** Significant at a 1% level
*Source* Computed Data

**Table 7** KMO and Bartlett's test for the challenges faced

| Kaiser–Meyer–Olkin measure of sampling adequacy | | 0.709 |
|---|---|---|
| Bartlett's test of sphericity | Approx. chi-square | 7360.293 |
| | df | 465 |
| | Sig | 0.000 |

*Source* Computed Data

**Table 8** Total variance explained

| Total variance explained | | | | | | |
|---|---|---|---|---|---|---|
| Component | Initial eigenvalues | | | Extraction sums of squared loadings | | |
| | Total | % of variance | Cumulative % | Total | % of variance | Cumulative % |
| 1 | 8.109 | 26.160 | 26.160 | 8.109 | 26.160 | 26.160 |
| 2 | 3.410 | 11.000 | 37.159 | 3.410 | 11.000 | 37.159 |
| 3 | 2.178 | 7.025 | 44.184 | 2.178 | 7.025 | 44.184 |
| 4 | 2.065 | 6.660 | 50.844 | 2.065 | 6.660 | 50.844 |
| 5 | 1.728 | 5.575 | 56.419 | 1.728 | 5.575 | 56.419 |

*Source* Computed Data

currencies and bitcoin. The R square (0.274) implies a moderate influence between two variables (Table 12).

Table 13 shows that there is a significant association ($p < 0.01$) between requirements and challenges faced by the respondents while deciding investments in digital currencies and bitcoin (Table 13).

Table 14 reflects that there is a positive impact between the variables. Further, the table expresses that a positive increase in the independent variable positively impacts the dependent variable (Table 14).

## 8 Suggestion

- The government should take the necessary steps to encourage the high-income group of people to invest more in Bitcoin, e.g., by providing tax concession
- Awareness programs on digital currency and bitcoin investment must be conducted in all the sectors of employment to overcome the unequal diversification in the portfolio management of respondents other than respondents working in the Finance and Banking sector
- Since younger individuals (below 25 years) showed a lower mean score, consider launching financial literacy programs focusing on cryptocurrency and traditional financial systems.
- Public sector employees showing a lower mean score, explore strategies or incentives that encourage these workers to engage with or invest in cryptocurrencies
- Policymakers can establish clear regulatory frameworks to reduce uncertainty and encourage investment in cryptocurrencies
- Financial Institutions can invest in enhanced security protocols and develop user-friendly platforms to simplify the investment process and boost confidence.

**Table 9** Factor analysis

| Rotated component matrix | Component | Factors |
|---|---|---|
| Transaction volume or stability | 0.734 | Transaction efficiency and stability |
| Lack of public awareness and utilization of blockchain technology | 0.716 | |
| Permanent record of transaction history | 0.707 | |
| Transaction fees involved in investments | 0.654 | |
| Network issues involved while executing transactions | 0.652 | |
| Time constraints involved in the verification of executed transactions | 0.650 | |
| Network capacity | 0.633 | |
| Privacy issues | 0.807 | Adoption barriers |
| Sel-risk, lack of ownership and power over digital currencies investments | 0.786 | |
| Investing in digital currencies is a complex process | 0.783 | |
| Cross-communication while using the wireless communication network | 0.739 | |
| Environmental issues (energy intensive and carbon footprints) | 0.657 | |
| Cultural barriers involved in investment | 0.653 | |
| Security for user accessibility/interface authentication | 0.746 | Security and resource management |
| Adequate capacity of resources for storage of transaction data | 0.737 | |
| Market fluctuations | 0.718 | |
| Lack of understanding of the working model of digital currency and bitcoin | 0.754 | Knowledge and compliance gaps |
| Increase in chances of money laundering activities | 0.754 | |
| Monopoly role of government | 0.789 | Regulatory challenges |
| Inability to self-regulate | 0.756 | |
| No refund policy in case of wrong transactions | 0.678 | |

**Table 10** Descriptive statistics

| Variables | Mean | Std. deviation | N |
|---|---|---|---|
| Requirements for upscaling of investment in digital currencies and bitcoin | 4.19 | 0.498 | 384 |
| Challenges faced in investing in digital currencies and bitcoin | 4.04 | 0.579 | 384 |

*Source* Computed Data

**Table 11** Correlations

| | | Requirements | Challenges |
|---|---|---|---|
| Pearson correlation | Requirements | 1.000 | 0.523 |
| | Challenges | 0.523 | 1.000 |
| Sig. (1-tailed) | Requirements | – | 0.000 |
| | Challenges | 0.000 | – |
| N | Requirements | 384 | 384 |
| | Challenges | 384 | 384 |

*Source* Computed Data

**Table 12** Model summary

| Model | R | R square | Adjusted R square | Std. error of the estimate |
|---|---|---|---|---|
| 1 | 0.523[a] | 0.274 | 0.272 | 0.425 |

[a] *Predictors* (Constant), Challenges
*Source* Computed Data

**Table 13** ANOVA

| Model | | Sum of squares | df | Mean square | F | Sig. |
|---|---|---|---|---|---|---|
| 1 | Regression | 26.021 | 1 | 26.021 | 143.849 | 0.000[b] |
| | Residual | 69.101 | 382 | 0.181 | | |
| | Total | 95.122 | 383 | | | |

[a] *Dependent Variable* Requirements
[b] *Predictors* (Constant), Challenges

**Table 14** Coefficients

| Model | | Unstandardized coefficients | | Standardized coefficients | t | Sig. |
|---|---|---|---|---|---|---|
| | | b | Std. Error | Beta | | |
| 1 | (Constant) | 2.368 | 0.153 | | 15.436 | 0.000 |
| | CAVG | 0.450 | 0.038 | 0.523 | 11.994 | 0.000 |

[a] *Dependent Variable* Requirements

# 9 Conclusion

The study has explored the multifaceted impact of Bitcoin diversification on traditional financial systems, identifying significant shifts in investment behavior and market dynamics. The findings indicate that as investors increasingly allocate resources to Bitcoin, traditional financial institutions face pressures to adapt, innovate, and integrate cryptocurrency into their offerings.

In addressing the challenges associated with digital currencies and blockchain networks, it highlighted key issues such as security vulnerabilities, regulatory uncertainties, and adoption barriers. These challenges necessitate concerted efforts from stakeholders including policymakers, financial institutions, and technology providers to create a secure and compliant environment that fosters trust and encourages broader adoption.

Moreover, the analysis reveals specific Bitcoin investors' requirements, such as enhanced security measures, clearer regulatory guidelines, and accessible educational resources. By meeting these demands, stakeholders can facilitate increased usage of Bitcoin and strengthen its position within the broader financial ecosystem. As the investment landscape evolves, understanding these dynamics and addressing the associated challenges will be crucial for both individual investors and traditional financial systems to thrive in a rapidly changing economic environment. The continued collaboration among all stakeholders is essential to harness the full potential of digital currencies while mitigating associated risks.

**Copyright Permission Evidence** The content used in this book chapter does not include any material from copyrighted documents. All data were sourced from publicly available, open-access documents, ensuring compliance with copyright regulations. All tables included in the chapter were created by the authors, based on the data generated from this study. This chapter incorporates findings from a quantitative study conducted using primary data collected from respondents. Prior to participation, informed consent was obtained from all respondents, who were fully informed about the purpose and nature of the data collection. To safeguard privacy, the anonymity of all respondents will be strictly maintained, ensuring that their identities remain confidential throughout the research and publication process.

## References

1. Aggarwal, S., Kumar, N.: The blockchain technology for secure and smart applications across industry verticals. Adv. Comput. **121**, 227–266 (2021). https://doi.org/10.1016/bs.adcom.2020.08.012
2. Agrawal, T., Kumar, V., Pal, R., Wang, L., Chen, Y.: Blockchain-based framework for supply chian traceability: a case example of textile and clothing industry. Comput. Ind. Eng. **154** (2021). https://doi.org/10.1016/j.cie.2021.107130
3. Anbugeetha, D., Nandhini, B.: Evolution of money: from barter system to digital money. New Era Digit. Payments 55–65. New Chennai Publications (2021)
4. Attaran, M., Gunasekaran, A.: Blockchain-enabled technology: the emerging technology set to reshape and decentralise many industries. Int. J. Appl. Decis. Sci. **12**(4), 424–444 (2019). https://doi.org/10.1504/IJADS.2019.102642
5. Aysan, A., Demir, E., Gozgar, G., Lau, C.: Effects of geopolitical risks on bitcoin returns and volatility. Res. Int. Bus. Financ. **47** (2019). https://doi.org/10.1016/j.ribaf.2018.09.011
6. Azar, P., Casillas, A., Farboodi, M.: The DeFi intermediation chain. Liberty Street Econ. (2024). https://libertystreeteconomics.newyorkfed.org/2024/08/the-defi-intermediation-chain/#:~:text=Decentralized%20Finance%2C%20or%20DeFi%2C%20is,borrowing%2C%20trading%2C%20and%20insurance
7. Banshal, D.: Smart Contracts: Game-changer for Streamlining Processes and Reshaping Industries. DAMCO (2024). https://www.damcogroup.com/blogs/smart-contracts-for-streamlining-processes-and-reshaping-industries

8. Baur, D., Hoang, L.: The bitcoin gold correlation puzzle. J. Behav. Exp. Financ. **32** (2021). https://doi.org/10.1016/j.jbef.2021.100561
9. Baur, D., Dimpfl, T.: The volatility of bitcoin and its role as a medium of exchange and a store of value. Empir. Econ. **61**, 2663–2683 (2021). https://doi.org/10.1007/s00181-020-01990-5
10. Bourveau, T., Brendel, J., Schoenfeld, J.: Decentralized finance (DeFi) assurance: early evidence. Rev. Acc. Stud. **29** (2024). https://doi.org/10.1007/s11142-024-09834-8
11. Bunjaku, F., Gjorgieva-Trajkovska, O., Miteva-Kacarski, E.: Cryptocurrencies—advantages and disadvantages. J. Econ. **2**(1) (2017)
12. Buterin, V.: A next generation smart contract & decentralized application platform. Ethereum White Pap. (2015). https://blockchainlab.com/pdf/Ethereum_white_paper-a_next_generation_smart_contract_and_decentralized_application_platform-vitalik-buterin.pdf
13. Chauriha, S. (2024). Are smart contracts really smart? Law Sch. Policy Rev. Kautilya Soc. https://lawschoolpolicyreview.com/2024/01/13/are-smart-contracts-really-smart/#:~:text=The%20idea%20behind%20smart%20contracts,transactions%20more%20efficient%20and%20trustworthy
14. Chao, C., Ting, I., Tseng, Y., Wang, B., Wang, S., Wang, Y.: The study of DeFi in virtual community and financial supervision. In: MISNC'22: Proceedings of the 9th Multidisciplinary International Social Networks Conference, pp. 74–80 (2022). https://doi.org/10.1145/3561278.3561297
15. Chen, Y.: Blockchain tokens and the potential democratization of entrepreneurship and innovation. Bus. Horiz. **61**(4), 567–575 (2018). https://doi.org/10.1016/j.bushor.2018.03.006
16. Cheney, C.: In Sierra Leone, New Kiva Protocol Uses Blockchain to Benefit Unbanked. Devex (2019). https://www.devex.com/news/in-sierra-leone-new-kiva-protocol-uses-blockchain-to-benefit-unbanked-95490
17. CRISIL: The Big Shift in Financialisation (2022). https://www.crisil.com/content/dam/crisil/our-analysis/reports/Research/documents/2022/12/the-big-shift-in-financialisation.pdf
18. Diaz, A., Esparcia, C., Huelamo, D.: Stablecoins as a tool to mitigate the downside risk of cryptocurrency portfolios. N. Am. J. Econ. Financ. **64** (2023). https://doi.org/10.1016/j.najef.2022.101838
19. Duan, K., Urquhart, A.: The instability of stablecoins. Financ. Res. Lett. **52** (2023). https://doi.org/10.1016/j.frl.2022.103573
20. Elsayed, A., Gozgor, G., Lau, C.: Risk transmissions between bitcoin and traditional financial assets during the COVID-19 era: the role of global uncertainities. Int. Rev. Financ. Anal. **81** (2022). https://doi.org/10.1016/j.irfa.2022.102069
21. Farell, R.: An analysis of the cryptocurrency industry. Whart. Res. Sch. **130** (2015). http://repository.upenn.edu/wharton_research_scholars/130
22. Hayes, A.: Blockchain Facts: What Is It, How It Works, and How It Can Be Used. Investopedia (2024). https://www.investopedia.com/terms/b/blockchain.asp
23. Hayes, A.: 10 Important Cryptocurrencies Other Than Bitcoin. Investopedia (2024). https://www.investopedia.com/tech/most-important-cryptocurrencies-other-than-bitcoin/
24. Herman, A.: The Economics of Bitcoin. Durham University (2014). https://cointhinktank.com/upload/The%20Economics%20of%20Bitcoin.pdf
25. Huynh-The, T., Gadekallu, T., Wang, W., Yenduri, G., Ranaweera, P., Pham, Q., Costa, D., Liyanage, M.: Blockchain for the metaverse: a review. Futur. Gener. Comput. Syst. **143** (2023). https://doi.org/10.1016/j.future.2023.02.008
26. IBM.: What is Blockchain? (2024). https://www.ibm.com/topics/blockchain
27. IBM.: Benefits of Blockchain (2024). https://www.ibm.com/topics/benefits-of-blockchain
28. Ikegwu, C.: Governance challenges faced by the bitcoin ecosystem: the way forward. Int. J. Cryptocurrency Res. **3**(2), 74–83 (2023). https://doi.org/10.51483/IJCCR.3.2.2023.74-83
29. Infosys.: Rise of DeFi: How Decentralized Finance is Revolutionizing the Landscape (2024). https://www.infosysbpm.com/blogs/finance-accounting/rise-of-defi-how-decentralized-finance-is-revolutionizing-the-landscape.html
30. Invest India: The Rise of Digital Currencies: Opportunities for Economies (2024). https://www.investindia.gov.in/team-india-blogs/rise-digital-currencies-opportunities-economies

31. Iwai, K.: Evolution of money. Ugo Pagano and Antonio Nicita eds. Evol. Econ. Divers. (1997). https://www.researchgate.net/publication/24135537_Evolution_of_Money
32. Javaid, M., Haleem, A., Singh, R., Suman, R., Khan, S.: A review of blockchain technology applications for financial services. BenchCouncil Trans. Benchmarks, Stan. Eval. **2**(3) (2022). https://doi.org/10.1016/j.tbench.2022.100073
33. Jose, M.: The Evolution of Cryptocurrencies: Decentralized Finance. LinkedIn (2024). https://www.linkedin.com/pulse/evolution-cryptocurrencies-part-3-decentralized-finance-jose-manuel-efrkc
34. Joseph, J.: Digital currencies and cross-border payments: an overview. U.S. Int. Trade Comm. (2023). https://www.usitc.gov/publications/332/executive_briefings/ebot_digital_currency.pdf
35. Kaspersky.: What is Cryptocurrency and How Does It Work? https://www.kaspersky.com/resource-center/definitions/what-is-cryptocurrency
36. Kimani, D., Adams, K., Attah-Boakye, R., Ullah, S., Frecknall-Hughes, J., Kim, J.: Blockchain, business, and the fourth industrial revolution: whence, whither, wherefore and how? Technol. Forecast. Soc. Chang. **161** (2020). https://doi.org/10.1016/j.techfore.2020.120254
37. Kriptomat.: A Short History of Cryptocurrencies. https://kriptomat.io/cryptocurrencies/history-of-cryptocurrency/
38. Kucheryavenko, M.P., Dmytryk, O., Golovashevych, O.: Cryptocurrencies: development, signs, and classification. Financ. Credit Act. Probl. Theory Pract. **3**(30), 371–374. https://doi.org/10.18371/fcaptp.v3i30.179737
39. Lorenz, G.: Regulating decentralized financial technology: a qualitative study on the challenges of regulating DeFi with a focus on embedded supervision. Standford J. Blockchain Law Policy (2024). https://stanford-jblp.pubpub.org/pub/regulating-defi/release/1
40. Luu, H., Nguyen, C., Nasir, M.: Implications of central bank digital currency for financial stability: evidence from the global banking sector. J. Int. Finan. Markets. Inst. Money **89** (2023). https://doi.org/10.1016/j.intfin.2023.101864
41. Maheshwari, R.: Popular Types of Cryptocurrencies & How Do They Work. Forbes (2023). https://www.forbes.com/advisor/in/investing/cryptocurrency/types-of-cryptocurrency/
42. Majeed, U., Khan, L., Yaqoob, I., Kazmi, A., Slah, K., Hong, C.: Blockchain for IoT-based smart cities: recent advances, requirements, and future challenges. J. Netw. Comput. Appl. **181** (2021). https://doi.org/10.1016/j.jnca.2021.103007
43. Mashhour, O., Aziz, A.: The era of cryptocurrencies: a study about the advantages and disadvantages. In: SOIS Conference on Arts and Humanities (SICAH 2019) (2019)
44. Mhlanga, D.: Blockchain technology for digital financial inclusion in the industry 4.0, towards sustainable development? Blockchain Econ. **6** (2023). https://doi.org/10.3389/fbloc.2023.1035405
45. Muniyandy, E., Radhika, V.S., Mohammad, S., Joyice, S., Dasari, T., Verma, A., Kiran, A.: Blockchain technology for secure and trustworthy decentralized applications. Int. J. Intell. Syst. Appl. Eng. **12**(21s), 875–883 (2024). https://ijisae.org/index.php/IJISAE/article/view/5484
46. Nakamoto, S.: Bitcoin: A Peer-to-Peer Electronic Cash System. Available at United States Sentencing Commission (2009). https://www.ussc.gov/sites/default/files/pdf/training/annual-national-training-seminar/2018/Emerging_Tech_Bitcoin_Crypto.pdf
47. Pele, D., Wesselhofft, N., Hardle, W., Kolossiatis, M., Yatracos, Y.: Classification of cryptocurrencies. SSRN (2020). https://doi.org/10.2139/ssrn.3548462
48. Pinkerton, J. (2024). The history of bitcoin. U.S.News (2024). https://money.usnews.com/investing/articles/the-history-of-bitcoin#:~:text=Bitcoin%20was%20launched%20in%20January,identity%20has%20never%20been%20verified.
49. Prasad, E.: The brutal truth about bitcoin. The New York Times (2021). https://www.nytimes.com/2021/06/14/opinion/bitcoin-cryptocurrency-flaws.html
50. Qaroush, Z., Zakarneh, S., Dawabsheh, A.: Cryptocurrencies advantages and disadvantages: a review. Int. J. Appl. Sci. Smart Technol. **4**(1) (2022)
51. Qiu, T., Zhang, R., Gao, Y.: Ripple vs SWIFT: transforming cross border remittance using blockchain technology. In: Procedia Computer Science. 2018 International Conference on

Identification, Information, and Knowledge in the Internet of Things, IIKI 2018, vol. 147, pp. 428–434 (2019). https://doi.org/10.1016/j.procs.2019.01.260
52. Ravikiran, A.S.: What is Blockchain Technology? How Does Blockchain Work? SimpliLearn (2024). https://www.simplilearn.com/tutorials/blockchain-tutorial/blockchain-technology#:~:text=A%20blockchain%20platform%20is%20a,consensus%20among%20the%20network%20participants
53. Ruhmann, F., Konda, S., Horrocks, P., Taka, N.: Can blockchain technology reduce the cost of remittances? OECD Dev. Co-Oper. Work. Pap. **73**. https://www.oecd-ilibrary.org/docserver/d4d6ac8f-en.pdf?expires=1727088820&id=id&accname=guest&checksum=69E43C0C03D7A491FCB79EBE59A66E29
54. Seth, S.: What is Central Bank Digital Currency? (CBDC). Investopedia (2024). https://www.investopedia.com/terms/c/central-bank-digital-currency-cbdc.asp#:~:text=Key%20Takeaways,of%20monetary%20and%20fiscal%20policies
55. Sharif, M., Ghodoosi, F.: The ethics of blockchain in organizations. J. Bus. Ethics 1009–1025 (2022). https://doi.org/10.1007/s10551-022-05058-5
56. Shyamasundar, R.K., Patil, V.: Blockchain: revolution in TRUST. Proceed. Ind. Natl. Sci. Acad. **96**(2) (2018). https://doi.org/10.16943/ptinsa/2018/49340
57. Spithoven, A.: Theory and reality of cryptocurrency governance. J. Econ. Issues **53**(2), 385–393 (2019). https://doi.org/10.1080/00213624.2019.1594518
58. Singh, G.: Blockchain Use Case: Revolutionizing real-world Asset Tokenization on the Bitcoin Blockchain. LinkedIn (2024). https://www.linkedin.com/pulse/blockchain-use-case-revolutionizing-real-world-asset-bitcoin-singh-wiygf
59. Sristy, A.: Blockchain in the Food Supply Chain—What Does the Future Look Like? (2021). https://tech.walmart.com/content/walmart-global-tech/en_us/blog/post/blockchain-in-the-food-supply-chain.html
60. Subramani, D.: Ethereum Blockchain for Decentralized Applications (History & Introduction)—Part 1. LinkedIn (2024). https://www.linkedin.com/pulse/ethereum-blockchain-decentralized-applications-part-1-subramani-vexac
61. Tan, E., Mahula, S., Crompvoets, J.: Blockchain governance in the public sector: a conceptual framework for public management. Gov. Inf. Q. **39** (2022). https://doi.org/10.1016/j.giq.2021.101625
62. Tang, Y.: Central bank digital currencies: a comprehensive study of characteristics, implications, and future perspectives. Adv. Econ. Manage. Polit. Sci. **65**(1), 232–238 (2023). https://doi.org/10.54254/2754-1169/65/20231641
63. Tripathi, S.: Smart Contracts Explained: Understanding Their Functionality on the Blockchain. Spydra (2024). https://www.spydra.app/blog/smart-contracts-explained-understanding-their-functionality-on-the-blockchain
64. Tschorsch, F., Scheuermann, B.: Bitcoin and beyond: a technical survey on decentralized digital currencies. IEEE Commun. Surv. Tutor. **18**(3), 2084–2123 (2016). https://doi.org/10.1109/COMST.2016.2535718
65. Vasudeva, S.: Cryptocurrency as an investment or speculation: a bibliometric review study. Bus. Anal. J. **44**(1), 34–50 (2023). https://doi.org/10.1108/BAJ-07-2022-0008
66. Zarrin, J., Phang, H., Saheer, L., Zarrin, B.: Blockchain for decentralization of internet: prospects, trends, and challenges. Clust. Comput. **24**, 2841–2866 (2021). https://doi.org/10.1007/s10586-021-03301-8
67. Zheng, Z., Xie, S., Dai, H., Chen, W., Chen, X., Weng, J., Imran, M.: An overview on smart contracts: challenges, advances and platforms. Futur. Gener. Comput. Syst. **105**, 475–491 (2020). https://doi.org/10.1016/j.future.2019.12.019

## *Websites*

68. https://en.wikipedia.org/wiki/History_of_money
69. https://www.pwc.com/us/en/industries/financial-services/library/cryptocurrency-evolution.html#:~:text=Cryptocurrencies%20carry%20groundbreaking%20potential%20to,history%20or%20a%20bank%20account
70. https://www.ibm.com/blockchain/industries/financial-services https://corporatefinanceinstitute.com/resources/cryptocurrency/types-of-cryptocurrency/#:~:text=Broadly%20speaking%2C%20we%20will%20classify,and%20Central%20Bank%20Digital%20Currencies.
71. https://byjus.com/current-affairs/cryptocurrency/
72. https://en.wikipedia.org/wiki/Bitcoin#:~:text=As%20a%20decentralized%20system%2C%20bitcoin,blockchain%20that%20records%20bitcoin%20transactions.
73. https://trustmachines.co/learn/what-is-segregated-witness-segwit/
74. https://utimaco.com/service/knowledge-base/blockchain/what-smart-contract-blockchain#:~:text=Definition%3A%20A%20smart%20contract%20is,a%20distributed%2C%20decentralized%20blockchain%20network.
75. https://consensys.io/blockchain-use-cases/law
76. https://www.ibm.com/topics/smart-contracts
77. https://webisoft.com/articles/blockchain-applications/

# Digital Transformation for Supplier Selection: A Pragmatic Constructivist Case Study at a Restaurant Chain in Vietnam

**Nguyen Thi Kim Oanh, Chi-Minh Tăng, and Tran Thi Thuy Anh**

**Abstract** This chapter presents a case study of digital transformation for supplier selection at a leading Vietnamese restaurant chain through a pragmatic constructivist perspective focusing on four dimensions: facts, possibilities, values, and communication. This perspective provides a theoretical framework for analyzing the success of digital transformation for supplier selection. The findings show that the restaurant chain faced challenges in operational efficiency (e.g., resource allocation) and control issues, which were derived from manual processes, personal relationship-based decisions, system shortages, and inconsistent evaluation criteria. The possibilities offer the restaurant chain solutions to overcome these challenges with employment of various technological infrastructure, including E-ticket, E-process for task management, E-bidding for supplier sourcing, the Delivery in Full on Time (Difot) for evaluating supplier performance, and SAP integration to manage supplier data. These digital elements enable the restaurant chain to standardize its supplier selection process and improve its monitoring procedures. The underlying value for this transformation is fair competition among suppliers, leading to better operational outcomes. Digital transformation enhances objectivity and transparency of the supplier selection process, resulting in higher operational efficiency and supplier quality. These three dimensions are realized through communication with two key factors: coordination across the restaurant chain's departments for reconciliation of diverse objectives and knowledge sharing through internal staff development and supplier engagement programs. This case study provides rich descriptions of digital transformation for supplier selection in the context of the restaurant chain through the lens of pragmatic constructivism.

**Keywords** Digital transformation · Supplier selection · Pragmatic constructivist · Restaurant chain

N. T. K. Oanh (✉) · C.-M. Tăng · T. T. T. Anh
Vietnam National University, Hanoi, Vietnam
e-mail: nguyenthikimoanh@vnu.edu.vn



N. D. K. Nguyen et al. (eds.), *Digital Transformation and Sustainability: Methods and Applications*, Information Systems Engineering and Management 47,
https://doi.org/10.1007/978-3-031-91278-8_7

# 1 Introduction

Digital transformation has increased its importance for organizations that seek to optimize their operations and maintain competitive advantage in today competitive business environment [1]. It leads to a fundamental change in the way the business operates and how it delivers value to customers through three organizational layers: organizational core, organizational periphery, and external environment [2]. Digital transformation brings huge benefits and positive outcomes for supply chain management, an important element of external environment in the restaurant industry. By forecasting demand more accurately, the restaurants can optimize inventory levels, minimize food waste, and ensure ingredient availability to meet customer needs [3–5]. Digital transformation with adoption of blockchain technology enables the restaurant to ensure food traceability, food safety by tracking food products from farms to table [6, 7], enhancing compliance with regulatory requirements and builds consumer trust [8]. Cloud-based supply chain management platforms integrate data from various sources to provide real-time visibility across the restaurant's network of suppliers, distributors, and logistics providers [9]. Improved visibility enables better coordination, faster issue resolution, and more effective collaboration with supply chain partners [4, 10]. Internet of Things (IoT)-enabled tracking devices and route optimization algorithms are helping restaurants better manage their delivery fleets and third-party logistics providers [11, 12]. This results in more efficient deliveries, reduced fuel consumption, and improved customer experience with on-time order fulfillment [11, 13]. Although digital transformation brings significant benefits to restaurants, successful implementation requires a strategic, cross-functional approach that aligns technology investments with the unique needs and challenges of the restaurant industry [14].

Supplier selection is a vital aspect of supply chain management, which has become increasingly critical for the restaurant industry, particularly as organizations seek to optimize operations and maintain competitiveness in a dynamic market environment [15]. This is because the costs of main food raw materials accounts for an average of 30–50% of total operating costs while that of tools, equipment, and IT infrastructure accounts for 15–30% of the investment cost [16]. Moreover, restaurants face unique challenges in their supplier selection due to the perishable nature of ingredients, varying quality requirements, and complex supply chain networks [17]. These challenges are magnified by the industry's stringent food safety regulations, fluctuating demand patterns, and the need for consistent quality across multiple locations [18, 19]. Embracing digital transformation enables restaurants to reduce costs [20, 21], achieve better financial performance [22, 23], improve environmental performance [24], and contribute to both operational and strategic strategy implementation [23, 25]. This importance has been increasing due to more intense competitiveness and more demanding customers. Therefore, selecting suppliers that ensure quality, price competitiveness, delivery time, and achievement of business targets—such as providing sufficient volume to business locations across different geographical areas—is crucial for most restaurant businesses [15].

In emerging markets like Vietnam, where the food service industry has experienced significant growth and modernization over the past decade [26, 27]. The Vietnamese restaurant industry exhibits intense competition across its diverse segments, ranging from casual eateries to fine dining establishments since a significant transformation in Vietnam's restaurant industry in early 2000s marked by the entry of international food chains like KFC, Pizza Hut, McDonald's, and Lotteria; barbecue and hot pot restaurants such as Gogi, Ashima, and Kichi Kichi; the emergence of prominent beverage chains such as Highlands Coffee, The Coffee House, Phuc Long, and Starbucks; cloud kitchens such as GrabFood, Baemin, and ShopeeFood. While historical competitive advantages focused on cuisine novelty, atmosphere, and brand recognition, contemporary competition emphasizes operational aspects, including food quality, service excellence, and efficiency. [28]'s competitive pressures shape the industry dynamics. The market has minimal entry barriers, attracting both international and domestic players. Buyer power is moderated by market segmentation across price points, from budget to premium offerings. High market entry rates create substantial substitution risks, with rapid establishment and replacement cycles evident across different segments. Supplier power varies by category. For imported food suppliers, sufficient volume remains challenging, with Vietnam's import volume relatively small compared to larger markets. Domestic suppliers generally have reduced bargaining power due to numerous alternatives, except for specific items where limited suppliers can meet large-scale demands. In this competition landscape, full-service restaurants have experienced an increase in market share in since 2017 and accounted for the largest market share in the near future [27].

This chapter presents an in-depth case study of digital transformation in supplier selection at one of Vietnam's leading restaurant chains, focusing on how the restaurant chain relates the reality of supplier selection processes to successfully drive digital transformation for supplier selection [29]. Currently, this growing restaurant chain that has numerous restaurants with a number of brands operating in multiple provinces in Vietnam improve efficiency of its supplier management processes operational efficiency by adopting digital solutions to manage complex operations [30–32]. Through the pragmatic constructivist framework proposed by Nørreklit [33], this study examines the digital transformation journey of the restaurant chain from traditional, relationship-based supplier selection to a sophisticated digital ecosystem. The analysis explores four interconnected dimensions: facts, possibilities, values, and communication [34]. Through nalysing interview data obtained from key stakeholders across different organizational levels, this chapter offers valuable insights into the challenges, solutions, and lessons learned during the digital transformation process. The findings are particularly relevant for organizations in emerging markets seeking to modernize their supplier selection processes while managing complex stakeholder relationships and organizational change [35].

## 2 Literature Review

### 2.1 Digital Transformation in the Restaurant Industry

Digital transformation refers to employment of digital technology to "create new or modify existing business models and processes or to support the transformation of organizational structure, resources, or relationships with internal and external actors" ([2], p. 1). This creates fundamental changes in multi-organizational layers, including organizational core, organizational periphery, and external environment. The first level of organizational core consists of three aspects: business models, organizational structures, and organizational culture; the second level of organizational periphery includes three components: organizational processes, digital platforms, and digital servitization. The third level of external environment encompass five features: competitive dynamics, digital ecosystems, cooperation with customers, digital ecosystems, supply chain management, and sustainability [2]. Digital transformation significantly influences the parameters of strategic business management, fundamentally reshaping operational processes and organizational interactions based on new technologies, representing changes in organizational thinking and operational methodologies [36]. This transformation leads to adjustments of organizations' management strategies to ensure alignment between technological capabilities and organizational objectives [37].

Digital transformation consists of three phases: digitization, digitalization, and digital transformation. Organizations must progress through the first two phases before reaching the third phase of transformation. Digitization refers to converting analog information into digital formats that can be processed and transmitted by computers. While digitization changes documentation processes, it does not create added value for organizations. However, it generates data and information that enables organizations to effectively automate tasks that previously required human intervention [38], resulting in reduced human errors, increased efficiency and accuracy in operational activities, and cost reduction [39]. This phase includes automation where processes are automated based on generated digital data [40]. This demonstrates that these two concepts are not only closely related but also mutually supportive in improving and optimizing operational processes. Digitalization refers to employing digital technologies to transform existing operational processes from production to firm-customer interactions, resulting in operational efficiency through productivity improvement and cost savings while creating added value for customers by enhancing their experiences [40]. The third phase represents the most pervasive change, coupled with rearranged operational processes and development of new business models to attain competitive advantages derived from new business logic regarding value creation and delivery to customers [40]. In the hospitality industry, the integration of digital transformation across various organizational levels enables firms to reduce operating costs, simplify workflows, and improve staff productivity, production efficiency [3], and firm-level productivity [41].

Digital transformation in restaurant chains' supplier selection processes plays a crucial role in improving operational efficiency and competitiveness, particularly in supply chain management. It is key to improving sustainability and operational efficiency in digital food supply chains [42], enhancing transparency, and improving supplier management quality [43]. The integration of new technology is a major limiting factor in digital transformation, affecting supplier selection process optimization [44]. Factors such as supplier digitalization capabilities [45], supplier-customer integration [46], and blockchain technology adoption [47] are vital in digital transformation for supplier selection processes. Optimization and efficiency of supplier selection can be enhanced through improved cybersecurity [48], or artificial intelligence (AI) in supplier search [49].

## 2.2 *Supplier Selection in Restaurant Chains*

Supplier selection is a complex process that involves evaluating and selecting suppliers based on a set of specific criteria [23]. In the restaurant industry, these criteria in the restaurant industry are often divided into many different categories, including price, product quality, delivery time, and supplier service ability [50]. Besides the criteria, sustainability and environmental impact are becoming more important than ever in supplier selection decisions. This is because customers are prioritizing restaurants with green strategies, thus, choosing environmentally responsible suppliers not only enhances brand image but also creates added value for customers [24]. Restaurants establish clear performance indicators and conduct periodic reviews is essential to not only maintain supplier relationships but also improve supply chain efficiency. This allows restaurants to make timely adjustments to factors that affect their quality and performance [51].

The supplier selection process follows a systematic seven-step cycle that ensures comprehensive evaluation and strategic decision-making. The process begins with recognizing the need for supplier selection, which may arise from business expansion, new product requirements, or performance issues with existing suppliers. This leads to identifying key sourcing requirements, where specific criteria including quality standards, volume needs, and delivery specifications are determined. The third step involves determining the sourcing strategy, encompassing decisions about single versus multiple sourcing or global versus local sourcing approaches. Following this, organizations conduct market research to identify potential supply sources that could meet their requirements. The fifth step involves limiting suppliers in the selection pool to a manageable number based on preliminary criteria and initial assessments. Organizations then determine their method of supplier evaluation and selection, establishing specific criteria, scoring systems, and evaluation procedures. The cycle culminates in selecting suppliers and reaching agreements, where final decisions are made and terms are negotiated. This cyclical process emphasizes the continuous nature of supplier selection, where each step influences subsequent ones and the completion of one cycle may initiate another as business needs evolve, ensuring organizations make informed decisions aligned with their strategic objectives [52]

The supplier selection process not only ensures that the supply sources meet the business's requirements in terms of quality, cost, and delivery time, but also covers the supply capacity, sustainability, and alignment with the restaurant's operational strategy [23], and strategic one [24]. Concerning operational strategy, supplier selection can have a profound impact on both service quality and customer satisfaction when competition is increasing. Implementing an effective supplier selection process not only helps to reduce risks but also enhances the performance of businesses in the food industry [53, 54]. Regarding strategic strategy, supplier selection can determine the long-term success of a restaurant through choosing green suppliers to improve the restaurant's environmental performance and the brand image [24], or develop long-term cooperative relationships with suppliers to bring long-term financial benefits and build customer trust [55].

## 2.3 The Pragmatic Constructivist Approach

Pragmatic constructivist has been used as one of theoretical lens to interpret social events/data in the field of business [29, 33, 56, 57]. It provides a comprehensive framework for understanding organizational practices and decision-making processes through four interconnected dimensions: facts, possibilities, values, and communication [58, 59]. Also, it is an evaluation method that can be used to illustrate how people could relate to their reality in a way that supports obtaining successful action in business [29]. According to Nørreklit, Raffnsøe-Møller et al. [57], this approach emphasizes the integration of factual observations with organizational values and possibilities while recognizing the crucial role of communication in implementing changes. The facts dimension focuses on the empirical reality of organizational processes and structures, examining concrete evidence and observable phenomena; facts could be socially constructed and not similar to things that actors face in the world. This dimension provides the foundation for understanding current practices and identifying areas requiring improvement. The possibilities should be grounded in factual knowledge, exploring potential solutions and opportunities available to organizations, considering technological capabilities, resource constraints, and market conditions. Values refer to motivations for actions, which examines the underlying principles and objectives that guide organizational decision-making, including both explicit and implicit value systems that influence strategic choices. This dimension is particularly crucial in understanding how organizations prioritize different objectives and evaluate success. Communication plays a vital role in realizing organizational action. While the integration of facts, possibilities, and values is essential, it alone is insufficient to create mutual understanding among stakeholders. Communication serves as a crucial mechanism through which individuals collaborate and understand each other's subjective values and reasoning across the other three dimensions. The communication dimension analyzes how information flows between stakeholders and how knowledge is shared and integrated across organizational levels [29, 33, 59, 60]. These four dimensions interact dynamically, creating a holistic framework for analyzing organizational practices and implementing change [59].

## 3 Methodology

This research employs a qualitative case study methodology to examine practices of digital transformation in supplier selection at a restaurant chain in Vietnam and to identify the company's successes, benefits, and challenges associated with these practices. This methodology enables practices of digital transformation in supplier selection to be examined within their real-life context of the restaurant chain through various sources of evidence [61].

Data were collected from a single case study through semi-structured interviews with managers and personnel involved in digital transformation for supplier selection at the restaurant chain about practices, benefits, and challenges that the restaurant chain encountered during its digital transformation for supplier selection [62]. By gathering data from interviewees across different departments, backgrounds, roles, and positions, the study obtained their insights into the digital transformation for supplier selection at the restaurant chain with rich descriptions and explorations [61]. The study involved semi-structured interviews based on an interview protocol were sent to these managers to collect data with eleven key personnel (8 males, 3 females) from various management levels and functional areas of the restaurant chain. The interviewees represent diverse perspectives on supplier selection and digital transformation processes, with experience ranging from 4 to 15 years in the industry (Table 1). The participants include senior management representatives: the Director of Internal Audit & Compliance (15 years of industry experience), Operation Manager (15 years), and Risk Management and Compliance Manager (15 years). These senior positions provide strategic insights into the organization's digital transformation initiatives and supplier selection processes. Mid-level management participants include a Compliance Team Leader (10 years' experience), Internal Audit Senior, Finance Controller Manager, and Brand Marketing Manager. Their roles offer operational perspectives on implementing digital transformation in supplier selection processes. The study also captured insights from operational-level management, including two Restaurant Managers and a Restaurant Supervisor, each with approximately 5 years of industry experience. Their involvement provides practical insights into the day-to-day implementation of supplier selection processes and digital transformation initiatives. The interviewees' tenure in their current positions' ranges from 1 to 10 years, with the Operation Manager and Compliance Team Leader having the longest tenure (10 years each) in their respective roles. This mix of experience levels and roles ensures a comprehensive understanding of both strategic and operational aspects of digital transformation in supplier selection processes.

Content analysis was employed to search for patterns that reflect the analytical framework of facts, possibilities, values, and communication. The interview

**Table 1** Interviewees' demographics

| No | Name | Gender | Position | Duration as current position | Experiences in industry |
|---|---|---|---|---|---|
| 1 | A | Male | Director of Internal audit & Compliance | 5 years | 15 years |
| 2 | B | Female | Internal Audit Senior | 2 years | 5 years |
| 3 | C | Male | Compliance Assurance Team Leader | 4 years | 8 years |
| 4 | D | Male | Restaurant manager | 2 years | 5 years |
| 5 | E | Female | Restaurant Supervisor | 1 year | 5 years |
| 6 | F | Female | Restaurant manager | 2 years | 5 years |
| 7 | G | Male | Brand Marketing Manager | 4 years | 4 years |
| 8 | H | Male | Risk Management and Compliance Manager | 2 years | 15 years |
| 9 | I | Male | Finance Controller Manager | 2 years | 7 years |
| 10 | J | Male | Operation Manager | 10 years | 15 years |
| 11 | K | Male | Compliance Team Leader | 10 years | 10 years |

*Source* Author's work

manuscripts were manually analyzed through two stages, considered appropriate for a small number of interviews. The first stage of analysis involved a coding technique to condense "a discrete phrase, sentence, or series of sentences which conveys one idea or one related set of perceptions" ([63], p. 113) into meaning units for each manuscript. These meaning units from 11 interviews were subsequently classified into themes based on the analytical framework [64].

## 4 Findings—A Pragmatic Constructivist Perspective on Digital Transformation for Supplier Selection

The restaurant chain with a decade of operations has emerged as a pioneering force in introducing the restaurant chain model to Vietnam. With a diverse portfolio encompassing five main casual dining cuisines—hotpot, BBQ, Asian, Western, and cafeteria—the company has rapidly expanded its footprint, currently owning about various brands and nearly hundreds of restaurants across provinces in Vietnam. Serving an impressive millions of customers annually, the restaurant chain's growth trajectory underscores the importance of efficient operations and supplier management.

***Facts dimension: pre-transformation challenges***

Prior to digital transformation, the restaurant chain's supplier selection process was characterized by significant operational inefficiencies and control challenges. Regarding operational inefficiencies, the restaurant chain lacked an integrated planning system and complete data for decision-making, leading to inefficient resource allocation and unrealized economies of scale: the restaurant chain struggled to optimize pricing, volume, and quality parameters, missing opportunities to reduce food costs and exploit seasonal supply advantages. Interview data consistently revealed a heavy reliance on manual processes and personal relationships in supplier selection, as one internal auditor noted: "Previously, supplier evaluation and selection were mainly based on intuition or personal relationships" (Interview 04). Communication was predominantly conducted through informal channels like email and chat, lacking standardization and transparency. Document management was paper-based, creating storage and retrieval inefficiencies.

Concerning control mechanisms, the restaurant chain struggled with conflicts of interest due to systematic process segregation: the procurement and purchase department oversees the entire supplier management lifecycle, from initial sourcing to ongoing evaluation. As highlighted by multiple interviewees, the process lacked systematic controls and data-driven evaluation methods, making it difficult to ensure fair and efficient supplier selection. This became more complicated by inconsistent monitoring processes, as noted by an operations manager: "From rudimentary activities without specific processes, it has become a regular and organized activity" (Interview 10). The absence of standardized evaluation criteria and centralized control mechanisms made it difficult to maintain consistency across the organization's extensive network. These challenges ultimately drove the organization's decision to pursue digital transformation in its supplier selection process.

***Possibilities dimension: technological solutions***

The restaurant chain employed various technological solutions during its digital transformation for supplier selection, including E-ticket, E-Process, E-bidding, SAP, and Difot systems. First, two task management systems, "E-ticket for handling unique, non-routine requests, and E-Process for managing repetitive tasks" (Interview 01), streamline interdepartmental workflows among purchasing, restaurant operations, and product development teams, enabling better measurement, management, and reduction of supplier selection timelines while ensuring process compliance across all stakeholders.

Second, the E-bidding platform and tender mailbox system transformed supplier communication and selection processes from manual email exchanges to an automated system. A significant improvement in control was achieved by separating the sourcing and evaluation functions through E-bidding automation, where suppliers submit bids independently through the system until closure. The supplier sourcing now begins with basic qualification criteria (business licenses, food safety certifications, and minimum one-year experience with major partners) for shortlisting. The evaluation phase incorporates detailed assessments of pricing, volume capabilities, and quality standards, including industry-specific metrics for product characteristics and logistics requirements. This automated process ensures transparency and eliminates potential bias in decision-making. The system maintains a comprehensive supplier database categorized by product types, facilitating supplier diversification during shortlisting processes and enabling "standardized evaluation, real-time monitoring, and data-driven decision-making" (Interview 02), enabling balanced supplier allocation to avoid over-dependency on any single vendor.

Third, the SAP T-code system improves supplier data management by maintaining essential supplier information (e.g., tax codes and legal representatives), preventing fictional supplier creation and ensuring accurate payment processing, effectively managing payment terms and conditions, and analyzing supplier performance over time, supporting strategic re-sourcing decisions. Fourth, Difot enables the restaurant chain to evaluate supplier performance through three criteria: quality, volume, and on-time delivery. Suppliers' performance is entered into the system to generate supplier scores, which are subsequently used to "filter future suppliers based on continuous improvement principles" (Interview 04). Data analytics allows the restaurant chain to analyze prices, identifying pricing trends and recognizing seasonal patterns, which "help optimize time, costs, has a comprehensive view from system analysis data, and provides an effective place to store and search for information" (Interview 02).

Through digital transformation, the restaurant chain evolved from a decentralized to a centralized operational model. This shift not only improved data integration and visibility but also substantially reduced agency problems and transaction costs. The investment in digital tools enabled better coordination across the restaurant chain's extensive network, allowing for more strategic supplier relationships and resource optimization. This systematic approach to supplier management has proven particularly valuable in achieving cost efficiencies through consolidated purchasing power and improved supply chain visibility.

***Values dimension: motivations and benefits***

The main motivation for the restaurant chain to adopt digital transformation in its supplier selection process is to create fair competition among suppliers for more efficient operations through enhancing transparency and objectivity of the supplier selection process. The restaurant chain experienced a shift from subjective to objective decision-making in its supplier selection process. The automated E-bidding system is based on "accurate and objective parameters like delivery performance, product quality, and price" rather than "personal relationships" (Interview 04). It also eliminated personal bias, focusing on merit-based partnerships rather than "individual benefits." The E-bidding system provides a transparent platform for supplier competition, ensuring equal opportunities for all suppliers and optimizing operational costs through fair price competition. This fairness and transparency contributed to sustainable supplier development through successfully attracting and retaining high-quality suppliers, leading to continuous improvement in supplier quality and performance, improving the restaurant chain's long-term operational efficiency. As Interviewee 08 noted, "E-bidding system promoted operational efficiency, reduces risk, and enhances transparency while enabling data-driven decision-making."

***Communication dimension: knowledge-sharing and coordination***

Communication is vital during digital transformation for supplier selection through departmental coordination and knowledge sharing. The coordination mechanism is necessary to resolve potential conflicts between different departments during digital transformation in the supplier selection process. For instance, Business Unit Managers who primarily focus on sales targets and short-term profitability of individual restaurants tend to select sub-optimal suppliers to save costs and increase profit in the short term instead of premium suppliers, as the Product Development team (Franchiser) prioritizes quality standards and product specifications. This conflict was solved through effective collaboration between these teams to reconcile diverse departmental objectives and ensure optimization of the whole restaurant chain. As a result, they agreed on criteria for selecting suppliers, which balances requirements for quality standards and operational efficiency across different departments throughout the entire restaurant chain network.

Knowledge sharing emerged as a critical success factor in the digital transformation of supplier selection, including both internal staff development and supplier engagement. Digital transformation has been coupled with difficulties in using new technological tools and communications across organizations. Thus, "training is needed in using specific technology applications in work, digital transformation culture communication," as Interviewee 08 emphasized. Internally, training on system operations and digital processes was provided to employees to ensure effective adoption of new technologies. These challenges were also faced by suppliers, as Interviewee 06 noted: "Some suppliers complaining about the complexity and time-consuming nature of bidding/tendering on the E-bidding system." Thus, the restaurant chain developed structured training programs for shortlisted suppliers, providing detailed guidance on e-bidding system usage, bidding procedures, and order management processes across the restaurant chain. This knowledge sharing fostered stronger supplier relationships since "sharing the benefits of the new system motivates supplier cooperation," as Interviewee 03 highlighted, creating mutual understanding and trust, transforming communication patterns to "quick, transparent communication" supported by systematic information management, as described by Interviewee 02. Internal and external knowledge sharing smoothed system operations and strengthened supplier relationships, encouraging suppliers to maintain high quality standards and respond promptly to restaurant chain feedback.

# 5 Discussion and Conclusion

## 5.1 Discussion

The findings illustrate changes in supplier selection as the outcome of digital transformation can support the restaurant chain to solve the problem of operational inefficiency and control challenges. The facts and possibilities dimensions are consistent with [40] regarding the typical progression from analog to digital operations following the sequence of digitization, digitalization, and digital transformation. The analysis reveals varying levels of digital transformation maturity across different stages of the supplier selection process [52]. The restaurant chain has successfully progressed through the digitization stage, establishing basic digital documentation and data collection systems across most processes as defined by Verhoef et al. [1]. The restaurant chain has also achieved significant advancement in digitalization, particularly evident in their integrated systems like E-bidding and Difot. This progress reflects digitalization as the use of digital technologies to alter existing operational processes, resulting in improved efficiency and cost savings [40].

However, true digital transformation—characterized by fundamental business model changes and complete process reinvention—remains a work in progress. The completion of two first stages of digital transformation becomes stable foundation for the restaurant chain to progress towards advanced analytics and AI for predictive demand forecasting [40]. The sourcing strategy and supplier identification stages demonstrate successful digitalization through E-bidding systems but could benefit from AI-driven market analysis tools [45]. The supplier pool limitation and evaluation stages show advanced digitalization through Difot and SAP integration. However, as suggested by [65], the integration of blockchain technology could enhance transparency and automate supplier verification. The final selection and agreement stage, while digitized, could evolve through the implementation of smart contracts and automated negotiation systems. To achieve comprehensive digital transformation, the organization should enhance its predictive capabilities through AI-driven analytics, deepen technology integration through blockchain and IoT solutions, and transform business models through innovative supplier relationship frameworks [12]. These improvements align with current research on digital transformation in supply chain management, suggesting that complete transformation requires not just technological adoption but fundamental changes in how value is created and delivered [40].

The pragmatic constructivist perspective highlights the critical role of organizational values and communication during digital transformation. Our findings indicate that while enhancing operational efficiency, digital transformation for supplier selection enables the restaurant chain to pursue values of fairness and transparency as a foundation to enhance operational performance. Moreover, communication through knowledge sharing and cooperation among stakeholders is essential for the restaurant chain to successfully transform to digital operations [29, 33, 59, 60].

## 5.2 Conclusion

This chapter has examined the digital transformation of supplier selection processes at a leading Vietnamese restaurant chain through a pragmatic constructivist lens. Through the examination of four interconnected dimensions—facts, possibilities, values, and communication—the study illustrates that the way an organization constructs its reality can support it to obtain success in digital transformation for supplier selection. The investigation through the pragmatic constructivist framework reveals the organization's comprehensive transformation journey. The facts dimension demonstrated the progression from manual, relationship-based processes to systematic, data-driven supplier selection. The possibilities dimension illustrated how integrated digital systems (E-ticket/E-Process, E-bidding, Difot, and SAP) created new opportunities for standardized evaluation and real-time monitoring. The values dimension highlighted how digital transformation established strong connections between fairness principles and operational outcomes, while the communication dimension emphasized the critical role of knowledge sharing and stakeholder coordination in successful implementation. The findings underscore that digital transformation is not merely a technological initiative but a comprehensive organizational change that requires careful consideration of facts, possibilities, values, and communication aspects to achieve sustainable success in supplier selection processes. While the organization has successfully achieved digitization and made significant progress in digitalization, complete digital transformation remains an ongoing journey. Theoretically, this case study successfully applies the pragmatic constructivist framework to evaluate the successful implementation of digital transformation for the restaurant chain, which is scarce in management literature, especially in digital transformation research. Practically, this case study suggests that successful digital transformation requires systematic integration of digital technologies, balance between automation and stakeholder engagement, strong focus on transparency and fairness in supplier relationships, and comprehensive communication frameworks. It is vital for organizations to recognize facts, identify possibilities, know how to connect them with values, and employ appropriate communication mechanisms to successfully implement organizational change.

## References

1. Verhoef, P.C., Broekhuizen, T., Bart, Y., Bhattacharya, A., Dong, J.Q., Fabian, N., Haenlein, M.: Digital transformation: A multidisciplinary reflection and research agenda. J. Bus. Res., **122**, 889–901. (2021). https://doi.org/10.1016/j.jbusres.2019.09.022
2. Plekhanov, D., Franke, H., Netland, T.H.: Digital transformation: A review and research agenda. Eur. Manag. J., **41**(6), 821–844. (2023). https://doi.org/10.1016/j.emj.2022.09.007
3. Peng, X., Zhu, J., Lee, S., Zhou, D., Song, W., Ying, T.: Digital transformation in the hospitality industry: A bibliometric review from 2000 to 2023. Int. J. Hosp. Manag., **120**, 103761 (2024). https://doi.org/10.1016/j.ijhm.2024.103761

4. Wang, G., Gunasekaran, A., Ngai, E. W. T., Papadopoulos, T.: Big data analytics in logistics and supply chain management: Certain investigations for research and applications. Int. J. Prod. Econ., **176**, 98–110. (2016). https://doi.org/10.1016/j.ijpe.2016.03.014
5. Wen, Z., Hu, S., De Clercq, D., Beck, M.B., Zhang, H., Zhang, H., Fei, F., Liu, J. Design, implementation, and evaluation of an Internet of Things (IoT) network system for restaurant food waste management. Waste Manag., **73**, 26–38. (2018). https://doi.org/10.1016/j.wasman.2017.11.054
6. Saberi, S., Kouhizadeh, M., Sarkis, J., Shen, L.: Blockchain technology and its relationships to sustainable supply chain management. Int. J. Prod. Res. **57**(7), 2117–2135 (2019)
7. Zhao, G., Liu, S., Lopez, C., Lu, H., Elgueta, S., Chen, H., Boshkoska, B.M.: Blockchain technology in agri-food value chain management: A synthesis of applications, challenges and future research directions. Comput. Ind. **109**, 83–99 (2019)
8. Cao, S., Johnson, H., Tulloch, A.:.Exploring blockchain-based traceability for food supply chain sustainability: towards a better way of sustainability communication with consumers. Procedia Comput. Sci., **217**, 1437–1445. (2023). https://doi.org/10.1016/j.procs.2022.12.342
9. Gammelgaard, B., Nowicka, K.: Next generation supply chain management: the impact of cloud computing. J. Enterp. Inf. Manag. **37**(4), 1140–1160 (2024). https://doi.org/10.1108/JEIM-09-2022-0317
10. Lin, M., Lin, C., Chang, Y.-S.: The impact of using a cloud supply chain on organizational performance. J. Bus. & Ind. Mark. **36**(1), 97–110 (2021). https://doi.org/10.1108/JBIM-04-2019-0154
11. Ding, Y., Jin, M., Li, S., Feng, D.: Smart logistics based on the internet of things technology: an overview. Int J Log Res Appl **24**(4), 323–345 (2021). https://doi.org/10.1080/13675567.2020.1757053
12. Taj, S., Imran, A. S., Kastrati, Z., Daudpota, S. M., Memon, R. A., Ahmed, J. (2023). IoT-based supply chain management: A systematic literature review. Internet Things, 24, 100982. https://doi.org/10.1016/j.iot.2023.100982
13. Rejeb, A., Simske, S., Rejeb, K., Treiblmaier, H., Zailani, S.: Internet of Things research in supply chain management and logistics: A bibliometric analysis. Internet Things, **12**, 100318. (2020). https://doi.org/10.1016/j.iot.2020.100318
14. Franco, C.W., Benitez, G.B., Sousa, P.R. d., Kliemann Neto, F.J., Frank, A.G.: Managing resources for digital transformation in supply chain integration: The role of hybrid governance structures. Int. J. Prod. Econ., **278**, 109428. (2024). https://doi.org/10.1016/j.ijpe.2024.109428
15. Saputro, T.E., Rosiani, T.Y., Mubin, A., Dewi, S.K., Baroto, T.: Green supplier selection under supply risks using novel integrated fuzzy multi-criteria decision making techniques. J. Clean. Prod., **449**, 141788. (2024). https://doi.org/10.1016/j.jclepro.2024.141788
16. Moreno-Gené, J., Daries, N., Cristobal-Fransi, E.: (2023). Effects of restaurant expenses on enhanced profitability: Do Michelin-starred restaurants perform differently? Int. J. Gastron. Food Sci., **34**, 100811. https://doi.org/10.1016/j.ijgfs.2023.100811
17. Kumar, A., Mangla, S. K., Kumar, P., Karamperidis, S.:.Challenges in perishable food supply chains for sustainability management: A developing economy perspective. Bus. Strat. Environ., **29**(5), 1809–1831. (2020). https://doi.org/10.1002/bse.2470
18. Abbas, H., Zhao, L., Gong, X., Faiz, N.: The perishable products case to achieve sustainable food quality and safety goals implementing on-field sustainable supply chain model. Socio-Econ. Plan. Sci., **87**, 101562 (2023) https://doi.org/10.1016/j.seps.2023.101562
19. Kaskela, J., Sund, R., Lundén, J.: Efficacy of disclosed food safety inspections in restaurants. Food Control., **123**, 107775. (2021). https://doi.org/10.1016/j.foodcont.2020.107775
20. Changalima, I.A., Ismail, I.J., McHopa, A.D.: Effects of supplier selection and supplier monitoring on public procurement efficiency in Tanzania: a cost-reduction perspective. Vilakshan - XIMB Journal of Management **21**(1), 55–65 (2024). https://doi.org/10.1108/XJM-04-2022-0077
21. Yang, M., Fu, M., Zhang, Z.: The adoption of digital technologies in supply chains: Drivers, process and impact. Technol. Forecast. Soc. Chang., **169**, 120795. (2021). https://doi.org/10.1016/j.techfore.2021.120795

22. Cho, M., Bonn, M. A., Giunipero, L., Divers, J.: Restaurant purchasing skills and the impacts upon strategic purchasing and performance: The roles of supplier integration. Int. J. Hosp. Manag., **78**, 293–303 (2019). https://doi.org/10.1016/j.ijhm.2018.09.012
23. Cho, M., Bonn, M.A., Giunipero, L., Jaggi, J.S.: Supplier selection and partnerships: Effects upon restaurant operational and strategic benefits and performance. Int. J. Hosp. Manag., **94**, 102781 (2021). https://doi.org/10.1016/j.ijhm.2020.102781
24. Shin, S., Cho, M.: Environmental cognition and environmental performance in restaurants: roles of green supplier selection and joint actions. Int. J. Contemp. Hosp. Manag., **35**(6), 1925–1942. (2023). https://doi.org/10.1108/ijchm-03-2022-0338
25. Cox, K.: Chapter 11—Digital purchasing and procurement systems: evolution and current state. In MacCarthy, B.L., Ivanov, D. (Eds.), The Digital Supply Chain (pp. 181–197). Elsevier. (2022). https://doi.org/10.1016/B978-0-323-91614-1.00011-3
26. Alt, R.: Digital transformation in the restaurant industry: current developments and implications. J. Smart Tour., **1**(1), 69–74 (2021). https://doi.org/10.52255/SMARTTOURISM.2021.1.1.9
27. Le, D.: Food & beverage industry in vietnam: market overview and investment opportunities (2024). In Corp Vietnam Retrieved 23/10/2024 from https://vietnam.incorp.asia/food-beverage-industry-in-vietnam/
28. Porter, M.E.: The Five Competitive Forces That Shape Strategy. Harv. Bus. Rev. **86**(1), 78–93 (2008)
29. Mitchell, F., Nørreklit, H., Nørreklit, L., Cinquini, L., Koeppe, F., Magnacca, F., Mauro, S.G., Jakobsen, M., Korhonen, T., Laine, T., Liboriussen, J.M.: Evaluating performance management of COVID-19 reality in three European countries: a pragmatic constructivist study. Account., Audit. & Account. J. **34**(6), 1345–1361 (2021). https://doi.org/10.1108/AAAJ-08-2020-4778
30. Chen, L., Kohpaiboon, A., Lee, C.: Introduction digital transformation in southeast asia. J. Southeast Asian Econ., **40**(1), 1–3. (2023). https://www.jstor.org/stable/27211222
31. Esposito, B., Sessa, M.R., Sica, D., Malandrino, O.: Service innovation in the restaurant sector during COVID-19: digital technologies to reduce customers' risk perception. The TQM Journal **34**(7), 134–164 (2022). https://doi.org/10.1108/TQM-01-2022-0016
32. Ha, H., Chuah, C.K.P.: Digital economy in Southeast Asia: challenges, opportunities and future development. Southeast Asia: Multidiscip. J. **23**(1), 19–35 (2023). https://doi.org/10.1108/SEAMJ-02-2023-0023
33. Nørreklit, H., Nørreklit, L., Mitchell, F.: Understanding practice generalisation—opening the research/practice gap. Qual. Res. Account. Manag. **13**(3), 278–302 (2016). https://doi.org/10.1108/QRAM-09-2015-0088
34. Baldvinsdottir, G.: The validity of management accounting language games – A pragmatic constructive perspective. Br. Account. Rev., **53**(6), 101039. (2021). https://doi.org/10.1016/j.bar.2021.101039
35. Zhu, X., Ge, S., Wang, N.: Digital transformation: A systematic literature review. Comput. & Ind. Eng., **162**, 107774 (2021). https://doi.org/10.1016/j.cie.2021.107774
36. Grab, B., Olaru, M., Gavril, R.M.: The impact of digital transformation on strategic business management. Ecoforum J., **8**(1) (2019)
37. Maksimenko, I., Vashko, T., Zdrestova-Zakharenkova, S.: Digital transformation and its challenges to the strategic management system. SHS Web of Conferences (2021)
38. Siderska, J.: Robotic Process Automation—a driver of digital transformation? Eng. Manag. Prod. Serv., **12**(2), 21–31. (2020). https://doi.org/10.2478/emj-2020-0009
39. Mugge, P., Abbu, H., Michaelis, T. L., Kwiatkowski, A., Gudergan, G.: Patterns of digitization: A practical guide to digital transformation. Res.-Technol. Manag., **63**(2), 27–35. (2020). https://doi.org/10.1080/08956308.2020.1707003
40. Verhoef, P.C., Broekhuizen, T., Bart, Y., Bhattacharya, A., Qi Dong, J., Fabian, N., Haenlein, M.: Digital transformation: A multidisciplinary reflection and research agenda. J. Bus. Res., **122**, 889–901. (2021). https://doi.org/10.1016/j.jbusres.2019.09.022
41. Lee, W., Jang, S., & Kim, H.S.: The effect of digital transformation: Boosting productivity in the restaurant industry. Int. J. Hosp. Manag., **123**, 103896. (2024). https://doi.org/10.1016/j.ijhm.2024.103896

42. Ortíz-Barrios, M. A., Madrid-Sierra, S. L., Petrillo, A., & Quezada, L. E. (2023). A novel approach integrating IF-AHP, IF-DEMATEL and CoCoSo methods for sustainability management in food digital manufacturing supply chain systems. J. Enterp. Inf. Manag.. https://doi.org/10.1108/jeim-04-2023-0199
43. Denesiuk, P.: International Supplier Quality Management in Global Supply Chains (based on "Dellini Restaurant" case) Private Higher Educational Establishment-Institute "Ukrainian-American …] (2024)
44. Orellano, M., Tiss, S.: Impacts of digital transformation on supply chain sustainability: a systematic literature review and expert assessment. In: Working Conference on Virtual Enterprises (2022)
45. Ricárdez-Estrada, J., Garay-Rondero, C. L., Romero, D., Wuest, T., Pinto, R.: Suppliers 4.0 digital capabilities development in support of digital supply chains: a scoping review. Int. J. Logist. Res. Appl., 1–28. (2024). https://doi.org/10.1080/13675567.2024.2391397
46. Li, J., Zhong, D., Ru, H., Jia, L.: Assessing the nexus between supplier and customer integration and environmental cost performance: insights into the role of digital transformation. Sustain., **16**(14), 5989. (2024). https://doi.org/10.3390/su16145989
47. Dehghani, M., Popova, A., Gheitanchi, S.: Factors impacting digital transformations of the food industry by adoption of blockchain technology. J. Bus. & Ind. Mark., **37**(9), 1818–1834. (2022). https://doi.org/10.1108/jbim-12-2020-0540
48. Kaur, H., Gupta, M., Singh, S.P.: Integrated model to optimize supplier selection and investments for cyber resilience in digital supply chains. Int. J. Prod. Econ., **275**, 109338 (2024). https://doi.org/10.1016/j.ijpe.2024.109338
49. Guida, M., Caniato, F., Moretto, A., Ronchi, S.:.Artificial intelligence for supplier scouting: an information processing theory approach. Int. J. Phys. Distrib. & Logist. Manag., **53**(4), 387–423. (2023). https://doi.org/10.1108/ijpdlm-12-2021-0536
50. Önder, E., Kabadayi, N.: Supplier selection in hospitality industry using ANP. Int. J. Acad. Res. Bus. Soc. Sci., **5**(1). (2015). https://doi.org/10.6007/ijarbss/v5-i1/1417
51. Vanbrabant, L., Verdonck, L., Mertens, S., Caris, A.: Improving hospital material supply chain performance by integrating decision problems: A literature review and future research directions. Comput. & Ind. Eng., **180**, 109235. (2023). https://doi.org/10.1016/j.cie.2023.109235
52. Monczka, R.M., Handfield, R.B., Crunipero, L.C., Patterson, J.L.: Purchasing and Supply Chain Management (7th ed.). Cengage Learning (2021)
53. Sureeyatanapas, P., Sriwattananusart, K., Niyamosoth, T., Sessomboon, W., Arunyanart, S.: Supplier selection towards uncertain and unavailable information: An extension of TOPSIS method. Oper. Res. Perspect., **5**, 69–79. (2018). https://doi.org/10.1016/j.orp.2018.01.005
54. Yazdani, M., Pamucar, D., Chatterjee, P., Torkayesh, A.E.: A multi-tier sustainable food supplier selection model under uncertainty. Oper. Manag. Res., **15**(1), 116–145. (2022). https://doi.org/10.1007/s12063-021-00186-z
55. Kim, B.Y.: The impact of supplier development on financial performance in the restaurant industry. Int. J. Hosp. & Tour. Adm., **7**(4), 81–103 (2006). https://doi.org/10.1300/j149v07n04_05
56. Lueg, R., Knapik, M.: Risk management with management control systems: a pragmatic constructivist perspective. Corp. Ownersh. Control. **13**, 72–81 (2016)
57. Nørreklit, H., Raffnsøe-Møller, M., Mitchell, F.: A pragmatic constructivist approach to accounting practice and research. Qual. Res. Account. Manag. **13**(3), 266–277 (2016). https://doi.org/10.1108/QRAM-05-2016-0039
58. Kelly, L.M., Cordeiro, M.: Three principles of pragmatism for research on organizational processes. Methodol. Innov. **13**(2), 2059799120937242 (2020). https://doi.org/10.1177/2059799120937242
59. Whitford, J., Zirpoli, F.: Pragmatism, practice, and the boundaries of organization. Organ. Sci., **25**(6), 1823–1839. (2014). http://www.jstor.org/stable/43663645
60. Mitchell, F., Nielsen, L.B., Nørreklit, H., Nørreklit, L.: Scoring strategic performance: a pragmatic constructivist approach to strategic performance measurement. J. Manage. Governance **17**(1), 5–34 (2013). https://doi.org/10.1007/s10997-012-9216-7

61. Yin, R.K.: Case study research—design and methods. Thousand Oaks, CA: (Sage Publications.) (2017)
62. Dolczewski M.:.Semi-structured interview for self-esteem regulation research. Acta Psychol., **228** (2022)
63. Burnard, P.: Searching for meaning: a method of analysing interview transcripts with a personal computer. Nurse Educ. Today, **14**(2), 111–117. (1994). https://doi.org/10.1016/0260-6917(94)90113-9
64. Malterud, K.: Systematic text condensation: A strategy for qualitative analysis. Scand. J. Public Health **40**, 795–805 (2012)
65. Vaezi, A., Rabbani, E., Yazdian, S.A.: Blockchain-integrated sustainable supplier selection and order allocation: A hybrid BWM-MULTIMOORA and bi-objective programming approach. J. Clean. Prod., **444**, 141216. (2024). https://doi.org/10.1016/j.jclepro.2024.141216

# Transforming Agri-Food Supply Chains with Blockchain: Insights from Australian Case Studies

**Imran Ali, Khoa Nguyen, Delwar Akbar, and Aniqa Mahmood Ali**

**Abstract** The integration of blockchain technology into agri-food supply chains offers transformative opportunities to enhance transparency, traceability and trust among stakeholders. This chapter delves into blockchain implementation and its benefits through four Australian case studies. These firms utilize blockchain to streamline documentation, enhance product evaluation and automate food recall processes, effectively addressing consumer demands for greater transparency and traceability. Transitioning from traditional to digital processes, along with the tokenization of assets and events, ensures the reliability and verifiability of supply chain data. However, challenges such as scalability, data privacy and supplier compliance remain crucial barriers. To address these issues, strategies include partnering with technology providers, integrating public blockchain networks and embedding conditional token minting within commercial workflows. Likewise, fostering collaboration among stakeholders and investing in scalable blockchain solutions are imperative for overcoming these challenges. This chapter underscores blockchain's pivotal role in transforming agri-food supply chains, highlighting its potential to enhance food safety and sustainability. It calls for continued exploration and widespread adoption of blockchain technology to ensure transparent and efficient agri-food systems.

**Keywords** Blockchain · Agri-food supply chains · Case studies · Australia

I. Ali (✉) · K. Nguyen · D. Akbar · A. M. Ali
School of Business and Law, Central Queensland University, Australia Bahria University, Islamabad, Pakistan
e-mail: i.ali@cqu.edu.au



N. D. K. Nguyen et al. (eds.), *Digital Transformation and Sustainability: Methods and Applications*, Information Systems Engineering and Management 47,
https://doi.org/10.1007/978-3-031-91278-8_8

# 1 Introduction

Agri-food supply chains—from 'farm to fork'—face immense pressure, exacerbated by the projected global population of 9.7 billion by 2050. This demographic expansion intensifies existing inefficiencies and exacerbates socioeconomic problems including food security, resource allocation and environmental sustainability [1, 2]. In response to these pressures, the advent of Industry 4.0, marked by the fusion of innovative technologies, is transforming supply chains on a global scale [3, 4]. Among these technologies, blockchain, initially conceived for cryptocurrencies, has evolved into a groundbreaking tool with broad applications across various industries. It provides a decentralized and immutable electronic ledger, ensuring transparency, traceability and efficiency in transaction recording among trading partners [5, 6]. In agri-food supply chains, blockchain can address pressing issues such as traceability, transparency, food waste and fraud [7, 8], thereby enhancing overall supply chain integrity.

Previous studies highlight blockchain's crucial potential to bolster supply chain resilience. For instance, Manzoor et al. [9] undertook a systematic review that underscored blockchain's capability to bolster supply chain robustness through improved visibility and information sharing. Concurrently, Ali and Aboelmaged [10] empirically explored the factors influencing the adoption of supply chain 4.0 within Australia's food and beverage sector, identifying key drivers such as organizational culture, technological infrastructure and perceived benefits. In a different geographical context, Saha et al. [11] investigated the adoption intentions of blockchain in Indian agri-food supply chains, revealing insights into regional challenges and opportunities. Further, Ali et al. [5] indicated how Malaysia successfully integrates blockchain with internet of things (IoT) for real-time monitoring, showcasing the synergy between these technologies in facilitating supply chain operations. Thus, this chapter delves into blockchain's transformative potential in agri-food supply chains in Australia, elucidating its practical benefits, challenges and viable solutions.

Through a critical examination of existing literature and practical case studies, the chapter demonstrates how blockchain enhances traceability, transparency, trust and efficiency across various stages of agri-food supply chains. The immutable nature of blockchain's transaction records reduces fraud risk, ensures product authenticity and streamlines operational processes [12]. Nevertheless, the adoption of blockchain in agri-food supply chains is not without challenges. The integration of blockchain with existing supply chain systems and other emerging technologies such as IoT can be burdensome for small and medium-sized enterprises (SMEs) regarding substantial investment in infrastructure and specialized expertise. Further, interoperability issues between different blockchain platforms and legacy systems can hinder seamless implementation. While blockchain's transparency is advantageous for traceability, it may conflict with competitive interests, as firms might be hesitant to disclose sensitive information on a decentralized network [7, 13].

Despite these hurdles, four Australian case studies illustrate successful blockchain implementations, leading to improved traceability and transparency, reduced food waste and mitigated fraud. Such case studies are part and parcel for several reasons. First, they reflect the unique challenges and opportunities within Australia's diverse agri-food sector, providing relevant and context-specific insights. Second, Australia's strong agricultural base and advanced technological infrastructure make it an ideal environment to demonstrate blockchain's practical applications and scalability. Third, these case studies highlight innovative solutions to regulatory and logistical barriers, offering valuable lessons for other regions with similar contexts. Finally, the success stories foster stakeholder confidence and encourage wide adoption by showcasing tangible benefits and return on investment.

This chapter equips industry professionals, researchers and policymakers with overarching insights into blockchain's benefits and adoption strategies in agri-food supply chains. By addressing sector-specific challenges, blockchain can pave the way for secure, sustainable and efficient food systems, revolutionizing agri-food supply chains to meet future demands.

## 2 Overview of Blockchain and Agri-Food Supply Chains

### *2.1 Blockchain Technology*

Blockchain, introduced as a decentralized ledger in the Bitcoin whitepaper, possesses the potential to revolutionize various sectors, economies and trust systems in the digital age. This advanced database technology facilitates transparent information exchange within firm networks by storing data in interconnected blocks [14]. Beyond its initial application in digital currencies, blockchain enables self-executing agreements, known as smart contracts, thereby eliminating intermediaries and broadening its use cases to areas such as supply chain management, healthcare and finance. Firms have rapidly recognized blockchain's transformative potential [15], leveraging its capabilities to enhance security, reduce costs and improve operational efficiency. With ongoing innovations that enhance scalability, interoperability and collaboration, blockchain is poised to make significant global impacts. As innovation and adoption accelerate, blockchain is expected to continue shaping trust frameworks, fostering greater transparency and driving progress within society [16]. This evolving technology promises to redefine traditional processes and establish new paradigms for digital interactions, contributing to more secure and efficient systems across various industries. Blockchain's capability to provide a trustworthy and decentralized infrastructure therefore positions it as a cornerstone for future digital transformation.

### 2.2 Agri-Food Supply Chains

The agri-food supply chains, spanning from 'field to fork', encompass food production, processing, packaging, transportation and consumption [14, 17]. Globalization has crucially increased the movement of products and information, intensifying demands on agri-food supply chains to meet evolving consumer needs [18, 19]. According to the FAO, nearly 30% of annual food production is lost globally, with 14% of farm products lost during transportation and storage post-harvesting [8, 20]. Contributing factors include limited storage infrastructure, mishandling and information gaps, which exacerbate food loss and inefficiencies. Traceability emerges as a critical mechanism to mitigate these losses, ensure food safety and prevent fraud within the inherently unpredictable agri-food supply chain [8]. Implementing robust traceability systems enhances transparency, improves accountability and streamlines operations, thereby addressing the multifaceted challenges posed by globalization and increasing demand. Similarly, effective traceability not only reduces food waste but also builds consumer trust by verifying the authenticity and safety of food products. Such a trust is inevitable for maintaining market stability and consumer loyalty in a highly competitive global market. Moreover, traceability systems facilitate compliance with international standards and regulations, thereby enabling smoother access to global markets and reducing the risk of costly recalls and reputational damage. Robust traceability systems in turn are indispensable for mitigating food loss, ensuring safety and fostering trust within agri-food supply chains. As globalization continues to evolve, the adoption of advanced traceability technologies will be pivotal in creating resilient, efficient and sustainable food systems capable of meeting future demands [8, 14, 17].

### 2.3 Blockchain in the Agri-Food Supply Chains

Blockchain technology is increasingly adopted across various industries, particularly within agri-food supply chains, due to its capability to enhance security, traceability and transparency in food production and distribution [21]. Numerous firms have implemented blockchain-based solutions while governments are actively developing legal frameworks to support this adoption. The inherent complexity of agri-food supply chains, coupled with the involvement of multiple stakeholders, often impedes transparency and efficiency [22]. Blockchain's decentralized and transparent ledger system effectively mitigates these challenges by providing immutable transaction records and fostering trust among participants. The technology's rapid expansion, evidenced by a projected compound annual growth rate (CAGR) of 49.75% between 2023 and 2030, underscores its significant transformative potential [23]. For instance, AgriDigital successfully completed its inaugural blockchain-based grain payment in December 2016 and has since processed over USD 360 million in farmer payments [24].

**Table 1** Top BaaS providers in agri-food supply chains (Author's work)

| | AWS | IBM | Oracle | SAP |
|---|---|---|---|---|
| Hosting platforms | Hyperledger Fabric, Ethereum, Quorum, Corda | Hyperledger Fabric | Hyperledger Fabric | Hyperledger Fabric |
| Blockchain type | Permissioned | Permissioned | Permissioned, consortium | Permissioned |
| Authentication & authorization services | Pay per use | Monthly subscription with free trial | Pay per use | – |
| Scalability | Provides API for quick node creation | IBM Smart Cloud only | – | – |

Table 1 outlines the top four leading firms offering Blockchain as a Service (BaaS), illustrating a significant trend where businesses adopt existing blockchain infrastructures via subscription-based cloud services [1, 25]. This shift toward BaaS and cloud-based solutions lowers entry barriers by eliminating the need for substantial upfront investment in infrastructure and technical expertise. It enables scalable deployment across diverse agri-food sectors, accommodating fluctuating demands and fostering rapid innovation. Cloud-based services enhance accessibility and interoperability, allowing seamless integration with existing systems and collaboration among supply chain stakeholders [26]. They also provide robust security measures and compliance support, crucial for handling sensitive data inherent in agri-food supply chains. The scalability and reliability of cloud infrastructures ensure that blockchain applications can expand alongside business needs without compromising performance. By leveraging BaaS through cloud services, agri-food businesses can revolutionize supply chains by enhancing operational efficiency, reducing fraud and ensuring product authenticity. This synergy positions blockchain as a transformative tool for achieving transparency, trust and sustainability in the agri-food industry.

## 3 Potential of Blockchain in Agri-Food Supply Chains

This section explores the trends and benefits of blockchain adoption in agri-food supply chains, the challenges hindering its implementation and strategies to overcome these obstacles.

## 3.1 *Trends and Benefits to Adoption: Transparency, Traceability, Trust*

Blockchain is increasingly indispensable in agri-food supply chains due to its ability to augment traceability, transparency and trust [27]. Its decentralized and immutable nature ensures secure transaction recording, providing stakeholders with accurate and tamper-proof data [14]. Enhanced transparency enables consumers to track food products from farm to fork, verifying their origins and handling, crucial for ensuring food safety and quality assurance. Improved traceability addresses issues such as contamination or fraud often obscured by traditional and fragmented data management systems in supply chains [21]. Blockchain provides a unified platform where every transaction is recorded in real-time and accessible to all participants, facilitating quick identification and resolution of problems [28]. Improved traceability supports compliance with regulatory requirements and certifications, increasingly demanded by global markets [29]. Trust is built through the transparency and traceability offered by blockchain [30]. In intricate agri-food supply chains with multiple stakeholders, trust is often compromised by information asymmetry and lack of accountability [13]. Blockchain's distributed ledger technology prevents any single entity from controlling the data, reducing fraud and manipulating risks [31]. This fosters greater confidence among consumers, producers and regulators, potentially leading to increased market access and consumer loyalty [16].

## 3.2 *Challenges to Adoption: Technological, Organizational, Regulatory*

Despite its potential, blockchain adoption in agri-food supply chains faces several challenges. Technological hurdles include the need for robust infrastructure and integrating blockchain with existing systems [32]. Many agri-food producers, especially small and medium enterprises (SMEs), can lack technical expertise and resources to adopt and maintain blockchain solutions [33]. Organizational challenges involve resistance to change and the need for collaboration among diverse stakeholders [34, 35].

Successfully implementing blockchain requires a cultural shift towards greater data sharing and cooperation, which can be difficult to achieve in an industry characterized by competition and siloed operations [36]. The complexity of the agri-food supply chains, with its numerous actors and processes, complicates the standardization and interoperability of blockchain applications [13]. Regulatory challenges also pose significant barriers to blockchain adoption. The legal and regulatory frameworks governing blockchain technology are still evolving, creating uncertainty for businesses considering its implementation [37]. Addressing security, data privacy and compliance with international standards is inevitable for the lawful and ethical application of blockchain in agri-food supply chains [38].

### *3.3 Strategies to Overcome Challenges for Successful Adoption*

To overcome these challenges, several strategies can be employed. From a technological perspective, investing in infrastructure and providing training and support for blockchain implementation are crucial [26]. Governments and industry associations can play facilitate this by providing subsidies, grants and educational programs to build necessary technical capabilities [39]. Organizationally, fostering a culture of collaboration and openness is essential. This can be achieved through developing industry-wide standards and best practices, encouraging data sharing and promoting the benefits of blockchain adoption through pilot projects and case studies. Engaging all stakeholders in the design and implementation process ensures that their concerns are addressed and their needs met [40]. Regulatory strategies should focus on developing clear and supportive legal frameworks that address the unique aspects of blockchain technology. Policymakers need to work closely with industry stakeholders to create regulations that protect consumers and businesses while promoting innovation. International cooperation is also important to ensure that blockchain applications in agri-food supply chains comply with global standards and facilitate cross-border trade [41]. As such, Fig. 1 depicts a digitized farm augmenting blockchain, where transaction data is shared and immutable, ensuring a secure, transparent and tamper-resistant supply chain accessible to authorized stakeholders [13]. It is noticeable that farmers record crop data, manufacturers log processing details, distributors track transportation and retailers monitor inventory in blockchain-basedagri-food supply chains. Each stage guarantees traceability, transparency and security through a shared and immutable ledger.

## 4 Practical Case Studies

Selecting four Australian case studies—Tyson Foods, Nestlé, Fresh Supply Co. and Walmart—is indispensable, as Australia's robust technological infrastructure and unique regulatory environment provide an ideal testing ground for blockchain applications. These firms demonstrate blockchain's ability to enhance traceability and transparency by addressing distinct challenges. Tyson Foods ensures product authenticity in complex supply chains; Nestlé achieves end-to-end visibility to boost consumer trust; Fresh Supply Co. optimizes freshness and reduces waste; and Walmart mitigates fraud and enhances food safety through rapid tracking. Such cases highlight benefits such as improved supply chain integrity and transparency while presenting challenges like interoperability and high adoption costs. Innovative solutions include stakeholder engagement, customized platforms and integration with cloud computing and IoT, offering insights for scalable blockchain adoption globally.

Tyson Foods, renowned for brands like Hillshire Farm and Jimmy Dean, is employing blockchain to bolster supply chain traceability and transparency in line

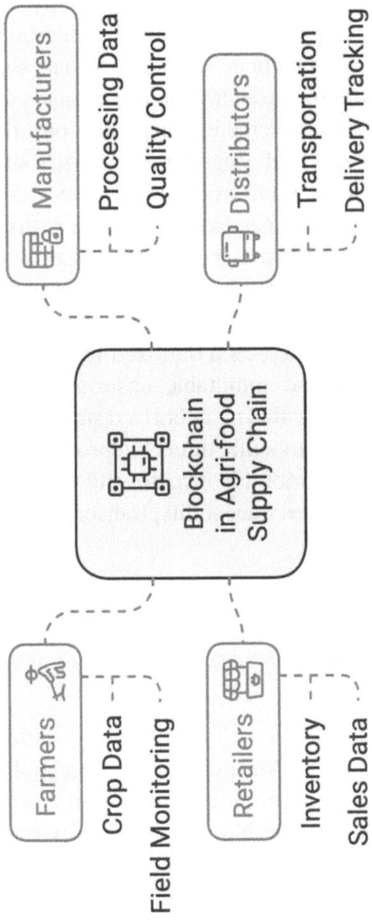


**Fig. 1** Blockchain in agri-food supply chains (*Source* Author's work)

with consumer expectations [42]. Blockchain is a secure and share ledger that links and stores transaction records in immutable blocks. Tyson Foods uses this technology to centralize supplier documentation, facilitate product evaluations and automate food recall actions. In 2018, Tyson Foods partnered with FoodLogiQ, a software developer specializing in food traceability, to pilot blockchain in its supply chain. This initiative is part of Tyson's broader digital transformation of its supplier management system [43]. Food safety has always been a priority at Tyson, yet some documentation remains analogue. Blockchain technology is expected to digitize these processes, enhancing the security and safety of food production. Tyson's food safety team learned about FoodLogiQ through Tyson Ventures, the venture capital arm of Tyson Foods. Impressed by FoodLogiQ's focus on mapping the global food supply chain, Tyson Ventures financed the company. Blockchain technology at Tyson Foods aims not only to reduce food waste and track contamination issues but also to meet the growing demand for more information about our food. As Tyson Foods continues to explore and implement blockchain technology, this innovation promises to transform the food industry [44, 45].

Nestlé, the world's largest food and beverage company, is at the forefront of using blockchain to improve supply chain transparency and traceability. While specific information about Nestlé's blockchain initiatives in Australia is not readily available, the company has made significant global strides in this area [46]. In 2019, Nestlé initiated a pioneering open blockchain pilot with OpenSC, enabling consumers to trace their food origins. Nestlé became the first major food and beverage company to pilot open blockchain technology with this initiative [47]. The program initially tracked milk from farms in New Zealand to Nestlé factories and warehouses in the Middle East. Subsequently, the technology was tested with palm oil sourced from the Americas, allowing Nestlé to assess its scalability [48]. Since 2017, Nestlé has also participated in the IBM Food Trust. Nestlé offered consumers access to blockchain data for the first time with Mousline purée in France in April 2019 [47]. Nestlé's use of blockchain technology goes beyond reducing food waste and tracking contamination issues. It addresses the growing demand for more information about the food we consume. As Nestlé continues to explore and implement blockchain technology, it is evident that this innovation has the potential to transform the food industry [45, 47].

Walmart, a global retail leader, has been a trailblazer in adopting blockchain technology to improve supply chain traceability and transparency [48]. In 2019, Walmart partnered with IBM to integrate blockchain into enhanced food safety requirements for suppliers. In the subsequent year, Walmart and Sam's Club required suppliers of leafy greens such as romaine lettuce and spinach to implement IBM's blockchain for global food traceability [49]. Walmart employed blockchain to address payment disputes with its 70 third-party freight carriers. Implementing a blockchain network not only resolved payment issues but also significantly improved operational efficiencies [50]. Furthermore, Walmart joined forces with nine other major companies in 2018 to collaborate with IBM on developing the Food Trust Blockchain. This initiative aims to create a unified decentralized platform for tracking food supplies globally. Although these initiatives are not specifically detailed for Australia, they underscore Walmart's commitment to transparency, sustainability and innovation on a global scale. For specific information about Walmart's blockchain initiatives in Australia, it is advisable to visit their official website or contact them directly [51, 52]. Walmart's proactive adoption of blockchain technology highlights its leadership in leveraging innovative solutions to address logistical challenges and enhance food safety across its extensive supply chain network [53].

Fresh Supply Co. (FSCO), an Australian company, is at the forefront of integrating blockchain technology in the agricultural sector. FSCO's mission is to enhance transparency and verifiability in the agricultural supply chain, achieving this by incorporating blockchain into their operations [54]. FSCO shifted from private Mastercard blockchain to the public Hedera network, facilitating asset and event tokenization across the agrifood supply chain. This shift digitizes food production and trade, tracking processes such as transforming cow to beef, milk to cheese, shipping, customs and order fulfillment with complete transparency. This transparency provides financiers with comprehensive data to validate the movement of goods and assets, thereby substantiating cash flows continuously [55]. FSCO has incorporated the Hedera Token Service (HTS) into its API tool, Continuity to monitor the payment lifecycles and approval processes. This integration allows organizations to embed conditional minting of tokens within commercial workflows. The upgraded Continuity API utilizes HTS for enhanced transparency and verifiability of payment transactions throughout the value chain. It allows users to track cash transaction conditions and create non-fungible tokens (NFTs) transferable between wallets. This capability enablers financial institutions to use data for confident credit decisions, expanding consumer trade financing opportunities [56, 57].

In a nutshell, Table 2 summarizes the benefits, challenges, functions and solutions of blockchain adoption by Tyson Foods, Nestlé, Fresh Supply Co. and Walmart. Each firm leverages blockchain to enhance traceability and transparency while addressing unique challenges and implementing innovative solutions.

## 5 Conclusion, Recommendations and Call to Action

Blockchain offers transformative potential for agri-food supply chains by enhancing traceability, transparency and trust. Key findings demonstrate that blockchain can significantly reduce inefficiencies and fraud, improve food safety and meet the growing consumer demand for detailed product information. Notable examples from industry leaders such as Tyson Foods, Nestlé, Walmart and FSCO underscore the diverse applications and benefits of blockchain across different segments of the supply chain.

Blockchain adoption offers industry professionals a robust solution to several prevalent issues in agri-food supply chains. Firms should invest in blockchain infrastructure and training to fully leverage its benefits. Pilot projects and collaborations with technology providers, namely those seen with IBM and SAP, can facilitate

**Table 2** A case study analysis (Author's work)

| Aspect | Tyson Foods | Nestlé | Walmart | FSCO |
|---|---|---|---|---|
| Benefits | Centralize supplier documents Facilitate product evaluation Automate food recall actions | Enhance transparency and traceability Provide consumers access to product history | Enhance traceability and transparency Improve operational efficien resolve payment disputes | Enhance transparency and verifiability Tokenize assets and events provide data for validating asset movement |
| Challenges | Transition from analogue to digital documentation Ensure scalability and interoperability | Scale pilot projects globally Ensure data privacy and security | Implement technology across a large network ensure supplier compliance | Transition from private to public Ensure complete digitalization of food production and trade processes |
| Functions | Track contamination issues reduce food waste | Track milk and palm oil participate in IBM food trust | Track leafy greens manage freight carrier payments | Track production processes monitor payment lifecycle mint transactional data as NTFs |
| Solutions | Partnered with FoodLogiQ | Collaborated with OpenSC | Partnered with IBM | Integrated hedera token service into continuity API |

a smoother transition and showcase the practical benefits of blockchain. Professionals should also foster a culture of collaboration and data sharing to maximize the technology's potential.

Researchers should continue investigating the scalability and interoperability of blockchain solutions across various segments of the agri-food supply chain. Exploring blockchain integration with emerging technologies namely internet of things and artificial intelligence could offer invaluable insights for generating more resilient and efficient supply chains. Additionally, researchers should examine the socio-economic influences of blockchain implementation, especially for SMEs and farming communities. Policymakers are paramount for broader blockchain adoption. Clear regulatory frameworks must be developed to address security, data privacy and international standard compliance. Policymakers should also engage with industry stakeholders to create supportive environments that encourage innovation while safeguarding consumer interests. Providing incentives, such as grants and subsidies, can also accelerate blockchain adoption, particularly among smaller players in the agri-food industry.

The transformative potential of blockchain in agri-food supply chains calls for a concerted effort from all stakeholders. Industry professionals are urged to adopt blockchain solutions and collaborate on best practices. Researchers are encouraged to delve deeper into the technology's implications and drive forward-thinking studies.

Policymakers should establish supportive regulations and incentives to foster innovation and adoption. Continued blockchain exploration and implementation in agri-food supply chains will not only enhance efficiency and safety but also contribute to building a more transparent and trustworthy food system. As the technology continues to evolve, it is imperative that all stakeholders remain engaged and proactive in leveraging blockchain to create a more effective, secure and sustainable food supply chain for the future.

## References

1. UNDP, https://www.undp.org/publications/blockchain-agri-food-traceability, last accessed 2024/06/29.
2. Schrobback, P., Rolfe, J., Akbar, D., Rahman, A., Kinnear, S., Bhattarai, S.: Horticulture producer's willingness to participate in contract-based supply chain coordination: A case study from Queensland (Australia). PLoS ONE **18**(5), e0285604 (2023)
3. Iftikhar, A., Ali, I., Arslan, A., Tarba, S.: Digital innovation, data analytics, and supply chain resiliency: a bibliometric-based systematic literature review. Ann. Oper. Res. (2022).
4. Nguyen, N.D.K., Ali, I., Gupta, S., Chen, R., Naresho, B.S.: Bridging the nexus between cloud ERP and enterprise resilience. J. Glob. Inf. Manag. **32**(1), 1–22 (2024)
5. Ali, M.H., Chung, L., Kumar, A., Zailani, S., Tan, K.H.: A sustainable Blockchain framework for the halal food supply chain: Lessons from Malaysia. Technol. Forecast. Soc. Chang. **170**, 120870 (2021)
6. Vern, P., Panghal, A., Mor, R. S., Kumar, V., Sarwar, D.: Unlocking the potential: leveraging blockchain technology for agri-food supply chain performance and sustainability. International Journal of Logistics Management (2024).
7. Food and Agriculture Organization of the United Nations, https://openknowledge.fao.org/server/api/core/bitstreams/9de595cd-9161-48b7-b68e-79a329a10c13/content, last accessed 2024/06/29.
8. Vern, P., Panghal, A., Mor, R. S., Kamble, S. S.: Blockchain technology in the agri-food supply chain: a systematic literature review of opportunities and challenges. Manag. Rev. Q. (2024).
9. Manzoor, R., Sahay, B. S., Singh, S. K.: Blockchain technology in supply chain management: an organizational theoretic overview and research agenda. Ann. Oper. Res. (2022)
10. Ali, I., Aboelmaged, M. G. S.: Implementation of supply chain 4.0 in the food and beverage industry: perceived drivers and barriers. Int. J. Prod. Perform. Manag., 71(4), 1426–1443 (2022)
11. Saha, A., Raut, R.D., Kumar, M., Paul, S. K., Cheikhrouhou, N.: The intention of adopting blockchain technology in agri-food supply chains: evidence from an Indian economy. J. Model. Manag. (2024).
12. Sahoo, S., Kumar, A., Mishra, R., Tripathi, P.: Strengthening Supply Chain Visibility with Blockchain: A PRISMA-Based Review. IEEE Trans. Eng. Manage. **71**, 1787–1803 (2024)
13. Menon, S., Jain, K.: Blockchain Technology for Transparency in Agri-Food Supply Chain: Use Cases, Limitations, and Future Directions. IEEE Trans. Eng. Manage. **71**, 106–120 (2024)
14. Nguyen, N. D. K., Chaudhuri, A., Ullah, A.: Achieving business sustainability with blockchain technology. Handb. Digit. Innov., Transform., Sustain. Dev. Post-Pandemic Era, **78** (2024).
15. Yousefi, S., Tosarkani, B.M.: Exploring the role of blockchain technology in improving sustainable supply chain performance: a system-analysis-based approach. IEEE Trans. Eng. Manage. **71**, 4389–4405 (2024)
16. Brusset, X., Kinra, A., Naseraldin, H., Alkhudary, R.: Increasing willingness to pay in the food supply chain: a blockchain-oriented trust approach. Int. J. Prod. Res. (2024).
17. Akbar, D., Nguyen, T.T.T., Rahman, A., Rolfe, J., Kinnear, S., Bhattarai, S.: Managing exogenous and endogenous risks in Australia's agricultural supply chains. Australas. J. Reg.Nal Stud. **28**(1), 74–93 (2022)

18. Ali, I., Arslan, A., Chowdhury, M., Khan, Z., Tarba, S.Y.: Reimagining global food value chains through effective resilience to COVID-19 shocks and similar future events: A dynamic capability perspective. J. Bus. Res. **141**, 1–12 (2022)
19. Cozzio, C., Viglia, G., Lemarie, L., Cerutti, S.: Toward an integration of blockchain technology in the food supply chain. J. Bus. Res. **162**, 113909 (2023)
20. FAO, I., UNICEF, WFP and WHO, https://openknowledge.fao.org/server/api/core/bitstreams/8b27c570-2f8b-4350-8d5a-8e82432e6db7/content, last accessed 2024/06/30
21. Zheng, K., Ding, K., Hui, J., Zhang, F., Lv, J., Chan, F.T.S.: Blockchain-based credible manufacturing data sharing for a collaborative manufacturing supply chain. Int. J. Prod. Res. **62**(6), 2233–2254 (2024)
22. Li, X., Du, J., Li, W., Shahzad, F.: Green ambitions: A comprehensive model for enhanced traceability in agricultural product supply chain to ensure quality and safety. J. Clean. Prod. **420**, 138397 (2023)
23. Verified Market Research, https://www.verifiedmarketresearch.com/product/blockchain-in-supply-chain-finance-market/, last accessed 2024/06/30
24. Akram, M.W., Akram, N., Shahzad, F., Rehman, K.U., Andleeb, S.: Blockchain technology in a crisis: Advantages, challenges, and lessons learned for enhancing food supply chains during the COVID-19 pandemic. J. Clean. Prod. **434**, 140034 (2024)
25. Nguyen, N.D.K., Ali, I.: Implementation of cloud customer relationship management in banking sector: strategies, benefits and challenges. International Journal of Electronics and Communication Engineering **15**(6), 242–247 (2021)
26. Ali, I., Nguyen, N.D.K., Gupta, S.: A multi-disciplinary review of enablers and barriers to Cloud ERP implementation and innovation outcomes. J. Enterp. Inf. Manag. **36**(5), 1209–1239 (2023)
27. Feng, H., Wang, X., Duan, Y., Zhang, J., Zhang, X.: Applying blockchain technology to improve agri-food traceability: A review of development methods, benefits and challenges. J. Clean. Prod. **260**, 121031 (2020)
28. John, E.P., Mishra, U.: Integrated multitrophic aquaculture supply chain fish traceability with blockchain technology, valorisation of fish waste and plastic pollution reduction by seaweed bioplastic: A study in tuna fish aquaculture industry. J. Clean. Prod. **434**, 140056 (2024)
29. Manzoor, R., Sahay, B.S., Singh, S.K.: Examining the factors that facilitate or hinder the use of blockchain technology to enhance the resilience of supply chains. IEEE Trans. Eng. Manage. **71**, 10626–10649 (2024)
30. Zhang, X., Feng, X., Jiang, Z., Gong, Q., Wang, Y.: A blockchain-enabled framework for reverse supply chain management of power batteries. J. Clean. Prod. **415**, 137823 (2023)
31. Zhu, S., Li, J., Wang, S., Xia, Y., Wang, Y.: The role of blockchain technology in the dualchannel supply chain dominated by a brand owner. Int. J. Prod. Econ. **258**, 108791 (2023)
32. Li, K., Lee, J.Y., Gharehgozli, A.: Blockchain in food supply chains: a literature review and synthesis analysis of platforms, benefits and challenges. Int. J. Prod. Res. **61**(11), 3527–3546 (2023)
33. Bhatia, M.S., Chaudhuri, A., Kayikci, Y., Treiblmaier, H.: Implementation of blockchainenabled supply chain finance solutions in the agricultural commodity supply chain: a transaction cost economics perspective. Production Planning and Control (2023).
34. Sousa, M.J., Rocha, Á.: Skills for disruptive digital business. J. Bus. Res. **94**, 257–263 (2019)
35. Nguyen, N. D. K., Ali, I.: Blockchain technology for sustainablie supply chain. In 2nd Australian International Conference on Industrial Engineering and Operations Management, (2023)
36. Klöckner, M., Schmidt, C.G., Fink, A., Flückiger, L., Wagner, S.M.: Exploring the physical–digital interface in blockchain applications: Insights from the luxury watch industry. Transp. Res. Part E: Logist. Transp. Rev. **179**, 103300 (2023)
37. Zhao, G., Liu, S., Lopez, C., Lu, H., Elgueta, S., Chen, H., Boshkoska, B.M.: Blockchain technology in agri-food value chain management: A synthesis of applications, challenges and future research directions. Comput. Ind. **109**, 83–99 (2019)
38. Martínez-Castañeda, M., Feijoo, C.: Use of blockchain in the agri-food value chain: State of the art in Spain and some lessons from the perspective of public support. Telecommun. Policy **47**(6), 102574 (2023)

39. Zhang, X., Li, Z., Li, G.: Impacts of blockchain-based digital transition on cold supply chains with a third-party logistics service provider. Transp. Res. Part E: Logist. Transp. Rev. **170**, 103014 (2023)
40. Menon, S., Jain, K.: Blockchain technology for transparency in agri-food supply chain: Use cases, limitations, and future directions. IEEE Trans. Eng. Manage. **71**, 106–120 (2021)
41. Kramer, M.P., Bitsch, L., Hanf, J.: Blockchain and its impacts on agri-food supply chain network management. Sustainability **13**(4), 2168 (2021)
42. Tyson Foods, https://www.tysonfoods.com/innovation, last accessed 2024/07/01.
43. Tyson Foods, https://www.tysonfoods.com/news/news-releases/2018/8/tyson-foods-selects-foodlogiq-connect-supply-chain, last accessed 2024/07/02.
44. Tyson Foods, https://www.tysonfoods.com/news/news-releases/2021/2/tyson-foods-accelerates-change-operate-speed-market, last accessed 2024/07/03
45. Nestlé, https://www.nestle.com/sites/default/files/2021-11/investor-seminar-2021-accelerating-digital-transformation.pdf, last accessed 2024/07/03
46. Nestlé, https://www.nestle.com/media/pressreleases/allpressreleases/nestle-open-blockchain-pilot, last accessed 2024/07/03
47. Nestlé, https://www.nestle.com/sites/default/files/asset-library/documents/media/news-feed/nestle-carrefour-mousline-puree-blockchain-pr-april-2019.pdf, last accessed 2024/07/03
48. Charlebois, S., Latif, N., Ilahi, I., Sarker, B., Music, J., Vezeau, J.: Digital traceability in agri-food supply chains: a comparative analysis of OECD member countries. Foods **13**(7), 1075 (2024)
49. Cointelegraph, https://cointelegraph.com/news/walmarts-foray-into-blockchain-how-is-the-technology-used, last accessed 2024/07/02
50. Potts, J.: Blockchain in agriculture. Available at SSRN 3397786 (2019).
51. Zdnet, https://www.zdnet.com/finance/blockchain/walmart-implements-ibms-blockchain-for-food-traceability/, last accessed 2024/07/02
52. Sharma, A., Sharma, A., Bhatia, T., Singh, R.K.: Blockchain enabled food supply chain management: A systematic literature review and bibliometric analysis. Oper. Manag. Res. **16**(3), 1594–1618 (2023)
53. Vitasek, K., Bayliss, J., Owen, L., Srivastava, N.: https://hbr.org/2022/01/how-walmart-canada-uses-blockchain-to-solve-supply-chain-challenges, last accessed 2024/07/02
54. Intelligent Investor, https://www.intelligentinvestor.com.au/investment-news/fresh-supply-co-brings-blockchain-to-the-farm/150751, last accessed 2024/07/03
55. HBAR Foundation, https://www.hbarfoundation.org/blog-post/fresh-supply-co-fsco-the-largest-user-of-mastercard-provenance-migrates-to-hedera-for-real-world-asset-tokenization, last accessed 2024/07/03
56. HBAR Foundation, https://www.hbarfoundation.org/blog-post/2023-a-year-in-review-for-the-hbar-foundation, last accessed 2024/07/03
57. PR Newswire, https://www.prnewswire.com/news-releases/fresh-supply-co-fsco-integrates-hederas-token-service-into-continuity-api-boosting-lending-confidence-through-verifiable-payments-302010936.html, last accessed 2024/07/03

# Strategic Approach to Facilitate Integration of IoT in the Rural Healthcare Management of Bangladesh

**Md. Adnan Hossain, Dewan Sabbir Ahammed Rayhan, and Iftekhar U. Bhuiyan**

**Abstract** The rural people of Bangladesh face a pronounced disparity in access to medical services compared to urban areas, exacerbated by an inadequate number of skilled medical professionals relative to the arrival of seriously ill patients seeking treatment. Leveraging the Internet of Things (IoT) offers a potential solution to these challenges, presenting noteworthy technological advancements with the capacity to revolutionize daily life. This study aims to strategically determine the viability of selecting an appropriate approach to enable IoT integration in Bangladesh's rural healthcare sector. To achieve this goal, the study first identified significant constraints and potential solutions to integrate IoT in rural healthcare through structured questionnaire interviews and then prioritized these factors using the Analytical Hierarchy Process (AHP) method. Additionally, a SWOT analysis was performed to evaluate the strengths, weaknesses, opportunities, and threats of the rural healthcare sector, incorporating expert insights. The outcomes of the SWOT analysis revealed IoT-based telemedicine as a trustworthy and appropriate opportunity in the rural healthcare sector of Bangladesh. IoT-based telemedicine, which integrates IoT with traditional telemedicine practices, offers a compelling method for delivering healthcare remotely. At the end of the study, two models are being constructed by considering the outcomes of the AHP method and SWOT analysis and named "Interaction of duty doctors with patients and IoT devices" and "Direct interaction of patients with IoT

Md. A. Hossain · D. S. A. Rayhan · I. U. Bhuiyan (✉)
Institute of Appropriate Technology, Bangladesh University of Engineering and Technology, Dhaka, Bangladesh
e-mail: iftekhar@iat.buet.ac.bd

D. S. A. Rayhan
e-mail: 0419292002@iat.buet.ac.bd

Md. A. Hossain
LM Ericsson Bangladesh Limited, Dhaka, Bangladesh

D. S. A. Rayhan
Department of Industrial & Production Engineering, National Institute of Textile Engineering and Research, Nayarhat, Savar, Dhaka, Bangladesh



N. D. K. Nguyen et al. (eds.), *Digital Transformation and Sustainability: Methods and Applications*, Information Systems Engineering and Management 47,
https://doi.org/10.1007/978-3-031-91278-8_9

devices". The proposed models were validated through the Delphi technique. Lastly, one model is being constructed from the suggested models to gain an appropriate and accepted approach to help integrate IoT into the management of Bangladesh's rural healthcare sector.

**Keywords** Rural healthcare management · Internet of things (IoT) · Analytical hierarchy process (AHP) · Delphi technique · SWOT analysis · Bangladesh

## 1 Introduction

During the COVID-19 pandemic, the rural people of Bangladesh have suffered a lot to get proper healthcare facilities [1, 2]. Generally, rural people go to urban medical centers to get a consultation from expert medical professionals [3, 4], but in the lockdown, people are not allowed to travel. The rural healthcare sector is still underdeveloped and it doesn't become reliable to the rural people. Most of the time it gets difficult to get the proper treatment in the first checkup with the available local doctors [3, 5, 6]. Many children are dying due to diarrhea, pneumonia, malaria, and so on [7, 8]. Women are suffering from pregnancy and delivery-related complications [9, 10]. The Internet of Things (IoT) can play a vital role in this case where there are many other opportunities in the rural healthcare sector. Remote treatment can be more efficient by integrating IoT and even critical patient handling in remote areas [3, 11]. With the integration of IoT into the healthcare, the rural healthcare organization may facilitate better treatment facilities for the rural people [12, 13]. IoT in healthcare services allows for saving time and improving customer experience. Some basic uses of IoT in healthcare sectors are Hospital Information Management System (HIMS), Electronic Healthcare Records System (EHR), and Mobile Healthcare (mHealth) [14–16]. HIMS collects and manages information such as medical records and data about doctors and patients. The system provides up-to-date information about appointments, hospital rooms, and so on [17–19]. EHR provides real-time information instantly and securely as well as shares data among various specialists to achieve better treatment results within a network [20]. The mHealth development gives a lot of opportunities like getting consultations online, scheduling offline consultations, and many more [21, 22]. Especially in COVID-19, there was massive usage of mobile app-based healthcare treatments [23, 24].

The IoT is the network of physical objects embedded with electronics, software, sensors, and network connectivity [25]. It assists in collecting, recording, exchanging, and analyzing new data streams faster and more accurately [26]. The IoT has the potential to provide solutions that significantly enhance many facets of daily life, including security, health, education, and energy efficiency (GSM [27]). IoT is creating a wealth of new applications that could enhance our quality of life. Researchers and practitioners worldwide have been paying close attention to IoT in recent years [28]. At present, the IoT has emerged as one of the twenty-first century's most potent communication platforms [29]. Globally, healthcare organizations are evolving into more user-centered, coordinated, and efficient systems. This trend suggests enhanced and simplified access to medical records and associated data, as well as more integrated and interoperable healthcare services [30]. Healthcare is becoming more and more dependent on IoT to enhance care quality, boost access to treatment, and most importantly lessen the cost of care [26]. Increasing the effectiveness of biomedical systems and healthcare infrastructures is one of the most difficult objectives of modern-day society [31]. Patient tracking, tracing, and monitoring may become simpler with the integration of IoT in healthcare [32].

Clinical care employs the IoT to monitor patients' physiological conditions via sensors and collect and analyze their information. Wearable devices with sensors are helpful not just for patients but also for normal individuals to check their health status [30]. For instance, Masimo Radical-7 remotely checks on the patient's condition and notifies the clinical staff of it [33]. IBM recently tracked hand washing at one of Ohio Health's hospitals using radio frequency identification (RFID) technology. Typically, when an accident occurs, someone has to notify the hospital to send an ambulance. However, with the IoT, wearable devices automatically signal the nearby Wi-Fi router whenever an accident occurs, and hospitals then send an ambulance based on the patient's health conditions, including heartbeat [25]. IoT-powered noninvasive monitoring can be used to continuously check on hospitalized patients whose physiological status requires careful attention. This kind of technology employs sensors to collect detailed physiological data, analyzes and stores the data using gateways and the cloud, and then wirelessly sends the data to caregivers for additional examination and analysis [25]. As an integrated information system, HIMS is made to record data from doctors as well as from computers, mobile devices, and IoT devices like medical analyzers and personal digital assistants. The IoT-based HIMS can collect a much larger and more comprehensive data repository than traditional HIMS without human efforts [34]. IoT applications are driving the creation of platforms for ambient assisted living (AAL) systems, which will provide services in the areas of improving safety and security, gaining access to medical and emergency systems, and helping people with everyday tasks, health, and activity monitoring. The IoT is crucial to healthcare applications, ranging from the prevention of disease to the management of chronic illnesses [35].

The Digital Bangladesh initiative of the government may offer the necessary impetus for the growth of the nation's IoT sector. The Bangladesh Telecommunication Regulatory Commission (BTRC) issued a regulation in April 2018 that allowed the IoT to lawfully enter Bangladesh. The country can deploy IoT technology in nine areas, according to the BTRC: smart buildings, industry automation, smart grids, water and waste management, smart agriculture, telecare, intelligent transportation systems, environment management, smart urban lighting, and smart parking [36]. Telemedicine was developed to treat patients who lived in remote locations, far from local medical facilities, or in areas where there was a shortage of medical personnel. Telemedicine is used well in Bangladesh. The IoT is improving telemedicine in many ways, enabling providers to provide care that may be superior to many in-person treatment techniques. For large populations residing in rural places, telemedicine may be a more affordable and convenient way to provide access to up-to-date healthcare services. Bangladesh has previously adopted the ideas of videoconferencing and teleconsultation. These infrastructures just require investment, upgrading, and the hiring of qualified workers [37]. IoT-based telemedicine has the potential to shorten patient wait times, lower operating costs for the health system, and enhance collaboration and communication between departments and hospitals. The primary objective of this study is to strategically identify the key challenges impeding the integration of IoT into the rural healthcare sector of Bangladesh, along with exploring potential solutions to mitigate these challenges. Furthermore, the second objective of this study is to develop comprehensive and tailored models that facilitate the effective incorporation of IoT into the rural healthcare management of Bangladesh. Upon validation, the proposed model is anticipated to enhance the management of the rural healthcare system through the application of IoT and telemedicine.

## 2 Materials and Methods

This study aims to find the viability of choosing a suitable strategy to facilitate the integration of IoT in the rural healthcare sector of Bangladesh. A brief description of the method to conduct this research is described in this section sequentially.

### 2.1 Research Process and Design

This research was done in the following phases:

***First Phase:*** In this phase, a survey was conducted by a structured questionnaire to collect qualitative data about the major constraints to integrating IoT in the rural healthcare sector of Bangladesh. Finally, quantitative data were collected and analyzed with the Analytic Hierarchy Process (AHP) method to prioritize and rank the significant constraints. Both data were collected by arranging face-to-face interviews. The interviewees were the management personnel of telecommunication industries and related organizations in Bangladesh.

***Second Phase:*** In this phase, the comments and feedback were gathered in the form of qualitative data on some possible solutions through Key Informant Interviews (KIIs), who is the authority and opinion leader of the industry with a structured questionnaire. Quantitative data were also collected from all key informants and analyzed with the AHP method to prioritize and rank the potential solutions to mitigate the major constraints.

***Third Phase:*** In this phase, a SWOT (Strengths, Weaknesses, Opportunities, and Threats) analysis was conducted to find out the environmental, social, economic, and technical strengths considering a suitable sector as an opportunity in the rural healthcare of Bangladesh.

***Model Development Phase:*** Based on the result of the first, second, and third phases, two models were proposed and validated by the Delphi technique. Finally, a final model was developed to facilitate the integration of IoT in the rural healthcare sector of Bangladesh.

## 2.2 *Data Collection and Questionnaire Design*

The data in the first phase was collected through face-to-face interviews with information technology (IT) expert personnel from the IT and Telecom sector of Bangladesh and some expert doctors from Bangabandhu Sheikh Mujib Medical University (BSMMU) and Dhaka Medical College (DMC). Responders have shared opinions based on their experience and expertise. The data collection in the first phase was concentrated on a total of twenty-two interviewees' judgments from the following organizations of Bangladesh: ERICSSON, BANGLALINK, GRAMEEN PHONE, HUAWEI, BJIT, EASYTRAX, UBITRIX, DATA SOFT, AUGMEDIX BD, SAMSUNG R&D, BSMMU, and DMC. The data of the second and third phases was collected from expert IT management professionals and public health personnel working in both the government and private sectors of Bangladesh. Also, the feedback was taken from professional associations. A total of ten personnel were interviewed for collecting data in the second and third phases. The persons from the following organizations were considered as the experts for the second and third phases: ICT Ministry Representatives, Business Support and Solution Head of Banglalink, IoT Head of Grameenphone, Senior IT Professionals of Government Organizations, and IT Head of Private Company. In the model development phase, three senior experts working in the IT and Healthcare sectors were invited to participate in the Delphi rounds. Hence, a total of thirty-seven persons' judgments were reflected in the outcomes of this research.

Three different sets of questionnaires were designed for the first, second, and third phases of the study. The first phase questionnaire was designed to find out qualitative data regarding the significant constraints that have an impact on the integration of IoT in the rural healthcare sector of Bangladesh. The second phase questionnaire was designed to find out the possible solutions that will be required to overcome the major constraints. The third phase questionnaire was designed to find out the strengths, weaknesses, opportunities, and threats for the integration of IoT in the rural healthcare sector of Bangladesh. To collect quantitative data, interviewees were requested to select any numeric value between 1 to 9 or 1 to 1/9 for marking priority of the comparisons and these data were required for AHP analysis in the first and second phases. The AHP rating scale was adopted from [38, 39]. The three different sets of questionaries' designed for this study are listed below:

| First phase questionnaire: |
|---|
| (a) How do you feel about the impact of IoT in healthcare?<br>(b) What are the current challenges of the healthcare sector in Bangladesh?<br>(c) What are the common problems you feel in adopting IoT considering Bangladesh 's industrial evolution?<br>(d) What are the common constraints to integrating IoT in the healthcare sector of Bangladesh?<br>(e) How do you feel about removing the constraints in facilitating IoT? |
| Second phase questionnaire: |
| (a) What will be the consequences of the integration of IoT in the healthcare sector of Bangladesh?<br>(b) How do you feel about the scopes and opportunities of IoT in healthcare?<br>(c) How possibly IoT can be integrated with the healthcare sector of Bangladesh?<br>(d) What are the prospects of IoT in the healthcare sector of Bangladesh?<br>(e) What can be the use of cases of IoT in the healthcare sector of Bangladesh? |
| Third phase questionnaire: |
| (a) What are the environmental, social, economic, and technical assets of IoT?<br>(b) What are the strengths of IoT in the Healthcare sector of Bangladesh?<br>(c) What are the weaknesses of IoT in the Healthcare sector of Bangladesh?<br>(d) What are the opportunities for IoT in the healthcare sector of Bangladesh?<br>(e) What are the threats of IoT in the healthcare sector of Bangladesh? |

## 2.3 *Research Tools*

In this study's first, second, and third phases, two technology assessment tools were used: the Analytic Hierarchy Process (AHP) and SWOT analysis. At the study's model and framework development phase, one technology forecasting tool, the Delphi technique, was implemented to validate the proposed models and develop the final model.

### 2.3.1 Analytic Hierarchy Process (AHP)

The analytic hierarchy process (AHP) is a Multi-criteria Decision Analysis (MCDA) tool used for organizing and analyzing complex decisions developed by Thomas Saaty in the 1970s [38, 39]. MCDA techniques help the decision-makers assess alternatives by explicitly considering multiple criteria or factors. MCDA techniques can be categorized into three types: value measurement models, outranking models, and goal, aspiration, or reference-level models. AHP is a value measurement model advantageous for technology assessment [40] and the most popular decision-making tool [41]. The unique feature of AHP is its adaptability to be integrated with various approaches, such as Fuzzy Logic, Linear Programming, and Quality Function Deployment (QFD). AHP is an Eigen value approach to pair-wise comparisons and offers a method for adjusting the numerical scale used to measure both quantitative and qualitative performance [42]. The procedure for using the AHP can be summarized below [42].

***Step-1:*** Modeling the problem in the form of a hierarchy that includes the decision goal, the alternatives for achieving it, and the criteria for assessing the alternatives.

***Step-2:*** Establish priority between the components of the hierarchy by evaluating each one in turn using pairwise comparisons.

***Step-3:*** Create a set of general priorities for the hierarchy by combining these assessments.

***Step-4:*** Make sure the judgments are consistent.

***Step-5:*** Determining a final choice based on the outcomes of this process.

### 2.3.2 SWOT Analysis

A structured strategic planning technique called a SWOT analysis is used to assess a project's or commercial venture's strengths, weaknesses, opportunities, and threats. Because SWOT analysis can guide subsequent planning processes to accomplish the goal, it is crucial to identify them. It helps us improve our strengths, solve our weaknesses, reduce risks, and seize as many opportunities for success as we can [43].

### 2.3.3 Delphi Method

The Delphi method is a forecasting procedure that uses the answers to several rounds of questionnaires distributed to a group of experts [44]. The model is very popular in the technological field and business sectors, especially when data are scarce for analysis [45]. Since the panel is informed of the group's opinions after several rounds of questioning, the Delphi method aims to arrive at the correct response by consensus [44, 46]. One of the major features of Delphi survey research is that participants are anonymous to each other, and responses are made in batches, so minimizing influence, in much the same way that the norms of brainstorming require participants to avoid judgment [47–49]. Validation is the process of determining whether a tool is appropriate for its intended use. It is crucial to examine the circumstances in which the validation process is conducted while evaluating the various types of validation processes [50]. The Delphi technique in this study focuses on the overall model structure and its applicability in rural healthcare organizations of Bangladesh. To validate the suggested models, the method asks the panel of experts to explain what they would do in certain circumstances. The Delphi process includes selecting the expert panel, designing the questionnaire and scoring methodologies, determining the number of iteration rounds, and analyzing the data.

In Delphi studies, the choice of an expert panel is typically based on the disciplinary areas of expertise needed to address the particular problem. The expert panel's merits, not its size, are used to evaluate representation with panel size. This is because the expert panels are not considered representative samples for statistical purposes by the Delphi technique [50]. Therefore, for this study, five experts were chosen based on their expertise in the healthcare industry and IoT. The number of iteration rounds varies and depends on the goal of the research. Skulmonski et al. [51] found that most research requires only two or three iterations. The goal of the first iteration is to pinpoint the general concerns relating to critical patient handling and remote treatments within the suggested models in the rural healthcare sector of Bangladesh. The first-round questionnaire is typically unstructured and seeks an open response. To maximize the effectiveness of data collecting and solicit specific information about the constructed models, open-ended questions were used. In the second phase, which was more focused, the questionnaire asked for a rating or ranking. The second questionnaire asked the experts to indicate whether they agreed or disagreed with the revised model. The questions are designed to measure the degree of agreement using a five-point Likert scale. There are five points on the scale, ranging from strongly disagree (1) to strongly agree (5). After every round, experts are presented with the findings in a Delphi study until there is agreement or consistency in the findings. In a particular round, consensus was defined as the experts' agreement on how to rate a given item. The minimal level of agreement on any given item was 75%. Consequently, when the experts rate each item with a score of more than 3.7, consensus is reached in this study.

### 2.3.4 Key Informant Interviews (KIIs)

Key informant interviews (KIIs) are often regarded as an essential component of qualitative research. KIIs frequently support other research methods like focus group discussions (FGDs) and surveys. Because of the prominence and experience of the key informant, KIIs may unintentionally be positioned within the hierarchy of research methods as yielding more valuable knowledge. Key informants are deemed to provide significant knowledge-more knowledge than might be obtained from "ordinary" persons in interviews [52]. KIIs are often conducted using one of the following two techniques: phone and in-person interviews. The reasons for conducting KIIs are listed below [53].

***Reason-1:*** To know about a serious community topic or problem from a small group of well-connected and knowledgeable community experts.

***Reason-2:*** Understanding the motivations and views of community members about a certain issue.

***Reason-3:*** To gather information from people with varied experiences and viewpoints while also asking in-depth and probing questions.

***Reason-4:*** To discuss sensitive matters, get candid responses, or obtain the necessary depth of knowledge. Individual or small group conversations (maximum two to three people) provide a comfortable setting for individuals to engage in frank and open in-depth discussions.

***Reason-5:*** To receive more open or in-depth responses. The focus group dynamic may prevent us from discussing sensitive topics or gathering the necessary knowledge. Some individuals may be unable to express their perspectives on sensitive topics due to group dynamics.

# 3 Results

## *3.1 Identification of Constraints to Facilitate IoT in Rural Healthcare Management*

In the initial phase of the study, questionnaire data was gathered through insightful, face-to-face interviews with experts to identify the constraints affecting the implementation of IoT in rural healthcare management. After meticulously reviewing the responses, the major constraints were categorized into three broad categories: "Adoption of IoT," "Public Awareness," and "Facilitating IoT in Healthcare". Each of these categories was then further divided into two sub-categories. For instance, the Adoption of IoT was fragmented into "Individual effort" and "Infrastructure" while Public Awareness was into "Resources" and "Management". Additionally, Facilitating IoT in rural healthcare was divided into "Technology provision" and "Economic provision." Throughout this process, four common constraints were addressed across all six sub-categories. To prioritize and rank the significant constraints, the Analytic Hierarchy Process (AHP) method was utilized. Before commencing the AHP analysis, the priority setting among the categories, sub-categories, and specific constraints was carefully established through comparative evaluations.

## *3.2 Prioritization of Constraints to Facilitate IoT in Rural Healthcare Management*

In the pursuit of fostering IoT integration in rural healthcare management, it is imperative to methodically prioritize constraints. In this study, for effective prioritization of constraints by AHP calculation, a comprehensive set of comparison ratings was obtained from twenty-two experts, with meticulous attention to ensuring consistency, resulting in a consistency ratio (CR) below 10%. Table 1 presents the comparison ratings provided by Expert-1 and the calculated priorities of major constraints categories and sub-categories. It also delineates the comparison ratings and calculated priorities for constraints falling under 'Individual effort,' 'Infrastructure,' 'Resources,' and 'Management', and the priorities of constraints pertaining to 'Technology Provision' and 'Economic Provision'.

The calculations of AHP outcomes for Expert-1 are shown in detail below:

Adequate IoT Resources = (0.691 × 0.667 × 0.421) + (0. 691 × 0.333 × 0.579) + (0.218 × 0.8 × 0.535) + (0. 218 × 0.2 × 0.58) + (0.091 × 0.5 × 0.404) + (0.091 × 0.5 × 0.55) = 0.4893.

Professional Association = (0. 691 × 0. 667 × 0.267) + (0. 691 × 0. 333 × 0.270) + (0. 218 × 0.8 × 0.270) + (0. 218 × 0.2 × 0.308) + (0.091 × 0.5 × 0.443) + (0.091 × 0.5 × 0.281) = 0.1963.

Operational Environment = (0. 691 × 0. 667 × 0.181) + (0. 691 × 0. 333 × 0.072) + (0. 218 × 0.8 × 0.120) + (0. 218 × 0.2 × 0.053) + (0.091 × 0.5 × 0.099) + (0.091 × 0.5 × 0.108) = 0.1281.

Strategic Planning = (0. 691 × 0. 667 × 0.121) + (0. 691 × 0. 333 × 0.078) + (0. 218 × 0.8 × 0.075) + (0. 218 × 0.2 × 0.059) + (0.091 × 0.5 × 0.054) + (0.091 × 0.5 × 0.062) = 0.095.

The AHP decision hierarchy for the first phase analysis, as evaluated by Expert-1, has been visually represented in Fig. 1. The AHP priorities from all twenty-two experts involved in the initial phase of analysis have been determined. The average priority percentages for the factors "adequate IoT resources," "professional association," "operational environment," and "strategic planning" are reported as 55%, 24%, 14%, and 9% respectively. In addition, the standard deviation (SD) figures for these percentages are recorded as 4%, 6%, 1%, and 3% respectively.

### 3.3 *Potential Solutions for the Integration of IoT in Rural Healthcare Management*

Interviewees participating in the first phase were requested to recommend possible solutions to overcome the constraints based on their opinions and real-world experiences. The outcomes of the first phase analysis were shown to the key informants participating in the second phase. With the help of the second phase questionnaire, ten individual key informants were asked to recommend possible solutions for mitigating the major constraints to facilitate the integration of IoT in the rural healthcare sector of Bangladesh. The responses of the interviewees were studied critically. Finally, six possible solutions for the integration of IoT in the rural healthcare sector of Bangladesh were found in the second phase which were addressed among the three responsible categories (Organization X, Organization Y, and Organization Z). The six possible solutions are: Need effective associations and resource empowerment among the organizations (A), Need proper professional training about new technology and industry attachment with academic institutes (B), Set up IoT incubator

**Table 1** Comparison rating provided by Expert-1 and AHP priorities

| *Major constraints categories and sub-categories* | | |
|---|---|---|
| Comparison between categories and Sub-categories | Ratings | AHP priorities |
| The Severity of 'Constraints related to Adoption of IoT' over 'Constraints related to Public Awareness' | 4 | Constraints related to the Adoption of IoT = 0.691 |
| The Severity of 'Constraints related to Adoption of IoT' over 'Constraints related to Facilitate IoT in Healthcare' | 6 | Constraints related to Public Awareness = 0.218 |
| The Severity of 'Constraints related to Public Awareness' over 'Constraints related to Facilitate IoT in Healthcare' | 3 | Constraints related to Facilitate IoT in Healthcare = 0.091 |
| The Severity of 'Constraints related to Individual effort' over 'Constraints related to Infrastructure' | 1 | Constraints related to Individual effort = 0.667, Constraints related to Infrastructure = 0.333 |
| The Severity of 'Constraints related to Resources' over 'Constraints related to Management' | 4 | Constraints related to Resources = 0.8, Constraints related to Management = 0.2 |
| The Severity of 'Constraints related to Technology Provision' over 'Constraints related to Economic Provision' | 1 | Constraints related to Technology Provision = 0.5, Constraints related to Economic Provision = 0.5 |
| *Constraints under sub-categories 'Individual effort' and 'Infrastructure'* | | |
| Comparison between Constraints | Ratings | AHP Priorities |
| The Severity of 'Lack of adequate IoT resources' over 'Lack of professional association' | 3 | • Lack of adequate IoT resources = 0.431,<br>• Lack of professional association = 0.267,<br>• Lack of operational environment = 0.181,<br>• Lack of strategic planning = 0.121 |
| The Severity of 'Lack of adequate IoT resources' over 'Lack of operational environment' | 2 | |
| The Severity of 'Lack of adequate IoT resources' over 'Lack of strategic planning' | 2 | |
| The Severity of 'Lack of professional association' over 'Lack of operational environment' | 2 | |
| The Severity of 'Lack of professional association' over 'Lack of strategic planning' | 3 | |
| The Severity of 'Lack of operational environment' over 'Lack of strategic planning' | 2 | |

(continued)

**Table 1** (continued)

| | | |
|---|---|---|
| The Severity of 'Lack of adequate IoT resources' over 'Lack of professional association' | 3 | • Lack of adequate IoT resources = 0.535,<br>• Lack of professional association = 0.270,<br>• Lack of operational environment = 0.120,<br>• Lack of strategic planning = 0.075 |
| The Severity of 'Lack of adequate IoT resources' over 'Lack of operational environment' | 4 | |
| The Severity of 'Lack of adequate IoT resources' over 'Lack of strategic planning' | 5 | |
| The Severity of 'Lack of professional association' over 'Lack of operational environment' | 3 | |
| The Severity of 'Lack of professional association' over 'Lack of strategic planning' | 4 | |
| The Severity of 'Lack of operational environment' over 'Lack of strategic planning' | 2 | |
| *Constraints under sub-categories 'Resources' and 'Management'* | | |
| Comparison between Constraints | Ratings | AHP Priorities |
| The Severity of 'Lack of adequate IoT resources' over 'Lack of professional association' | 1 | • Lack of adequate IoT resources = 0.404,<br>• Lack of professional association = 0.443,<br>• Lack of operational environment = 0.099,<br>• Lack of strategic planning = 0.054 |
| The Severity of 'Lack of adequate IoT resources' over 'Lack of operational environment' | 4 | |
| The Severity of 'Lack of adequate IoT resources' over 'Lack of strategic planning' | 7 | |
| The Severity of 'Lack of professional association' over 'Lack of operational environment' | 5 | |
| The Severity of 'Lack of professional association' over 'Lack of strategic planning' | 8 | |
| The Severity of 'Lack of operational environment' over 'Lack of strategic planning' | 2 | |

(continued)

**Table 1** (continued)

| | | |
|---|---|---|
| The Severity of 'Lack of adequate IoT resources' over 'Lack of professional association' | 2 | • Lack of adequate IoT resources = 0.579,<br>• Lack of professional association = 0.270,<br>• Lack of operational environment = 0.072,<br>• Lack of strategic planning = 0.078 |
| The Severity of 'Lack of adequate IoT resources' over 'Lack of operational environment' | 8 | |
| The Severity of 'Lack of adequate IoT resources' over 'Lack of strategic planning' | 8 | |
| The Severity of 'Lack of professional association' over 'Lack of operational environment' | 4 | |
| The Severity of 'Lack of professional association' over 'Lack of strategic planning' | 3 | |
| The Severity of 'Lack of operational environment' over 'Lack of strategic planning' | 1 | |
| *Constraints under sub-categories 'Technology Provision' and 'Economic Provision'* | | |
| Comparison between Constraints | Ratings | AHP Priorities |
| The Severity of 'Lack of adequate IoT resources' over 'Lack of professional association' | 3 | • Lack of adequate IoT resources = 0.58,<br>• Lack of professional association = 0.308,<br>• Lack of operational environment = 0.053,<br>• Lack of strategic planning = 0.059 |
| The Severity of 'Lack of adequate IoT resources' over 'Lack of operational environment' | 8 | |
| The Severity of 'Lack of adequate IoT resources' over 'Lack of strategic planning' | 8 | |
| The Severity of 'Lack of professional association' over 'Lack of operational environment' | 9 | |
| The Severity of 'Lack of professional association' over 'Lack of strategic planning' | 5 | |
| The Severity of 'Lack of operational environment' over 'Lack of strategic planning' | 1 | |

(continued)

**Table 1** (continued)

| | | |
|---|---|---|
| The Severity of 'Lack of adequate IoT resources' over 'Lack of professional association' | 2 | • Lack of adequate IoT resources = 0.55,<br>• Lack of professional association = 0.281,<br>• Lack of operational environment = 0.108,<br>• Lack of strategic planning = 0.062 |
| The Severity of 'Lack of adequate IoT resources' over 'Lack of operational environment' | 5 | |
| The Severity of 'Lack of adequate IoT resources' over 'Lack of strategic planning' | 9 | |
| The Severity of 'Lack of professional association' over 'Lack of operational environment' | 3 | |
| The Severity of 'Lack of professional association' over 'Lack of strategic planning' | 4 | |
| The Severity of 'Lack of operational environment' over 'Lack of strategic planning' | 2 | |

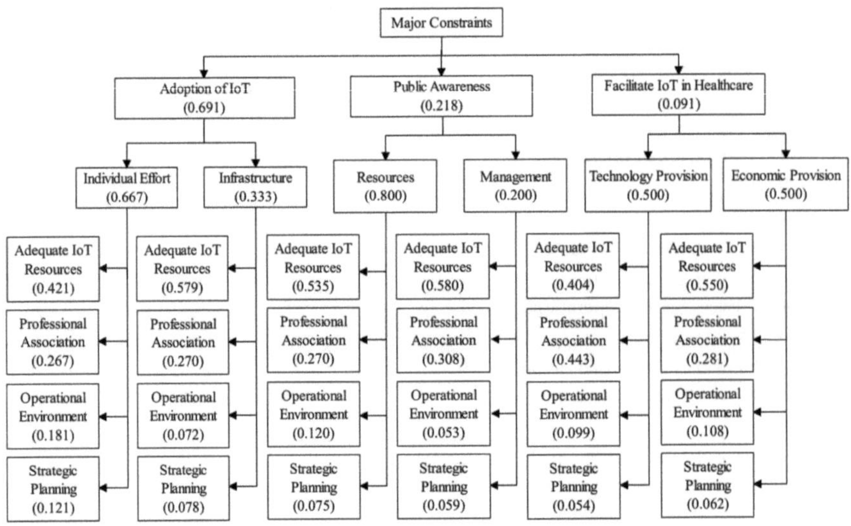


**Fig. 1** AHP decision hierarchy of major constraints to facilitate IoT in rural healthcare management reflected by Expert-1 in the first phase analysis (*Source* Author's Work)

with stable power, high-speed internet, easy communication facility, etc. (C), Proactive actions from Public health organizations to get familiar and used to about the new technology equipment (D), Implement a future-oriented ICT policy to promote IoT in Healthcare sector (E), and Encourage foreign and local investors by reducing tax and improving loan facilities, etc. (F). Following a comprehensive analysis, the solutions were subsequently prioritized and ranked by using the AHP method. Before starting the AHP analysis, the priority setting among categories and possible solutions were made by comparison.

## *3.4 Prioritizing Solutions to Facilitate IoT in Rural Healthcare Management Using AHP*

For the AHP calculation in the second phase, comparison ratings were collected from all participating ten key informants (KIs) in the second phase questionnaire data collection. The CR of the comparisons was verified and found to be below 10%. Table 2 presents the comparison ratings provided by KI-1 and the calculated priorities of the responsible categories. Additionally, it shows the comparison rating provided by KI-1 and priorities of potential solutions with respect to 'Organization X', 'Organization Y', and 'Organization Z' respectively.

The outcomes of AHP analysis in the second phase for KI-1 are calculated as below:

Need effective associations and resource empowerment among the organizations (A) $= 0.073 \times 0.255 + 0.166 \times 0.229 + 0.761 \times 0.195 = 0.2050$.

Need proper professional training about new technology and industry attachment with academic institutes (B) $= 0.073 \times 0.238 + 0.166 \times 0.246 + 0.761 \times 0.244 = 0.2265$.

Set up IoT incubator with stable power, high-speed internet, easy communication facility (C) $= 0.073 \times 0.198 + 0.166 \times 0.243 + 0.761 \times 0.230 = 0.1895$.

Proactive actions from Public health organizations to get familiar and used to the new technology equipment (D) $= 0.073 \times 0.095 + 0.166 \times 0.178 + 0.761 \times 0.194 = 0.1841$.

Implement a future-oriented ICT policy to promote IoT in the Healthcare sector (E) $= 0.073 \times 0.128 + 0.166 \times 0.057 + 0.761 \times 0.056 = 0.0614$.

Encourage foreign and local investors by reducing tax and improving loan facilities etc. (F) $= 0.073 \times 0.085 + 0.166 \times 0.046 + 0.761 \times 0.043 = 0.0389$.

The AHP decision hierarchy for KI-1 of the second phase analysis has been provided in Fig. 2. The AHP priorities for all ten KIs of the second phase analysis have been determined. The average priority values and standard deviations for A, B, C, D, E, and F are 0.230, 0.233, 0.215, 0.177, 0.069, and 0.051 respectively, and 3%, 1%, 7%, 2%, 1%, and 1% respectively.

## *3.5 SWOT Analysis to Integrate IoT in the Rural Healthcare Sector of Bangladesh*

The SWOT analysis was conducted with thirty-two interviewees from the first and second phases to determine the technological, social, economic, and environmental

**Table 2** Comparison rating provided by KI-1 and AHP priorities

| *The Categories* | | |
|---|---|---|
| Comparison between Categories | Ratings | AHP priorities |
| Importance of 'Organization X' over 'Organization Y' | 3 | Organization X = 0.073 |
| Importance of 'Organization X' over 'Organization Z' | 8 | Organization Y = 0.166 |
| Importance of 'Organization Y' over 'Organization Z' | 6 | Organization Z = 0.761 |
| *Possible solutions with respect to 'Organization X'* | | |
| Comparison of Possible Six Solutions | Ratings | AHP Priorities |
| Importance of 'A' over 'B' | 2 | Need effective associations and resource empowerment among the organizations (A) = 0.255,<br>Need proper professional training about new technology and industry attachment with academic institutes (B) = 0.238,<br>Set up IoT incubator with stable power, high-speed internet, easy communication facility, etc. (C) = 0.198,<br>Proactive actions from Public health organizations to get familiar and used to the new technology equipment (D) = 0.095,<br>Implement a future-oriented ICT policy to promote IoT in the Healthcare sector (E) = 0.128,<br>Encourage foreign and local investors by reducing tax and improving loan facilities etc. (F) = 0.085 |
| Importance of 'A' over 'C' | 1 | |
| Importance of 'A' over 'D' | 3 | |
| Importance of 'A' over 'E' | 2 | |
| Importance of 'A' over 'F' | 2 | |
| Importance of 'B' over 'C' | 3 | |
| Importance of 'B' over 'D' | 2 | |
| Importance of 'B' over 'E' | 1 | |
| Importance of 'B' over 'F' | 3 | |
| Importance of 'C' over 'D' | 2 | |
| Importance of 'C' over 'E' | 3 | |
| Importance of 'C' over 'F' | 2 | |
| Importance of 'D' over 'E' | 1 | |
| Importance of 'D' over 'F' | 1 | |
| Importance of 'E' over 'F' | 2 | |

(continued)

**Table 2** (continued)

| *Possible solutions with respect to 'Organization Y'* | | |
|---|---|---|
| Comparison of Possible Six Solutions | Ratings | AHP Priorities |
| Importance of 'A' over 'B' | 1 | • Need effective associations and resource empowerment among the organizations (A) = 0.229,<br>• Need proper professional training about new technology and industry attachment with academic institutes (B) = 0.246,<br>• Set up IoT incubator with stable power, high-speed internet, easy communication facility, etc. (C) = 0.243,<br>• Proactive actions from Public health organizations to get familiar and used to the new technology equipment (D) = 0.178,<br>• Implement a future-oriented ICT policy to promote IoT in the Healthcare sector (E) = 0.057,<br>• Encourage foreign and local investors by reducing tax and improving loan facilities etc. (F) = 0.046 |
| Importance of 'A' over 'C' | 2 | |
| Importance of 'A' over 'D' | 1 | |
| Importance of 'A' over 'E' | 4 | |
| Importance of 'A' over 'F' | 3 | |
| Importance of 'B' over 'C' | 1 | |
| Importance of 'B' over 'D' | 3 | |
| Importance of 'B' over 'E' | 4 | |
| Importance of 'B' over 'F' | 3 | |
| Importance of 'C' over 'D' | 3 | |
| Importance of 'C' over 'E' | 4 | |
| Importance of 'C' over 'F' | 5 | |
| Importance of 'D' over 'E' | 4 | |
| Importance of 'D' over 'F' | 8 | |
| Importance of 'E' over 'F' | 2 | |

(continued)

**Table 2** (continued)

| *Possible solutions with respect to 'Organization Z'* | | |
|---|---|---|
| Comparison of Possible Six Solutions | Ratings | AHP Priorities |
| Importance of 'A' over 'B' | 1 | Need effective associations and resource empowerment among the organizations (A) = 0.234,<br>Need proper professional training about new technology and industry attachment with academic institutes (B) = 0.244,<br>Set up an IoT incubator with stable power, high-speed internet, easy communication facility, etc. (C) = 0.230,<br>Proactive actions from Public health organizations to get familiar and used to the new technology equipment (D) = 0.194,<br>Implement a future-oriented ICT policy to promote IoT in the Healthcare sector (E) = 0.056,<br>Encourage foreign and local investors by reducing tax and improving loan facilities etc. (F) = 0.043 |
| Importance of 'A' over 'C' | 2 | |
| Importance of 'A' over 'D' | 1 | |
| Importance of 'A' over 'E' | 4 | |
| Importance of 'A' over 'F' | 3 | |
| Importance of 'B' over 'C' | 1 | |
| Importance of 'B' over 'D' | 2 | |
| Importance of 'B' over 'E' | 5 | |
| Importance of 'B' over 'F' | 4 | |
| Importance of 'C' over 'D' | 2 | |
| Importance of 'C' over 'E' | 4 | |
| Importance of 'C' over 'F' | 6 | |
| Importance of 'D' over 'E' | 4 | |
| Importance of 'D' over 'F' | 8 | |
| Importance of 'E' over 'F' | 2 | |

effects of integrating IoT in Bangladesh's rural healthcare system. The third phase questionnaire was asked to get their feedback on key points (Strengths, Weaknesses, Opportunities, and Threats). SWOT analysis found the following strengths: Low-wage workers (S-1), Talented individual efforts (S-2), Mass people interaction with the internet and telecom devices (S-3), Strong telecommunication network (S-4), The inclination to the IoT sector (S-5), Increasing trend of foreign investments (S-6), Appropriate and instant treatment in remote areas (S-7), Sufficient Health and IT background resources (S-8); the following weaknesses: Less technical skill and quality workers (W-1), Lack of high-speed internet facilities (W-2), Inefficient power

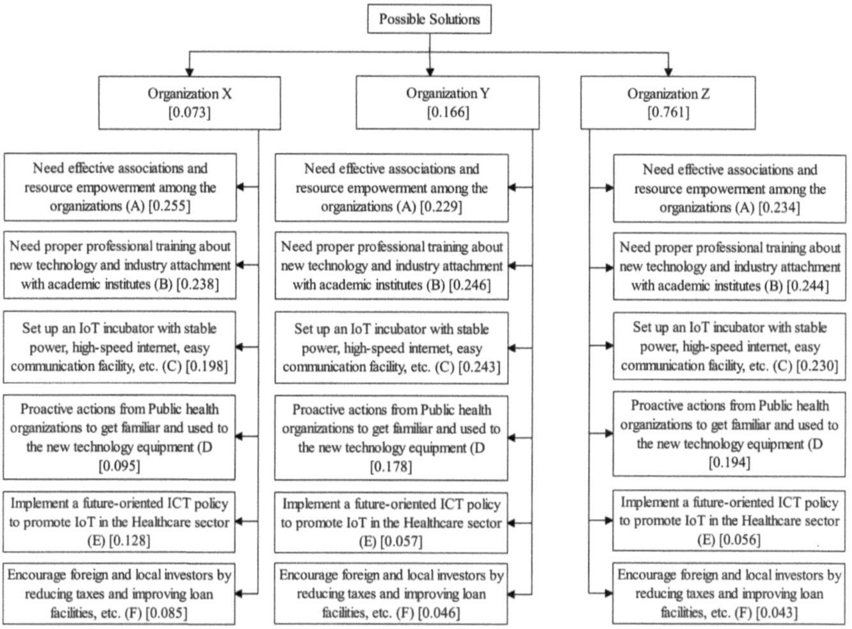


**Fig. 2** AHP decision hierarchy of proposed solutions for overcoming constraints in integrating IoT in rural healthcare management reflected by KI-1 in the second phase analysis (*Source* Author's Work)

system (W-3), Lack of appropriate planning (W-4); the following opportunities: IoT for patient monitoring (O-1), Economic growth (O-2), IoT based telemedicine (O-3), Remote patient monitoring (O-4), Preventive healthcare (O-5), Healthcare vehicle system (O-6); and the following threats: Weak association with government policy (T-1), Competitive other emerging countries (T-2), Bureaucracy (T-3), Political instability (T-4).

Then consensus were gathered from experts and counted the votes of each listed point of SWOT analysis. The inclination to the IoT sector (S-5) received 80% of the vote, while the inefficient power system (W-3) received 80%, IoT-based telemedicine (O-3) received 90%, and political instability received (T-4) 80%. The SWOT analysis indicates that the following are the environmental strengths of integrating IoT in Bangladesh's rural healthcare sector: the large number of people interacting with internet and telecom devices, as well as the availability of sufficient health and IT background resources; the social strengths, which include professional individual efforts and prompt, appropriate treatment; the economic strengths, which include low-wage workers and the growing trend of foreign investments; and the technical strengths, which include a robust telecommunication network and a preference for the IoT sector. Several opportunities emerged from the expert's feedback to incorporate IoT in the rural healthcare sector of Bangladesh. However, the majority of experts suggest that IoT-based telemedicine is a good option among them.

## *3.6 Proposed Models for IoT-Based Tele-Medicine in Rural Healthcare of Bangladesh*

There are numerous government and private local healthcare centers and medicals at the district and thana levels, but the majority of them employ recently hired MBBS physicians and medical officers. They always make an effort to provide the best treatment possible, but if a critical condition or unforeseen circumstance arises, they refer patients to specific specialists. To overcome this circumstance, IoT can play a vital role in obtaining or obtaining immediate assistance and input from knowledgeable medical professionals. In some circumstances, due to a lack of medical officers and resources, it becomes difficult to provide proper care to all patients immediately. Patients must therefore wait in the queue as a result. Treatment can be provided via IoT devices with the aid of artificial intelligence (AI), indirect communication, or without the direct participation of on-call physicians. Only duty doctors can able to oversee that situation. If the patients are not in such a critical state, this will allow for the simultaneous care of several patients.

### 3.6.1 Study of Treatment Scenarios

Ear, Nose, and Throat (ENT), one of the important departments of the medical sector, was taken into account, aiming to construct comprehensive models to facilitate the integration of IoT in the rural healthcare sector of Bangladesh. For that purpose, consultations were conducted with two medical officers working remotely in a government health organization. Based on their feedback on ENT, the following treatment scenarios are found.

**Scenario-1**

Problem statement: A mother of a 4-year-old boy complains that the boy had repeated an attack of upper respiratory tract infection, mouth breathing, nasal obstruction, and deafness.

Probable diagnosis: Otitis media with effusion due to enlarged adenoid.

Treatment: Conservative treatment for 15 days, referred to an ENT specialist for further guidelines and surgery.

**Scenario-2**

Problem statement: A 25-year-old lady complains about bilateral nasal obstruction with a runny nose for 1 year.

Probable diagnosis: Ethmoidal polyp.

Treatment: Initial treatment was given to remove pain immediately and referred to an ENT specialist for further guidelines and a detailed checkup.

**Scenario-3**

Problem statement: A 35-year-old male complains about right-sided non-offensive ear discharge for six years with hearing loss and tinnitus.

Probable diagnosis: Right chronic supportive otitis media.

Treatment: Initial treatment was given to remove pain immediately and referred to an ENT specialist for further guidelines and a detailed checkup.

**Scenario-4**

Problem statement: A 40-year-old lady complains of a single mass on the left side of her neck.

Probable diagnosis: solitary thyroid module on the left side.

Treatment: Initial treatment was given to remove pain immediately and referred to an ENT specialist for further guidelines and a detailed checkup.

**Scenario-5**

Problem statement: A fair-looking lady 30 years old complains about bilateral conductive hearing loss.

Probable diagnosis: Bilateral otosclerosis.

Treatment: Initial treatment was given to remove pain immediately and referred to an ENT specialist for further guidelines and a detailed checkup.

**Scenario-6**

Problem statement: A 14-year-old boy complains about severe unprovoked nasal bleeding with bilateral nasal obstruction.

Probable diagnosis: Juvenile nasopharyngeal angiofibroma.

Treatment: Initial treatment was given to remove pain immediately and referred to an ENT specialist for further guidelines and a detailed checkup.

Based on the scenarios, it is observed that if there is a communication channel available between the local MBBS doctors and the experts then the treatment can be given locally for the critical situations.

### 3.6.2 Proposed Models

To accomplish the aim of this research, two distinct models have been proposed. Based on IoT devices with and without AI, the models' flow is divided into two parts. The proposed model-1 is shown in Fig. 3(a), where the duty doctor interacts with patients and IoT devices. According to this model, the duty doctor of the health organization will ask general questions to the coming patients about health-related problems. The on-call doctor will next determine if the treatment may be provided locally or if expert professional advice is not required. Patients will receive treatment in their local healthcare center if it is feasible. Information about the healthcare issue will be sent to the central server via IoT devices and channels if the treatment cannot be provided locally. Expert medical specialists will select the health issue from the central database. A qualified medical expert will evaluate the case and offer suggestions or advice on how to proceed with the treatment. The on-call physician will take note of the comments and respond appropriately to treat the patients.

The proposed model-2 is shown in Fig. 3(b), where incoming patients directly interact with IoT devices with AI and the duty doctor only supervises the patients. According to the model, the patients will share their issues with IoT devices. With the aid of AI, the IoT device will examine the generic pattern of the issue. This information will be sent to the central server via IoT devices and a channel if the problem's pattern cannot be identified or if professional assistance is required. The available medical professionals will select this case from the central database. Professionals will analyze the information and make the recommendations required to treat the

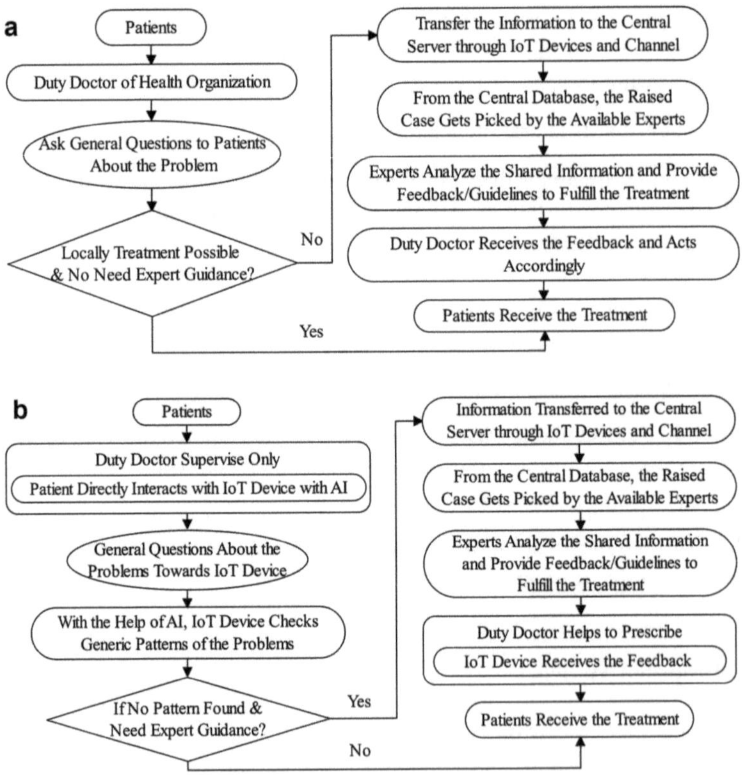


**Fig. 3** Proposed models **a** Model-1 (Interaction of the duty doctors with patients and IoT devices), **b** Model-2 (Direct interaction of patients with IoT devices) (*Source* Author's Work)

patients. Ultimately, the duty doctor will prescribe after the IoT device has received the feedback.

### 3.6.3 Architecture Diagram of Proposed Models

Figure 4(a) shows the whole process of the architecture diagram of Model-1. The duty physicians, according to the architecture diagram, will participate directly and engage directly with the patients and the IoT device. Using an IoT device, doctors will collect the necessary information from the patients and transmit it to a central server. After assigning the case to an available specialist or expert and receiving comments, the central server will respond to the duty physician. The duty physician will administer the treatment following receipt of the feedback.

According to the architecture diagram of proposed model-2, duty doctors can merely monitor with the aid of the AI embedded into the IoT device; they are not required to be directly involved. Like a chatbot, it will communicate with the patient. In this instance, machine learning and deep learning will be integrated into IoT devices, allowing for the possibility of all patterns for the first treatments depending on the patient's symptoms. Patients will provide the device with detailed information about all symptoms and issues. IoT devices will automatically get and evaluate the data. The IoT device will respond appropriately if the case is characterized as an inherent common scenario or treatment. If not, the data will be sent to the system or centralized server. After receiving input, the central server will respond to the duty physician and allocate the case to an available specialist or expert. The duty physician will administer the treatment following receipt of the feedback. This whole process of the architecture diagram is shown in Fig. 4(b).

## 3.7 *Validation of Proposed Models Using Delphi Technique*

The content analysis technique was employed in this study to determine the key themes that emerged from the first unstructured questionnaire of the Delphi study. The findings are then converted into a revised model and a structured questionnaire for the following round. Because the results are quantitative, they are assessed using ranking and rating procedures, and then the final model is presented. In the first round, each expert was interviewed. During this session, the experts were invited to provide qualitative input on the models' components. The results of the first round were utilized to improve the proposed models and to create the questionnaire for the second round of the Delphi study. The experts have provided three major suggestions to improve the models. The first recommendation from the expert panel is to think about emergencies. The element is therefore included in the revised models. The second point raised by the experts is the importance of putting logic into the assessment of whether or not the patient is in critical condition. In the event of a catastrophic condition, it is recommended to forego the IoT-based process. As a result, a new

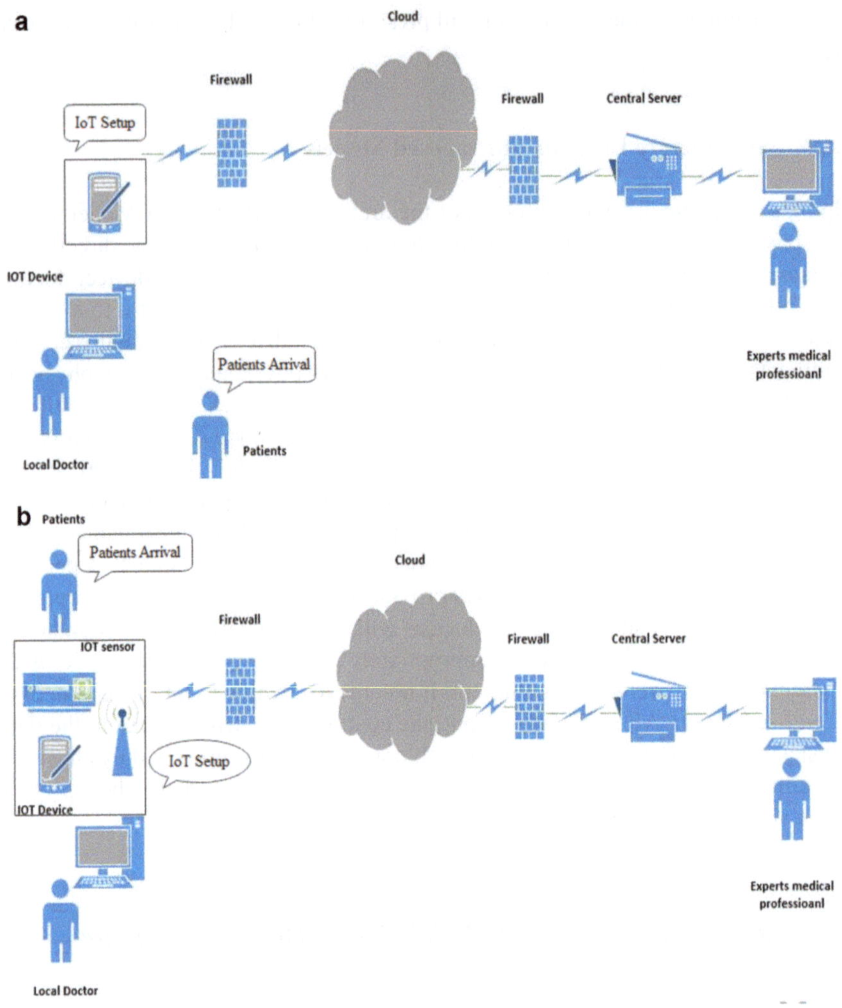


**Fig. 4** Architecture diagrams for proposed models [*Source* Author's Work]

circumstance arises. The third key adjustment to the models was the addition of an element for remote treatment, which was ensured through professional video conferencing. Following the first round, each expert received a questionnaire. The round two results of the Delphi study were stable, with each item scoring more than 3.7, except for one. The mean of experts rating for 'covers all the major aspects of remote treatments' and 'gives both options for remote treatment whether there is a queue or not' was 4.3, and for 'critical patients handling' was 3.3. Therefore, it is optional to convey a third round. Based on the results from this Delphi study, a single final model has been developed to integrate the two proposed models, as demonstrated in Fig. 5.

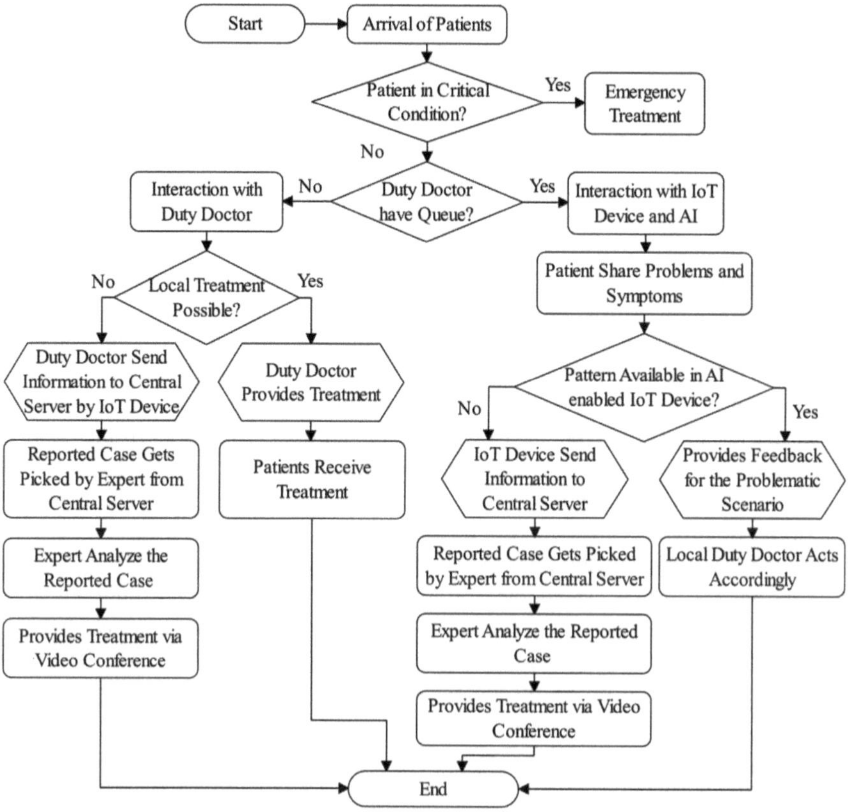


**Fig. 5** Final model to integrate IoT in the rural healthcare sector of Bangladesh (*Source* Author's Work)

## 3.8 Discussions

This research has developed and suggested a workable and acceptable method for enabling the integration of IoT in Bangladesh's rural healthcare system. Based on the experts' input, four primary obstacles to IoT integration in Bangladesh's rural healthcare system were found in the first phase. The opinions of the experts informed each of the four challenges. While there were a few other issues that were discussed in the in-person interviews, four primary issues emerged based on the majority of the comments. The resources needed to run IoT devices and associated technology are referred to as IoT resources. IoT devices have many features and options, and they are quite complex and important to manage. For resources to function end-to-end from the sensor to the cloud, they must be trained. IoT devices are related to the cloud and sensors. Data that occurs in the cloud is gathered and examined by sensors. The existence of professional associations necessitates that health organizations have doctors and IT/technical personnel with the necessary training to handle IoT-related devices and capabilities, which is currently lacking. Physicians must also be appropriately linked to or involved in the new technological uses. To have this environment, a proper operational environment needs to be created where IT and health professionals can work together, which is quite challenging in Bangladesh due to a lack of proper planning and policy-making, affecting such integration. Policy-makers should be able to make a strategic plan occasionally to facilitate such integration; otherwise, the advantages of such new technologies cannot be adequately obtained. The AHP analysis addresses two main challenges: "lack of adequate IoT resources" and "lack of proper professional associations" based on the priority weight. Strategic planning has gotten the least priority among the four major constraints.

There are six possible solutions obtained from this study to overcome the challenges. The two primary potential solutions identified by the AHP analysis have been prioritized as the "requirement of effective associations and resource empowerment among the organizations (A)" and the "necessity of proper professional training about new technology and industry attachment with academic institutes (B)", which have a significant effect on eliminating the addressed challenges. Encouraging foreign and domestic investors by shrinking taxes and improving loan facilities, etc. (F) has gotten the least priority among all six solutions. IoT and IoT-enabled services are new to rural and even urban people of Bangladesh in terms of new technology adaptation. Therefore, the current resource needs will be met via professional training and industrial attachment with academic institutions. Furthermore, it will contribute to the development of future resources. The primary requirement for overcoming the paucity of resources is resource empowerment within the organizations. Some laboratory environments, such as IoT incubators, must also overcome the need for an operational environment. Public health organizations should take proactive actions to become familiar with and use the latest technological tools to meet the standards of the professional association. To enhance IoT in the rural healthcare sector, a forward-thinking ICT policy must be put into place with the aid of sound strategic planning. By reducing travel and saving time, IoT in telemedicine offers great financial benefits for both patients and enterprises, according to the SWOT analysis. This kind of technological adoption and use will enable the next generation to swiftly adjust to future benefits. The two primary technological advantages are the robust telecommunications network and the fact that the majority of Bangladeshis use the internet efficiently these days. Finally, two models are built concerning ENT, a significant medical sector department and it has been shown that the help of IoT devices is creating an impact in remote healthcare organizations to improve patient care. Also, a final model is proposed based on the two constructed models to facilitate IoT in the rural healthcare sector of Bangladesh.

### *3.9 Recommendations*

Along with the major telecom and IT businesses, the government of Bangladesh has already taken the initiative to establish IT parks and IoT-focused IT incubators. There will be several chances to enhance healthcare advancements and the IoT under this umbrella that can benefit one another. All that is required is cooperation between the health sectors and these organizations. Some health sectors are already participating, nevertheless, necessitating more efforts by important role players and leaders in the public and private sectors. Even if there are prospective native investors, there are chances to draw in foreign investors to the IoT due to regional advancements in the healthcare industry. The following actions can be performed in light of Bangladesh's possible development:

- Reduce the complexity of government policymaking.
- Provide everyone with an equal chance to access high-quality healthcare facilities.
- Organize local fairs and exhibits to highlight these new developments in technology.

Additionally, to develop a productive workforce, the health technology-based education system and workplace must be improved. Bangladesh has a sizable number of international medical specialists working there. The primary cause of this is the lack of qualified domestic medical personnel. The government ought to take the lead in modernizing Bangladesh's educational requirements, industrial ties, and workplace conditions to foster exceptional professionalism. There is a need for more research in other areas, such as smart cities. IoT in healthcare needs more effective channels for data transfer and communication, where telecom companies might be helpful. Healthcare equipment's remote sensor data and patient data need to be analyzed at the other end, where IT companies are directly involved in data analysis and cloud computing. The security of patient data that is transferred and shared via Internet channels might be further investigated. Global data security and integrity are now more important than ever. As well, research can be done to create the algorithms or technologies needed to expand the use of IoT in Bangladesh's rural healthcare system.

## 4 Conclusions

Finding a workable and suitable strategy to support IoT integration in Bangladesh's rural healthcare system is the aim of this study. The entire study project is broken down into four phases to accomplish this purpose. Using AHP and SWOT technology evaluation tools, the necessary analysis was carried out based on the interviewees' input. The results of the first phase investigation address two primary challenges "lack of adequate IoT resources" and "lack of proper professional associations". To overcome these obstacles, two significant potential solutions are discussed as "requirement of effective associations and resource empowerment among the organizations" and the "necessity of proper professional training about new technology and industry attachment with academic institutes". These potential solutions, along with the findings of the SWOT analysis, show that the rural healthcare sector is a good fit for telemedicine integration. Telemedicine is seen as a dependable industry with a beneficial influence and numerous prospects in Bangladesh, given the state of the country's healthcare system. In light of the telemedicine industry, two models have been constructed to improve Bangladesh's rural healthcare system. Finally, a model is constructed by combining these two proposed models with the help of the Delphi technique. It is feasible to lower the patient death rate in rural and remote parts of Bangladesh by using this model in the healthcare system. In addition, persons who live in remote places can receive treatment from expert medical professionals within the shortest possible time. People will no longer have to go from door to door in search of better care. Every sector is greatly impacted by IoT, including healthcare, and it is becoming more and more essential to meet the demands of the younger generation and the latest technology advancements. This technological growth and its rapid adaptation are the main concerns of the industrialized nations. The regulatory commission of Bangladesh has already taken some action and approved specific IoT use cases that are already evident in the industrial and telecommunications sectors.

**Acknowledgements** Institute of Appropriate Technology, Bangladesh University of Engineering and Technology is acknowledged for providing the research facility.

**Author Contributions** All authors contributed equally to preparing this manuscript.

**Funding** This study did not receive financial support or funding.

**Data Availability Statement** The data supporting the study's findings will be made available to the manuscript's corresponding author upon request.

**Competing Interests** No conflicting interests are disclosed by the authors.

## References

1. Al-Zaman, Md.S.: Healthcare Crisis in Bangladesh during the COVID-19 Pandemic. Am. J. Trop. Med. Hyg. **103**(4), 1357–1359 (2020). https://doi.org/10.4269/ajtmh.20-0826
2. Rahman, A., Abdulla, F., Karimuzzaman, M., Hossain, M.M.: Burden of COVID-19 on health and wellbeing, education, and economy of Bangladesh. Clin. Case Rep. **10**(11), e6639 (2022). https://doi.org/10.1002/ccr3.6639
3. Islam, A., Biswas, T.: Health system in bangladesh: challenges and opportunities. Am. J. Health Res., **2**(6), 366. (2014). https://doi.org/10.11648/j.ajhr.20140206.18
4. Joarder, T., Rawal, L.B., Ahmed, S.M., Uddin, A., Evans, T.G.: Retaining doctors in rural bangladesh: a policy analysis. Int. J. Health Policy Manag., **7**(9), 847–858. (2018). https://doi.org/10.15171/ijhpm.2018.37
5. Hasan, M.: Quality of primary health care services in rural bangladesh: Patients' perspectives. OIDA Int. J. Sustain. Dev. **3**(8), 69–78 (2012)
6. Khandakar, M.S.A.: Rural health care system and patients' satisfaction towards medical care in Bangladesh: An empirical Study. J. Bus. Stud. **35**(2), 83–102 (2014)
7. Hossain, Md.I., Islam, Md.R., Saleheen, A.A.S., Rahman, A., Zinia, F.A., Urmy, U.A.: Determining the risk factors of under-five morbidity in Bangladesh: A Bayesian logistic regression approach. Discov. Soc. Sci. Health **3**(1), 21 (2023). https://doi.org/10.1007/s44155-023-00052-2
8. Qazi, S., Aboubaker, S., MacLean, R., Fontaine, O., Mantel, C., Goodman, T., Young, M., Henderson, P., Cherian, T.: Ending preventable child deaths from pneumonia and diarrhoea by 2025. Development of the integrated Global Action Plan for the Prevention and Control of Pneumonia and Diarrhoea. Arch. Dis. Child., **100**(Suppl 1), S23–S28 (2015).
9. Hossain, A.T., Siddique, A.B., Jabeen, S., Khan, S., Haider, M.M., Ameen, S., Tahsina, T., Chakraborty, N., Nahar, Q., Jamil, K., El Arifeen, S., Rahman, A.E.: Maternal mortality in Bangladesh: Who, when, why, and where? A national survey-based analysis. J. Glob. Health **13**, 07002 (2023). https://doi.org/10.7189/jogh.13.07002
10. World Health Organization.: Success factors for women's and children's health: Bangladesh. (2015). https://apps.who.int/iris/bitstream/handle/10665/178623/9789241509091_eng.pdf
11. Mannan, M.A.: Access to public health facilities in Bangladesh: A study on facility utilisation and burden of treatment. Bangladesh Dev. Stud., 25–80 (2013)
12. Ministry of Health and Family Welfare, G. of the P. R. of B.: Bangladesh digital health strategy 2023—2027. (2023). https://dghs.portal.gov.bd/sites/default/files/files/dghs.portal.gov.bd/page/4124d18a_ab99_40e2_8fef_ff4052948739/2024-04-23-07-09-48541d4dd55108137e50961ebcba0477.pdf
13. Zahid, A., Yadav, U.N., Mistry, S.K.: Digital health services in Bangladesh—The need for a sustainable design framework. World Med. & Health Policy **15**(4), 682–695 (2023). https://doi.org/10.1002/wmh3.580
14. Altynpara, E.: The role of IoT in healthcare industry: Benefits and use cases. IoT Healthc. (2023). https://www.cleveroad.com/blog/iot-in-healthcare/
15. Javaid, M., Haleem, A., Singh, R.P., Rab, S., Ul, M.I., Raina, A.: Internet of Things in the global healthcare sector: Significance, applications, and barriers. Int. J. Intell. Netw. **3**, 165–175 (2022). https://doi.org/10.1016/j.ijin.2022.10.002
16. Pisuwala, U.: Internet of things in healthcare: applications, benefits, and challenges. healthcare, IoT. (2023). https://www.peerbits.com/blog/internet-of-things-healthcare-applications-benefits-and-challenges.html
17. Anwar, K., Fikry, M., Natania Ibrahim, S., Nurlita, F., Sasmita Sari, A., Marlinda, N.: Literature review: Assessing the success factors of Hospital Management Information System (HMIS) implementation using the HOT-FIT method in Indonesia. New Appl. Stud. Manag., Econ. & Account. **6**(2), 26–38 (2023)
18. Mahla, M., Talati, S., Gupta, A., Agarwal, R., Tripathi, S., Bhattacharya, S.: The acceptance level of Hospital Information Management System (HIMS) among the nursing officials working

in a teaching hospital. J. Educ. Health Promot. **10**(1), 452 (2021). https://doi.org/10.4103/jehp.jehp_1410_20
19. Wadhwa, S., Saxena, A., Wadhwa, B.: Hospital information management system: An evolutionary knowledge management perspective. Int. J. Electron. Healthc. **3**(2), 232 (2007). https://doi.org/10.1504/IJEH.2007.013103
20. Menachemi, N., Collum.: Benefits and drawbacks of electronic health record systems. Risk Manag. Healthc. Policy, **47** (2011). https://doi.org/10.2147/RMHP.S12985
21. Imteaj, A., Hossain, M.K.: A smartphone based application to improve the health care system of Bangladesh. 2016 International Conference on Medical Engineering, Health Informatics and Technology (MediTec), 1–6. (2016). https://doi.org/10.1109/MEDITEC.2016.7835358
22. Ventola, C.L.: Mobile devices and apps for health care professionals: Uses and benefits. P & T: Peer-Rev. J. Formul.Y Manag. **39**(5), 356–364 (2014)
23. Jin, X., Yuan, Z., Zhou, Z.: Understanding the Antecedents and Effects of mHealth App Use in Pandemics: A Sequential Mixed-Method Investigation. Int. J. Environ. Res. Public Health **20**(1), 834 (2023). https://doi.org/10.3390/ijerph20010834
24. Sun, L., Buijsen, M.: Mobile health in China: Does it meet availability, accessibility, acceptability and quality standards? Health Policy Technol. **11**(3), 100660 (2022). https://doi.org/10.1016/j.hlpt.2022.100660
25. Badugu, S., Srikanth, K., Inampudi, L.N.: IoT for Healthcare. Int. J. Sci. Res. (IJSR) **5**, 2319–7064 (2016)
26. Kulkarni, A., Sathe, S.: Healthcare applications of the Internet of Things: A Review. Int. J. Comput. Sci. Inf. Technol. **5**(5), 6229–6232 (2014)
27. GSM Association. Understanding the Internet of Things (IoT) (pp. 1–13). (2014). https://www.gsma.com/solutions-and-impact/technologies/internet-of-things/wp-content/uploads/2014/08/cl_iot_wp_07_14.pdf
28. Gubbi, J., Buyya, R., Marusic, S., Palaniswami, M.: Internet of Things (IoT): A vision, architectural elements, and future directions. Futur. Gener. Comput. Syst. **29**(7), 1645–1660 (2013). https://doi.org/10.1016/j.future.2013.01.010
29. Gope, P., Hwang, T.: BSN-Care: a secure IoT-Based modern healthcare system using body sensor network. IEEE Sens. J. **16**(5), 1368–1376 (2016). https://doi.org/10.1109/JSEN.2015.2502401
30. Fernandez, F., Pallis, G.C.: Opportunities and challenges of the Internet of Things for healthcare: Systems engineering perspective. 2014 4th International Conference on Wireless Mobile Communication and Healthcare-Transforming Healthcare through Innovations in Mobile and Wireless Technologies (MOBIHEALTH), 263–266. (2014). https://doi.org/10.1109/MOBIHEALTH.2014.7015961
31. Catarinucci, L., De Donno, D., Mainetti, L., Palano, L., Patrono, L., Stefanizzi, M.L., Tarricone, L.: An IoT-Aware architecture for smart healthcare systems. IEEE Internet Things J. **2**(6), 515–526 (2015). https://doi.org/10.1109/JIOT.2015.2417684
32. Chiuchisan, I., Costin, H.-N., Geman, O.: Adopting the Internet of Things technologies in health care systems. International Conference and Exposition on Electrical and Power Engineering (EPE) **2014**, 532–535 (2014). https://doi.org/10.1109/ICEPE.2014.6969965
33. Eraslan, İ, Sarıoğlan, B.T., Ünsal, S., Erdoğan, G.: Diagnostic utility of perfusion index in identifying radial artery embolism during hyperthermic intraperitoneal chemotherapy procedure: A case report. J. Clin. Monit. Comput. **38**(2), 553–555 (2024). https://doi.org/10.1007/s10877-023-01102-5
34. Xie, C., Yang, P., Yang, Y.: Open knowledge accessing method in IoT-Based hospital information system for medical record enrichment. IEEE Access **6**, 15202–15211 (2018). https://doi.org/10.1109/ACCESS.2018.2810837
35. Chen, Y.-K.: Challenges and opportunities of internet of things. 17th Asia and South Pacific Design Automation Conference, 383–388. (2012). https://doi.org/10.1109/ASPDAC.2012.6164978
36. Hossain, K., Rahman, A.: Challenges before IoT in Bangladesh. The Financial Express (FE). 18 June 2024. (2019). https://thefinancialexpress.com.bd/views/columns/challenges-before-iot-in-bangladesh-1532880691

37. Ferdous, A.M.: Scopes and challenges of implementing Telemedicine in a developing country like Bangladesh [Master's Thesis, UiT Norges arktiske universitet]. (2017). https://munin.uit.no/handle/10037/11275
38. Saaty, T.L.: The analytic hierarchy process (AHP). J. Oper. Res. Soc. **41**(11), 1073–1076 (1980)
39. Saaty, T.L.: Decision making for leaders: The analytic hierarchy process for decisions in a complex world. RWS publications (2001)
40. Thokala, P., Duenas, A.: Multiple criteria decision analysis for health technology assessment. Value Health **15**(8), 1172–1181 (2012). https://doi.org/10.1016/j.jval.2012.06.015
41. Nejad, M.C., Mansour, S., Karamipour, A.: An AHP-based multi-criteria model for assessment of the social sustainability of technology management process: A case study in banking industry. Technol. Soc. **65**, 101602 (2021). https://doi.org/10.1016/j.techsoc.2021.101602
42. Vaidya, O.S., Kumar, S.: Analytic hierarchy process: An overview of applications. Eur. J. Oper. Res. **169**(1), 1–29 (2006). https://doi.org/10.1016/j.ejor.2004.04.028
43. Ayub, A., Razzaq, A., Aslam, M.S., Iftekhar, H.: A conceptual framework on evaluating SWOT analysis as the mediator in strategic marketing planning through marketing intelligence. Eur. J. Bus. Soc. Sci. **2**(1), 91–98 (2013)
44. Ostrom, L.T.: Impact of societal norms on safety, health, and the environment: case studies in society and safety culture. John Wiley & Sons (2022)
45. Giménez-Medina, M., Enríquez, J.G., Olivero, M.A., Domínguez-Mayo, F.J.: The innovation challenge in Spain: A Delphi study. Expert Syst. Appl. **230**, 120611 (2023). https://doi.org/10.1016/j.eswa.2023.120611
46. Kezar, A., Maxey, D.: The Delphi technique: An untapped approach of participatory research. Int. J. Soc. Res. Methodol. **19**(2), 143–160 (2016). https://doi.org/10.1080/13645579.2014.936737
47. Hirschhorn, F.: Reflections on the application of the Delphi method: Lessons from a case in public transport research. Int. J. Soc. Res. Methodol. **22**(3), 309–322 (2019). https://doi.org/10.1080/13645579.2018.1543841
48. Turoff, M.: An alternative approach to cross impact analysis. Technol. Forecast. Soc. Chang. **3**, 309–339 (1971). https://doi.org/10.1016/S0040-1625(71)80021-5
49. Turoff, M.: Delphi conferencing: Computer-based conferencing with anonymity. Technol. Forecast. Soc. Chang. **3**, 159–204 (1971). https://doi.org/10.1016/S0040-1625(71)80012-4
50. Nordin, N., Md Deros, B.M.D., Abdul Wahab, D., Ab. Rahman, M.N.: Validation of lean manufacturing implementation framework using delphi technique. Jurnal Teknologi, **59**(2). (2012). https://doi.org/10.11113/jt.v59.1596
51. Skulmoski, G.J., Hartman, F.T., Krahn, J.: The Delphi method for graduate research. J. Inf. Technol. Educ.: Res. **6**(1), 1–21 (2007)
52. Lokot, M.: Whose voices? whose knowledge? a feminist analysis of the value of key informant interviews. Int J Qual Methods **20**, 1–8 (2021). https://doi.org/10.1177/1609406920948775
53. UCLA Center for Health Policy Research.: Section 4 key informant interviews, health DATA program–data, advocacy and technical assistance. (2023). https://healthpolicy.ucla.edu/sites/default/files/2023-08/tw_cba23.pdf
54. Rahman, Q.M., Sikder, Md.T., Talha, Md.T.U.S., Banik, R., Pranta, M.U.R.: Perception regarding health and barriers to seeking healthcare services among rural rickshaw pullers in Bangladesh: A qualitative exploration. Heliyon **8**(10), e11152 (2022). https://doi.org/10.1016/j.heliyon.2022.e11152
55. Uddin, M.N., Hamiduzzaman, M.: Problems and prospects of decentralized rural health care system in Bangladesh. P & T: Peer-Rev. J. Formul.Y Manag. **2**(1), 47–75 (2009)

# From X to Z: Engaging Different Generations of Customers in Refurbishing Business Models in Electronics

**Omar Mohammed, Fatima Khitous, and Pauliina Ulkuniemi**

**Abstract** The electronic waste (E-waste) crisis is escalating, with an annual global production exceeding 50 million tons; transitioning to a circular economy (CE) model offers a potential solution, emphasizing resource efficiency through recycling, reuse, repair, refurbishing, and remanufacturing. However, a critical knowledge gap exists regarding customer engagement in circular business models, especially those underpinned by refurbishment. This study addresses part of the gap by exploring how companies could engage customers from different generations (Z, Y, and X) in their circular business models by understanding the differences in the customers' perceived value. Employing qualitative methods, including 52 in-depth semi-structured interviews, the findings show that Gen Z sees refurbished tech as an affordable option for diverse needs requiring a more cognitive engagement strategy. Gen Y opts for transparency and clear communication, allowing the company to engage them through an affective engagement strategy. Being more financially stable, Gen X values device longevity and after-sales service, which relies more on a behavioral engagement strategy.

**Keywords** Refurbishing business models · Refurbished electronics · Electronics waste · Circular economy · Customer engagement · Across-generation engagement

---

O. Mohammed (✉) · F. Khitous · P. Ulkuniemi
Oulu Business School, University of Oulu, Oulu, Finland
e-mail: omar.mohammed@oulu.fi

F. Khitous
e-mail: Fatima.Khitous@oulu.fi

P. Ulkuniemi
e-mail: pauliina.ulkuniemi@oulu.fi



N. D. K. Nguyen et al. (eds.), *Digital Transformation and Sustainability: Methods and Applications*, Information Systems Engineering and Management 47,
https://doi.org/10.1007/978-3-031-91278-8_10

## 1 Introduction

Approximately 50 million tons of electronic waste (E-waste) is generated globally annually [3, 27]. Our daily reliance on various electronic devices, with shorter product lifespans due to rapid technological advancements, contributes to this problem [31]. E-waste contains valuable materials and hazardous substances [12], creating two significant challenges: resource scarcity in the electronics industry [14] and environmental and health risks associated with unauthorized recycling [59]. As raw material sources deplete and demand for new products grows, recycling and prolonging the lifespan of the items becomes increasingly vital [47]. A circular economy (CE) reduces resource consumption, waste, and environmental impact [29]. In contrast to the linear economy, the circular economy promotes closing, slowing, and narrowing loops through circular business models (CBMs) [6]. CBMs enable using resources for as long as possible through recycling, reuse, repair, refurbishing, remanufacturing, and recycling [5, 13, 34]. The goal is to create a regenerative system that maximizes the value and utility of materials and products throughout their lifecycle. Instead of focusing on the "end-of-life" phase, refurbishing electronics business models have been introduced as a slowing-loop solution to aid in the transition towards a circular economy. Hence, how customers perceive refurbished electronics plays a pivotal role in shaping the demand for these products within the circular economy transition [28]. Refurbishment business models appear when used products or components are repaired, updated cosmetically, and restored to like-new quality, providing value to end users and B2B customers [13, 34, 54].

Existing research emphasizes understanding customer engagement in the circular economy [22]. Nevertheless, customer engagement within the circular business model must be addressed better [19]. Customer engagement, focused on building trust and enduring relationships through superior value delivery, is challenging due to customer disinterest and limited awareness of circular products, hindering the shift to a circular economy [29]. Although the conceptualization of customer engagement is somewhat debated [25], the stream of research focuses on defining it as customer-company interaction [8], which appears in three dimensions: cognitive (thoughts and understanding), affective (emotional connections), and behavioral (customer actions) [24]. In designing customer engagement strategies, understanding the customer's perceived value is critical [23]. Customer engagement is a vital strategy for enhancing company performance, including sales, competitiveness, and profitability [7]. For refurbishing business models, Wang et al. [16] suggest that companies prioritize improving perceived value to enhance customer engagement. Perceived value, as defined by [58], represents customers' assessments of a product or service's effectiveness based on perceived benefits and acquisition costs.

However, not all customers have that same perceived value. Individuals from the same era, with shared circumstances and experiences, influence each other [4]. Additionally, each generation possesses distinct characteristics, lifestyles, and values [56], which can impact the perceived value of refurbished electronics. Therefore, understanding these variations is crucial for companies' strategies to engage diverse customer groups in efficiently refurbishing business models. Hence, a cross-generational study enables a better understanding of customer engagement and value perception toward refurbishing business models, especially since the literature lacks insights into perceived value variations among different generations in the refurbished electronics context. Therefore, this study examines the customer-perceived value of different generations concerning refurbishing business models. It aims to answer the following research question: "*How can companies engage different generations of customers in their refurbishing business model by understanding differences in customer perceived value?*".

Empirically, this study focuses on exploring generational differences in Finland, a country known for its sustainability and circularity practices (Ministry of Economic Affairs and Employment Ministry of the Environment, 2021). In terms of the generations, the study focuses on generations Z, Y, and X. The study employs a qualitative approach, and data is composed of 52 in-depth semi-structured interviews of customers representing the different generations of customers. Theoretically, the study employs literature on customer-perceived value, customer engagement, and refurbishment CBMs to examine the value perceptions of the different generations and to form customer engagement strategies for companies with refurbishing business models. The study contributes to general theoretical knowledge of circular and refurbishing business models, particularly. It also provides a managerial contribution to companies developing effective marketing strategies to promote refurbished electronics consumption and remain competitive.

The paper is structured so that the following section provides the theoretical framework based on existing knowledge on the role of customers in refurbishing business models, characteristics of different generations, customer engagement, and customer perceived value. After that, the empirical research methods are presented. This is followed by findings from the empirical analysis presenting the identified differences in customer perceived value and forming the engagement strategies based on these. The last section presents the conclusions.

## 2 Literature Review

### *2.1 Role of Customers in Refurbishing Business Model*

Companies must adopt circular business models (CBMs) to meet the growing demand for sustainable products [18, 26]. Circular business model strategies allow the extended product lifespan through design, repair, refurbishing, and recycling to create value while aligning with stakeholders and actors for environmental, social, and economic benefits [41], thereby reducing the environmental damage while creating economic value. Among CBMs, there is a refurbishing business model, which appears when companies extend the life of used products or components that are repaired, updated cosmetically, and restored to like-new quality, providing value to end users and B2B customers. However, the success of a refurbishing business model hinges on engaging customers effectively [22]. Strikingly, prior research has not adequately addressed the role of customers in this pivotal shift toward a refurbishing business model [19, 29]. Customer preferences surrounding refurbished electronics exhibit significant diversity, with some studies, such as [9], emphasizing quality and reliability, while others, like [27], highlight price and extended warranties. Lieder et al. [32] also contend that customers increasingly favor access over ownership.

Interestingly, Wang et al. [54] suggest that the environmental impact may not be the primary incentive if customers lack trust in refurbished electronics. In this context, Kerber et al. [28] emphasize the pivotal role of customers' value perception in driving the demand for refurbishing CBMs. Furthermore, Bocken et al. [6] argue that circular business models must not only offer value to customers but also demonstrate a positive environmental impact. In light of these considerations, it becomes evident that refurbishing electronics business models plays an indispensable role in advancing the circular economy. However, they encounter significant challenges stemming from customer perceived value [36]. Hence, a deep understanding of the customers' differences in the perceived value must be explored.

## 2.2 A Cross-Generational Perspective on Customer Perceived Value

Generations, representing individuals born during specific eras, exhibit shared characteristics shaped by their times [42]. These shared traits, influenced by significant historical events, contribute to developing distinct behaviors, attitudes, and values that impact purchasing decisions. Generation categories, as defined by Berkup [4], include Generation Z (1996–2010), Generation Y (1980–1995), and Generation X (1965–1979). Understanding these generational distinctions is crucial for tailoring marketing strategies and facilitating the transition to a circular economy [56].

Each generation requires unique approaches to accelerate the shift toward sustainable consumption practices. Research highlights significant generational differences in buying behavior, emphasizing the importance of understanding how various cohorts perceive and prioritize value [56]. For example, Rahulan et al. [42] identified differences in purchasing compression sportswear, while Djafarova and Bowes [16] explored how Gen Y and Gen Z perceive value in shaping purchasing decisions. This understanding is vital for adapting circular business models to meet generational needs, particularly regarding preferences for refurbished electronics.

Historical and cultural contexts deeply influence generational distinctions [30]. Generation Z, born during the internet boom, values individual expression, environmental responsibility, and rapid access to information [38]. Millennials, transitioning from offline to online, prioritize experiences, purpose, and technology for communication. Events such as the 2008 financial crisis have shaped their financial habits and career choices [40]. Generation X emphasizes work-life balance, self-sufficiency, and hands-on learning. Growing up during the rise of personal computing instilled skepticism and an entrepreneurial spirit in this cohort [4].

Perceived value is pivotal in driving engagement with circular business models, particularly for refurbished electronics. Defined as a customer's comprehensive assessment of a product's effectiveness based on perceived benefits and acquisition costs [58], perceived value significantly influences purchase and repurchase intentions [37]. High perceived value fosters satisfaction, loyalty, and willingness to pay a premium, while low perceived benefits result in dissatisfaction, reduced repurchase intentions, and negative word-of-mouth [55].

The 'consumption value' theory [33, 45] offers a multidimensional perspective on perceived value, encompassing four key dimensions: monetary, functional, social, and environmental. Monetary value reflects customers' willingness to pay, functional value pertains to a product's features meeting customer needs, and social value considers its impact on customers' social lives or perceived status. Environmental value, as explored in the environmental value attitude system model [15], links individual values and attitudes to environmentally conscious behavior [45]. For instance [13, 44] emphasize the role of environmental and green value perceptions in shaping purchase intentions, highlighting concepts like ecological integrity, stewardship, resource preservation, and concern for future generations.

Combining insights into generational characteristics and perceived value dimensions provides a foundation for developing tailored engagement strategies. By aligning these strategies with generational preferences and value perceptions, companies can effectively engage diverse customer groups, fostering long-term loyalty and accelerating the transition to a circular economy.

## 2.3 Customer Engagement

Even though customer engagement has been discussed prominently in academic literature, there still needs to be more debate on conceptualizing it [25]. In line with Bruce and Shelly, 2010, this paper defines customer engagement as an involvement or interaction between the customer and the company. There are three dimensions of customer engagement: cognitive engagement, which pertains to how a customer thinks about an organization or their level of understanding and absorption of corporate communication, affective engagement, which refers to the emotional connection or attachment that customers have towards a company; and behavioral engagement, which refers to the actions of the customers with the company [24]. In interactive and ever-evolving business landscapes, customer engagement is crucial for improved company performance, encompassing increased sales, a more decisive competitive edge, and greater profitability. Brodie et al. [7] proposed that customer engagement is crucial in marketing activities, especially viral marketing. Additionally, companies can strategically improve value co-creation by strengthening customer engagement. Hence, the company must enhance customer engagement in its refurbishing business model.

### *2.4 Theoretical Framework*

This study explores how companies could engage customers from different generations (Z, Y, and X) in their circular business models by understanding the differences in the customers' perceived value. Analyzing monetary, functional, environmental, and social values will guide managers and researchers in tailoring engagement strategies for efficient customer involvement. Considering cognitive, affective, and behavioral dimensions in engagement strategies is crucial, but each generation may emphasize one dimension more than the others, requiring targeted focus from companies for effective engagement. This understanding accelerates the shift toward a circular economy, aiding in addressing the E-waste dilemma (Fig. 1).

## 3 Research Methodology

This research adopts a qualitative research design [57], aiming to comprehensively explore the perceived value of refurbished electronics among customers of the diverse generational cohorts of X, Y, and Z. The qualitative approach was chosen to examine the participants' views, beliefs, and lived experiences [46]. The data were collected in Finland and selected because of Finland's higher sustainability approach and the circular economy program that has to be achieved by 2030, according to the Ministry of Economic Affairs and Employment Ministry of the Environment (2021). Hence, the context is well-sustainable and fertilized to be investigated and understood. There are second-hand stores everywhere, which makes it a bit of the norm in Finland to buy from second-hand shops or form a refurbishing company, unlike elsewhere [21]. Hence, this context allows for an in-depth understanding of the studied phenomenon by shedding light on customers' perceived value towards refurbished electronic devices, as articulated in their own words.

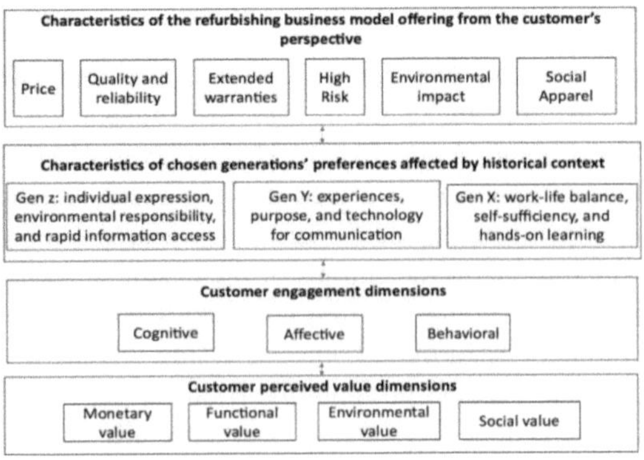


**Fig. 1** Customer perceived value and customer engagement dimensions in the refurbishing business models (*Source* Author's work)

Data was collected through semi-structured interviews, a well-suited method to facilitate an open-ended exploration of participants' diverse experiences and multi-faceted perceptions of refurbishing business models [35]. Interviews provide rich and detailed information, capturing the essence of the customer's perceived value of refurbished electronics. To ensure the effectiveness of the interview process, the study began with a pilot phase consisting of two interviews [10]. These pilot interviews allowed for refining the interview protocol, enhancing the study's overall rigor. Various strategies were employed to recruit participants to ensure a balanced representation of generational groups: informative flyers at the university campus and the snowballing technique [2]. Financial incentives in the form of prizes were introduced to attract participants further. Participants were selected based on their experience buying refurbished electronics for themselves or others. This purposive sampling allows the gathering of rich data from experienced customers and users [49]. The final sample comprised 52 interviewed individuals residing in Finland, whose ages ranged from 19 to 62 years old.

Thirty-six interviews were conducted face-to-face, while 16 were conducted online using Microsoft Teams. The duration of the interviews varied from 28 to 90 min, with an average length of 45 min. This allowed participants to share their experiences and viewpoints in-depth and fostered a holistic understanding of their perceived values regarding refurbishing business models. In total, 36 interviews were collected, totaling approximately 43 h of audio recordings.

The interviews were transcribed verbatim to analyze data [57]. The data underwent a systematic thematic analysis of our large data set [11], facilitated by Nvivo software. We started by preliminarily reading the interviews separately for each generation. Then, we categorized the data under four theoretical categories: Monetary, Functional, Environmental, and Social. After that, we compared all different generations to understand their differences regarding the perceived value and the customer engagement strategies of the refurbished electronics and refurbishing business models (Table 1).

## Findings

The following presents the findings of the empirical analysis, showing how the monetary, functional, environmental, and social perceived value differs between generations Z, Y, and X. It also shows the most appropriate engagement strategies that the refurbishing business model needs to focus on for each generation.

**Table 1** Presents the informants' details and their coding numbers

| Participants | GenZ | GenY | GenX | Total |
|---|---|---|---|---|
| Female | 5 | 9 | 5 | 19 |
| Male | 10 | 12 | 11 | 33 |
| Total | **15** | **21** | **15** | **52** |
| Participants | 1–15 | 16–36 | 37–52 | – |

## 3.1 Monetary Value

- **Generation Z**: Monetary value remains a critical driver for Gen Z. They are highly price-sensitive and pragmatic, often comparing refurbished electronics with new ones to ensure they get the best deal. Gen Z often perceives the decision to purchase refurbished electronics as a reflection of their financial savvy, allowing them to access technology that might otherwise be out of reach.

  *"The price is everything. If it is not cheaper than a new one, why would I bother with refurbished?"(8)*

  *"I am a student, so saving money matters. Refurbished gives me a good deal without losing out on quality." (12)*

- **Generation Y**: For Millennials, monetary value includes not just price but also the broader experience, such as after-sales support and the reputation of the refurbishing company. Gen Y appreciates refurbished electronics that balance cost savings and premium experiences, including access to high-end brands at reduced prices.

  *"I don't mind paying extra if the company provides a good warranty. It shows they stand by their products."(19)*

  *"I bought refurbished because it allowed me to afford a premium brand. That made me feel good about my decision." (25)*

- **Generation X**: Gen X values the practicality of refurbished electronics, particularly for their children's use. They consider the overall cost–benefit ratio, factoring in longevity and potential repair costs.

  *"Refurbished makes sense for my kids. They break things often, so there's no point in spending too much." (46)*

  *"If the refurbished device lasts long enough, it's worth it. Otherwise, I'd rather spend more on a new one." (41)*

## 3.2 Functional Value

- **Generation Z**: Gen Z prioritizes the functionality of refurbished products over appearance. They accept minor cosmetic flaws if the product meets their performance expectations. They also appreciate the opportunity to access advanced features and specifications at a lower cost through refurbished options.

  *"I don't care about scratches or dents. As long as it works well, I'm happy."* (14)

  *"Refurbished is just as good as new if it's tested properly. I trust brands that show me how they have fixed things."* (9)

- **Generation Y**: Millennials are more critical of functional value, requiring performance and aesthetics. They expect detailed technical reports and precise documentation of the refurbishing process. Aesthetic flaws, even minor ones, can deter this group from purchasing, as they often equate outward appearance with quality.

*"I need to know what's been replaced or repaired before I buy. Transparency builds trust." (20)*

*"If it looks and works like new, I'll consider refurbished. Otherwise, I'll pass." (18)*

- **Generation X**: For Gen X, functionality and reliability are non-negotiable. They expect refurbished products to perform at the same level as new ones and place significant trust in after-sales support.

*"Functionality is key. I want to feel like I'm buying a product that will last."* (42)

*"Customer service matters to me. If I have issues, I need to know the company will resolve them quickly."* (47)

### *3.3 Environmental Value*

- **Generation Z**: Gen Z is highly environmentally conscious and views refurbished electronics as a way to align their consumption habits with their sustainability goals. They often seek out companies that provide transparent information about the environmental benefits of refurbished products.

*"Every time I buy something refurbished, I feel like I'm doing my part to reduce waste."* (3)

*"It's not just about saving money. It's about saving the planet too."* (6)

- **Generation Y**: Millennials recognize the environmental benefits of refurbished electronics but often prioritize convenience and quality over sustainability. They are drawn to brands that integrate sustainability into their core values without compromising product performance.

*"Sustainability is important, but I still want a product that meets my needs. It's about finding the right balance."* (24)

*"If a refurbished device is sustainable and high-quality, that's a win-win for me."* (23)

- **Generation X**: Environmental considerations are not a primary motivator for Gen X but are becoming more relevant, often due to their children's influence.

*"I never thought about e-waste until my kids started talking about it. Now I try to make better choices." (49)*

*"I don't focus on sustainability when buying electronics, but it's a nice bonus if it's there." (45)*

### 3.4 Social Value

- **Generation Z**: Social value is highly significant for Gen Z. They associate refurbished electronics with ethical consumption and enjoy sharing their experiences with peers to promote these values. Gen Z actively advocates for refurbished electronics to challenge planned obsolescence and support circular economy principles.

  *"I convinced my friends to buy refurbished because it's cheaper and better for the environment." (5)*

  *"Knowing I'm supporting ethical practices makes me feel good about my purchase." (2)*

- ***Generation Y***: Millennials value peer opinions and social validation when purchasing refurbished electronics. They are more likely to buy a product if it aligns with current trends or enhances their social status.

  *"I checked with my friends before buying refurbished, and they all recommended it. That made my decision easier." (26)*

  *"It's great when I can afford something that looks good and performs well without breaking the bank." (22)*

- ***Generation X***: Social value plays a minor role for Gen X, as they prioritize functionality and reliability. However, their children often influence their attitudes toward their purchases' social and environmental impacts.

  *"My kids care a lot about the environment, and that rubs off on me. I try to make choices they'd approve of." (51)*

  *"I wouldn't buy refurbished for the social impact, but it's a nice bonus if it's there." (48)*

## 4 Discussions

The findings of cross-generational customer engagement dimensions strategies are based on the customer's perceived value in refurbishing business models in the electronics industry (Table 2).

The findings indicate that a cognitive engagement strategy may help engage Gen Z customers by ensuring customer satisfaction with pricing and functionality. This extends [27] argument that price is the most crucial factor when buying refurbished electronics. Gen Z consumers appear receptive to reasonable pricing strategies aligned with their financial situation. Gen Z customers prioritize functionality over aesthetics and understand that refurbished products can offer good quality. Their familiarity with refurbished device quality makes them more likely to be engaged with the refurbishing business model. This aligns with what [52] said, that the perceived quality of the refurbished electronics depends on the knowledge and awareness of the customer about it.

**Table 2** A summary of the findings of cross-generational customer engagement dimensions strategies based on the customer perceived value in refurbishing business models in the electronics industry (*Source* Author's work)

| Generations/ perceived value | Gen Z | Gen Y | Gen X |
|---|---|---|---|
| Monetary value | See refurbished electronics as an affordable solution to access a wider variety of devices, given their economic constraints. So, pricing strategies are a vital key to consider here | Millennials see refurbished devices as fitting their budget and social standing. They value high-end brands at reduced prices, and the value-to-feature ratio guides their choices | Financially stable Gen X customers valued device longevity and convenience, gravitating towards new electronics for efficiency and reduced risk |
| Functional value | Despite aesthetic differences, it prioritizes functional equivalence in refurbished electronics, seeking devices with advanced specifications and features due to budget constraints | Millennials value refurbished electronics for their cost-effectiveness and functionality, relying on detailed technical reports for assurance and emphasizing aesthetics in their choices | The older generation prioritizes product quality over price, while Gen X perceives new electronics to have superior functionality compared to refurbished ones |
| Environmental value | Appreciates the environmental advantages of refurbished electronics, with growing environmental awareness driving a preference for such options | Gen Y values the eco-benefits of refurbished electronics but seeks corporate transparency, expects quality, and advocates for company-produced, durable, repairable devices | Gen X focuses on price, quality, and aftersales over e-waste concerns, but their environmental interest is rising, influenced in part by their children's education |

(continued)

**Table 2** (continued)

| Generations/ perceived value | Gen Z | Gen Y | Gen X |
|---|---|---|---|
| Social value | Advocates for refurbished electronics due to social concerns like child labor in mineral fields, leading them to favor brands with positive societal and environmental footprints. So, this generation is discussing these aspects and advocates for each other to buy refurbished electronics | Gen Y emphasizes collective action for global challenges and greatly values peer opinions, reflecting a collectivist approach to social values | Gen X, please do not consider the social status when buying refurbished electronics; the most important thing for them rather than the status is that the devices have to perform well |
| Customer engagement strategies for each generation | Strategic communication plan for Gen Z engagement:<br>– Consistent emphasis on product prices and their societal and environmental impact<br>– Content alignment with company vision and values<br>*Cognitive engagement* | Community development for Gen Y engagement:<br>– Establishment of communication groups catering to their collectivistic nature<br>– Implementation of loyalty and rewards programs<br>– Prioritization of transparent and honest communication<br>*Affective engagement* | Customer journey development for Gen X engagement:<br>– Personalize content and interactions to resonate with their preferences<br>– Implementation of a tailored after-sales service<br>– Emphasis on active listening to understand and address their needs effectively<br>*Behavioral engagement* |

Gen Z and Y customers prioritize environmental consciousness when purchasing refurbished electronics. Engaging them requires a cognitive and affective strategy emphasizing environmental impacts, aligning with studies by Wang et al. [54] and [44] note they are proactive in resource conservation. Ethical practices and transparency in the supply chain enhance engagement, particularly in addressing concerns about mineral sourcing and human rights abuses associated with electronics production. Sharing such information fosters cognitive and behavioral engagement.

Based on the analysis, Gen Y customers can be more effectively engaged through an *affective engagement*, valuing a solid emotional connection. The data shows that Gen Y customers are willing to be engaged when the companies act more transparently and honestly. The prior studies show that the higher the knowledge about the environmental impact and education that companies give to the customers, the more customers are willing to buy the refurbished electronics [52]. However, the results of the present study show that the customers also have high expectations for after-sales service and extended warranties, extending the monetary value beyond the product to the service they receive, which also requires some behavioral engagement. This finding aligns with [39], who confirm that after-sales service is a significant factor in the perceived quality of refurbished electronics. Based on our data, Gen Y customers' busy lifestyle demands convenience and effortless transactions when purchasing refurbished electronics. This aligns with [1, 55], who argued that customers expect the same quality from refurbished electronics as the new ones. Gen Y customers expect the same quality of the transaction as new products. Gen Y customers appear to thrive on clear communication from the company for functional value. Gen Y customers prioritize reliability and rely on peer opinions, emphasizing the need for positive word of mouth. Gurtner and Soyez [20] argued that Gen Y transferred their trust in green and ecological products through word of mouth between each other. Gen Y customers appear to place significant importance on social status and image, requiring a wide product variety to represent them among their social group. Thus, our study suggests that a higher level of affective engagement is necessary for social perceived value.

Our findings suggest that engaging Gen X customers requires *behavioral engagement,* as these customers appear to differ in their perception of monetary value. Customization is essential, as customers of this generation often buy refurbished electronics for their children or themselves. Companies must tailor strategies to convince them in both scenarios. Based on the data, Gen X customers buy refurbished electronics more frequently for their children due to cost considerations and aim to use devices as long as they last. Our data indicates that the functional value is critical for Gen X customers. This aligns with [9], according to which functionality is the most crucial factor for refurbished electronics. Gen X customers perceive refurbished electronics as having inferior quality, placing more importance on functionality than monetary value. Ensuring product quality, addressing concerns about functionality, and providing excellent customer service are vital for engaging the customers of this generation. Thus, extensive education from the company side is necessary to correct the functional perceived value for Gen X customers. Prior research also emphasizes that education and advertising are essential for remanufactured and refurbished electronics [1].

Additionally, companies should focus on communicating the environmental and social value to Gen X customers who may need higher awareness in these areas. Bocken et al. [6] argued that companies must ensure the delivery of environmental value to their customers. Our study found that Gen X customers often adopt environmental and social values from their children, influencing their parents' choices and lifestyles. Our study suggests that targeting Gen X customers requires high behavioral engagement. Understanding these generational differences is essential for customer satisfaction and engagement in the refurbishing business models (Fig. 2).

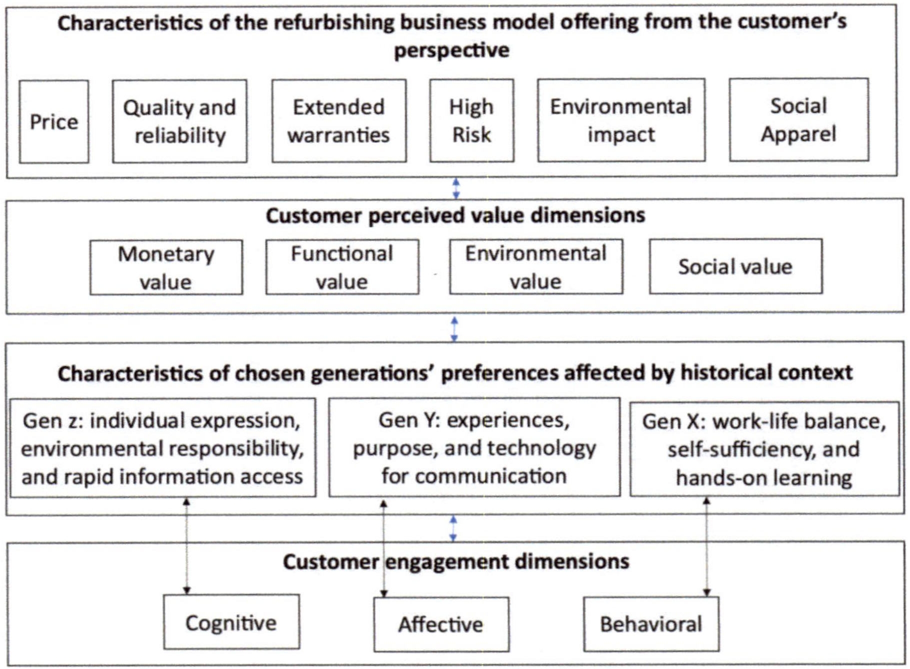


**Fig. 2** Customer engagement dimensions and strategies for different generations in refurbishing business models in the electronics sector (*Source* Author's work)

## 5 Conclusions

The study explores Gen Z, Y, and X's value perceptions regarding refurbished electronics, revealing differences in preferences. Gen Z prioritizes specifications; Gen Y seeks cost-effectiveness and transparency, and Gen X values product quality and longevity. Customized engagement strategies are recommended, with cognitive engagement for Gen Z, affective engagement for Gen Y, and behavioral engagement for Gen X.

The findings offer three main theoretical contributions. Firstly, the present study focuses on the vital actor in the circular economy transition, which is the customer. While few studies focus on the customer role in the circular economy, there still needs to be more knowledge regarding the customer role in the circular economy literature, particularly in the customer engagement literature [19]. In addition, prior studies focused more on customers' acceptance or switching behavior toward the circular economy [22]. Hence, our study indicates how crucial it is to understand the differences in the customers' perceived value in different generations. Indeed, there are previous studies, such as [17], which addressed customer engagement, however, despite its focus on the dynamic capabilities besides the customer engagement, it was more focused on the SMEs side rather than understanding the customer's differences per se, while the present study takes the customer perspective which added to our knowledge how refurbishing business models engage different generational customers by using different engagement strategies that fit better for each generation.

Secondly, by identifying the engagement strategies for different generations of customers, the present study strengthens the existing literature on marketing strategies of companies engaged in the circular economy, mainly refurbishing business models. Indeed, only some studies address marketing strategies for refurbishing business models. For example, Esmaeilian et al. (2021) provided different marketing strategies for refurbished products; however, the study focused more on the pricing (Monetary value) rather than on understanding other perceived values in the refurbished electronics. Van Weelden et al. [55] have interestingly discussed the barriers that prevent refurbished electronics from being considered by customers [51], however, the study should have highlighted the differences between the customers and proposed customized engagement strategies for each. Hence, the present study adds to this knowledge by providing a more comprehensive understanding of perceived value to understand better how refurbishing companies could better design their marketing and engagement strategies.

Thirdly, by examining the particular business model of refurbishing, especially within the electronics industry, and by depicting the unique characteristics of customers' perceived value in this context, the present study contributes to the literature on circular business models in general and refurbishing business models in particular. The rare prior refurbishment research focuses more on the business, such as operations, supply chain, reversal supply chain, and human resources. The present study offers insights into the customer side, which is the main engine of refurbishing businesses. The present study suggests that the refurbishing business models must consider how people from different generations will change their perceived value of the refurbished electronics and engage in the circular economy transition.

The study also offers managerial implications through the strategies the refurbishing companies adopt to engage customers representing different generations. Managers must build brand communities and enhance trust with those interested in environmental and social issues. Also, managers need to understand that the customer perceives some risks from buying refurbished electronics. Thus, they must balance this equilibrium by ensuring an outstanding and reliable after-sales service. Reliability and transparency are the vital keys to sustaining this refurbishing business model. Customers need to feel safe to buy when their expectations meet reality; hence, they are intensely engaged with the company. Therefore, viral marketing can be seen as one avenue because of this.

In the spirit of transparency, it is imperative to admit the limitations inherent in this study. Future studies could also examine Gen Alpha (people born between 2010 and 2023) customers as they will be the majority in the future, and it could give us an imagination of what the future looks like in the electronic industry. Besides, the present study represents the first exploration of the topic, and the data used in the present study was collected in one city in Finland. Indeed, more empirical studies are needed to examine the differences between countries and cultural contexts.to allow researchers and practitioners to understand how to engage customers in refurbishing business models in different cultural contexts. We also suggest that a study on the customer role in the circular reverse supply chain is highly needed. We noticed that the refurbishing business is based on the reverse supply chain when customers sell or trade in their electronics [1].

## References


1. Abbey, J.D., Kleber, R., Souza, G.C., Voigt, G.: The role of perceived quality risk in pricing remanufactured products. Prod. Oper. Manag. **26**(1), 100–115 (2017)
2. Alshenqeeti, H.: Interviewing as a data collection method: A critical review. Engl. Linguist. Res. **3**(1), 39–45 (2014)
3. Althaf, S., Babbitt, C.W., Chen, R.: Forecasting electronic waste flows for effective circular economy planning: resources. Conserv. Recycl. **151**, 104362 (2019)
4. Berkup, S.B.: Working with generations X and Y in generation Z: Management of different generations in business life. Mediterr. J. Soc. Sci. **5**(19), 218 (2014)
5. Bocken, N.M., Olivetti, E.A., Cullen, J.M., Potting, J., Lifset, R.: Taking the circularity to the next level: a special issue on the circular economy. J. Ind. Ecol. **21**(3), 476–482 (2017)
6. Bocken, N.M., De Pauw, I., Bakker, C., Van Der Grinten, B.: Product design and business model strategies for a circular economy. J. Ind. Prod. Eng. **33**(5), 308–320 (2016)
7. Brodie, R.J., Ilic, A., Juric, B., Hollebeek, L.: Customer engagement in a virtual brand community: An exploratory analysis. J. Bus. Res. **66**(1), 105–114 (2013)
8. Bruce, P., Shelley, R.: Assessing stakeholder engagement. Commun. Journal of New Zealand. **11**(2), 30–48 (2010)
9. Bundgaard, A.M., Huulgaard, R.D.: Luxury products for the circular economy? A case study of Bang & Olufsen. Bus. Strateg. Environ. **28**(5), 699–709 (2019)
10. Castillo-Montoya, M.: Preparing for interview research: The interview protocol refinement framework. Qual. Rep., **21**(5), (2016)

11. Castleberry, A., Nolen, A.: Thematic analysis of qualitative research data: Is it as easy as it sounds? Curr. Pharm. Teach. Learn. **10**(6), 807–815 (2018)
12. Chen, M., Ogunseitan, O.A., Wang, J., Chen, H., Wang, B., Chen, S.: Evolution of electronic waste toxicity: Trends in innovation and regulation. Environ. Int. **89**, 147–154 (2016)
13. Chen, Y.S., Chang, C.H.: Enhance green purchase intentions: the roles of green perceived value, green perceived risk, and green trust. Manag. Decis. **50**(3), 502–520 (2012).
14. Cucchiella, F., D'Adamo, I., Koh, S.L., Rosa, P.: Recycling of WEEEs: An economic assessment of present and future e-waste streams. Renew. Sustain. Energy Rev. **51**, 263–272 (2015)
15. Dembkowski, S., Hanmer-Lloyd, S.: The environmental value-attitude-system model: A framework to guide the understanding of environmentally-conscious customer behaviour. J. Mark. Manag. **10**(7), 593–603 (1994)
16. Djafarova, E., Bowes, T.: 'Instagram made Me buy it': Generation Z impulse purchases in fashion industry. J. Retail. Cust. Serv. **59**, 102345 (2021)
17. Elf, P., Werner, A., Black, S.: Advancing the circular economy through dynamic capabilities and extended customer engagement: Insights from small sustainable fashion enterprises in the UK. Bus. Strateg. Environ. **31**(6), 2682–2699 (2022)
18. Ferasso, M., Beliaeva, T., Kraus, S., Clauss, T., Ribeiro-Soriano, D.: Circular economy business models: The state of research and avenues ahead. Bus. Strat. Environ., **29**(8), 3006–3024 (2022)
19. Ghisellini, P., Cialani, C., Ulgiati, S.: A review on circular economy: the expected transition to a balanced interplay of environmental and economic systems. J. Clean. Prod. **114**, 11–32 (2016)
20. Gurtner, S., Soyez, K.: How to catch the generation Y: Identifying consumers of ecological innovations among youngsters. Technol. Forecast. Soc. Chang. **106**, 101–107 (2016)
21. Harrabin, R.: Finland: Where second-hand comes first. BBC News. (2018). https://www.bbc.com/news/science-environment-46099418
22. Hazen, B.T., Mollenkopf, D.A., Wang, Y.: Remanufacturing for the circular economy: An examination of customer switching behavior. Bus. Strateg. Environ. **26**(4), 451–464 (2017)
23. Hollebeek, L.D., Chen, T.: Exploring positively-versus negatively-valenced brand engagement: a conceptual model. J. Prod. & Brand. Manag. **23**(1), 62–74 (2014)
24. Hollebeek, L.D., Glynn, M.S., Brodie, R.J.: Customer brand engagement in social media: Conceptualization, scale development and validation. J. Interact. Mark. **28**(2), 149–165 (2014)
25. Hollebeek, L.D., Sharma, T.G., Pandey, R., Sanyal, P., Clark, M.K.: Fifteen years of customer engagement research: a bibliometric and network analysis. J. Prod. & Brand. Manag. **31**(2), 293–309 (2022)
26. Hopkinson, P., Zils, M., Hawkins, P., Roper, S.: Managing a complex global circular economy business model: Opportunities and challenges. Calif. Manage. Rev. **60**(3), 71–94 (2018)
27. Hunka, A.D., Linder, M., Habibi, S.: Determinants of customer demand for circular economy products. A case for reuse and remanufacturing for sustainable development. Bus. Strat. Environ., **30**(1), 535–550 (2021)
28. Kerber, J.C., de Souza, E.D., Bouzon, M., Cruz, R.M., Govindan, K.: Customer behaviour aspects towards remanufactured electronic products in an emerging economy: Effects on demand and related risks. Resour. Conserv. Recycl. **170**, 105572 (2021)
29. Kirchherr, J., Reike, D., Hekkert, M.: Conceptualizing the circular economy: An analysis of 114 definitions. Resour. Conserv. Recycl. **127**, 221–232 (2017)
30. Kolnhofer-Derecskei, A., Reicher, R.Z., Szeghegyi, A.: The X and Y generations' characteristics comparison. Acta Polytechnica Hungarica **14**(8), 107–125 (2017)
31. Kumar, A., Holuszko, M., Espinosa, D.C.R.: E-waste: An overview on generation, collection, legislation and recycling practices. Resour. Conserv. Recycl. **122**, 32–42 (2017)
32. Lieder, M., Asif, F.M., Rashid, A., Mihelič, A., Kotnik, S.: A conjoint analysis of circular economy value propositions for customers: Using "washing machines in Stockholm" as a case study. J. Clean. Prod. **172**, 264–273 (2018)
33. Lin, P.C., Huang, Y.H.: The influence factors on choice behavior regarding green products based on the theory of consumption values. J. Clean. Prod. **22**(1), 11–18 (2012)

34. Lüdeke-Freund, F., Gold, S., Bocken, N.M.: A review and typology of circular economy business model patterns. J. Ind. Ecol. **23**(1), 36–61 (2019).
35. Magaldi, D., Berler, M.: Semi-structured interviews. Encyclopedia of personality and individual differences, pp. 4825–4830 (2020)
36. Matsumoto, M., Chinen, K., Endo, H.: Remanufactured auto parts market in Japan: Historical review and factors affecting green purchasing behavior. J. Clean. Prod. **172**, 4494–4505 (2018)
37. Morar, D.D.: An overview of the customer value literature–perceived value, desired value. Mark. Inf. Decis. **6**, 169–186 (2013)
38. Munsch, A.: Millennial and generation Z digital marketing communication and advertising effectiveness: A qualitative exploration. J. Glob. Scholars Market. Sci. **31**(1), 10–29 (2021)
39. Nasiri, M.S., Shokouhyar, S.: Actual consumers' response to purchase refurbished smartphones: Exploring perceived value from product reviews in online retailing. J. Retail. Consum. Serv. **62**, 102652 (2021)
40. Naumovska, L.: Marketing communication strategies for generation Y–millennials. Bus. Manag. Strat. **8**(1), 123–133 (2017)
41. Nußholz, J.: Circular business models: defining a concept and framing an emerging research field. Sustain. **9**(10), 1810 (2017). http://www.mdpi.com/2071-1050/9/10/1810
42. Rahulan, M., Troynikov, O., Watson, C., Janta, M., Senner, V.: Customer behavior of generational cohorts for compression sportswear. J. Fash. Mark. Manag. **19**(1), 87–104 (2015)
43. Sashi, C.M.: Customer engagement, buyer-seller relationships, and social media. Manag. Decis. **50**(2), 253–272 (2012)
44. Severo, E.A., de Guimarães, J.C.F., Dorion, E.C.H.: Cleaner production and environmental management as sustainable product innovation antecedents: A survey in Brazilian industries. J. Clean. Prod. **142**, 87–97 (2017).
45. Sheth, J.N., Newman, B.I., Gross, B.L.: Why we buy what we buy: A theory of consumption values. J. Bus. Res. **22**(2), 159–170 (1991)
46. Sofaer, S.: Qualitative methods: what are they and why use them? Health Serv. Res. **34**(5 Pt 2), 1101 (1999)
47. Tanskanen, P.: Electronics waste: recycling of mobile phones. Post-Cust Waste Recycl Optim Prod, 129–150 (2012)
48. Terry, G., Hayfield, N., Clarke, V., Braun, V.: Thematic analysis. SAGE Handb. Qual. Res. Psychol. **2**, 17–37 (2017)
49. Tongco, M.D.C.: Purposive sampling as a tool for informant selection. (2007)
50. Vafadarnikjoo, A., Mishra, N., Govindan, K., Chalvatzis, K.: Assessment of customers' motivations to purchase a remanufactured product by applying Fuzzy Delphi method and single valued neutrosophic sets. J. Clean. Prod. **196**, 230–244 (2018)
51. Wahjudi, D., Gan, S.S., Tanoto, Y.Y., Winata, J.: Drivers and barriers of consumer purchase intention of remanufactured mobile phones: a study on Indonesian consumers. Int. J. Integr. Supply Manag. **13**(2–3), 178–191 (2020)
52. Wang, Y., Hazen, B.T.: Consumer product knowledge and intention to purchase remanufactured products. Int. J. Prod. Econ. **181**, 460–469 (2016)
53. Wang, J., Wang, S., Xue, H., Wang, Y., Li, J.: Green image and consumers' word-of-mouth intention in the green hotel industry: The moderating effect of Millennials. J. Clean. Prod. **181**, 426–436 (2018)
54. Wang, Y., Zhu, Q., Krikke, H., Hazen, B.: How product and process knowledge enable consumer switching to remanufactured laptop computers in circular economy. Technol. Forecast. Soc. Chang. **161**, 120275 (2020)
55. Van Weelden, E., Mugge, R., Bakker, C.: Paving the way towards circular consumption: exploring customer acceptance of refurbished mobile phones in the Dutch market. J. Clean. Prod. **113**, 743–754 (2016)
56. Williams, K.C., Page, R.A.: Marketing to the generations. J. Behav. Stud. Bus. **3**(1), 37–53 (2011)
57. Yin, R.K.: Qualitative research from start to finish. Guilford publications, (2015)

58. Zeithaml, V.A.: Customer perceptions of price, quality, and value: a means-end model and synthesis of evidence. J. Mark. **52**(3), 2–22 (1988)
59. Zeng, X., Xu, X., Zheng, X., Reponen, T., Chen, A., Huo, X.: Heavy metals in PM2. 5 and in blood, and children's respiratory symptoms and asthma from an e-waste recycling area. Environ. Pollut. **210**, 346–353 (2016)

# Navigating the AI Landscape for Sustainable Business Growth

**J. C. Pavithra and V. Mohana Sundari**

**Abstract** Artificial intelligence (AI) revolves around software programs that draw knowledge from data, identify patterns, and automatically make decisions on their own tasks that formally require human intelligence. This chapter examines how businesses can effectively traverse the AI landscape to drive eternal growth for the organization while bolstering wider employee performance. It focusses on AI-driven approaches for optimizing operations, improving performance, and spurring individuality. Businesses that incorporate AI into their business processes tend to boost productivity, mitigate environmental impact, and position themselves for long-term success. Through data analysis and insights, this study analyzes how AI technologies promote efficiency in organizations, communication and decision-making. The initial objective of this chapter is to examine how AI drives employee engagement and the performance of an employee. This study uses a mixed-methodology approach that incorporates both qualitative and quantitative methods. Simple random sampling was conducted to pick participants from the target demographic of employees in several organizations. A random number generator is employed to select the appropriate number of employees to ensure that every potential employee has an equal chance of being selected. Findings reveal that employee engagement and adoption of AI are positively correlated, which emphasizes increased productivity, performance and pursuit of happiness. Applications of this research include educating organizational leaders on the potential benefits and limitations of incorporating AI and enabling them to develop strategic decisions that will increase employee engagement and effectively accomplish employee performance.

**Keywords** Artificial intelligence · Employee engagement · Organization · Employee performance · Communication · Employees and decision making

---

J. C. Pavithra (✉) · V. M. Sundari
Faculty of Management, SRM Institute of Science and Technology, Kattankulathur, Chengalpattu District, Tamil Nadu, India
e-mail: pj0815@srmist.edu.in

J. C. Pavithra
Faculty of Management, SRM Institute of Science and Technology, Kattankulathur, India



N. D. K. Nguyen et al. (eds.), *Digital Transformation and Sustainability: Methods and Applications*, Information Systems Engineering and Management 47,
https://doi.org/10.1007/978-3-031-91278-8_11

# 1 Introduction

AI has emerged as an essential transformation driver in the digital era, delivering unbeatable prospects for organizational expansion and accelerating business innovation[35]. AI covers everything from process improvement to upgrading the experiences of customers, and it affords companies an assortment of probabilities for accomplishing sustainable success. It is no longer merely an interdisciplinary idea. AI is reorganizing everything from customer interactions to operational efficiency and it is imperative that companies comprehend how to implement AI for sustainable growth if they are to prosper in this revolutionary setting [11, 25]. Understanding how to thrive in this rapidly evolving world where AI presents both opportunities and challenges is becoming increasingly essential for companies worldwide. Bughin et al. [7], assert that for companies aiming to make use of AI, this chapter serves as a comprehensive guide, and it evaluates the multifaceted nature of integrating AI and emphasizes the vitality of a strategic approach that correlates AI efforts with a broader company's objectives. The pathway towards persistent growth through AI is a tangled one, encompassing not solely the incorporation of emerging technologies but also an enormous change in how companies operate, make decisions, and engage with those who are involved. Researchers [12, 16], consider the most important aspects that lead to AI integration and efficient adoption, revealing an approach towards companies striving for success for a prolonged period of time.

**The AI Revolution: Possibilities and Difficulties**

The incorporation of AI into company operations has been one of the largest advancements in technology of the twenty-first century. Davenport and Ronanki [14], claim that AI has the capabilities to establish whole novel company structures, maximize the utilization of resources and readily accessible new efficiencies. Other than these potentialities AI additionally poses considerable challenges, companies must continuously evolve to it in order to stay fiercely competitive, as AI is advancing at an extremely rapid pace and serves as tools that can accelerate growth in formerly inconceivable ways, from machine learning algorithms that strengthen customer-personalized services to predictive analytics that can predict market developments [35]. Lee and Roberts [26], assert that implementing an integrated approach that prioritizes long term consequences over immediate benefits is necessary to achieve sustainable growth through the use of AI. In a comparable way, Chen and Patel [10] emphasized the fact that effective AI integration relies upon fostering a culture that encourages innovative thinking, ethical standards, and continuous learning.

## *1.1 Success Components in Navigating AI for Growth*

Creating a workplace where AI can contribute and prosper to the businesses ambitious objectives is one of the most important aspects of a successful AI integration, rather than simply implementing new tools. This implies meticulous oversight that is consistent with the strategic objective of the company with an assortment of different substantial factors [6, 14]. The success of AI initiatives depends significantly on four crucial standards, of which we will explore in this chapter: appropriate teams, organizational culture, training and development, and leadership.

**Appropriate Teams: Building the Right Expertise**

The emergence of appropriate teams is the cornerstone of successful AI integration. Multidisciplinary teams with an equilibrium of business competence and expertise in technology are imperative for AI projects. Wilson and Daugherty [35], claim that assembling up teams with an appropriate mix of capabilities is extremely important for the success of AI initiatives. Data scientists, machine learning engineers, domain experts, and business strategists are often included in these teams. With an objective of ensuring that AI technologies are not developed in isolation but are alternatively integrated into the fundamental operations of the business, the authors emphasize that collaboration across disciplines must be essential for success to reconcile AI efforts with business objectives. Furthermore, Davenport and Ronanki [14] believe that when organizations concentrate solely on their technical knowledge, omitting to additionally take into consideration the bigger commercial operations circumstances, AI projects typically fail. They think that employees who have the ability to bridge the gap between advanced technology and business needs, along with those who recognize the organization's strategic objectives, constitute critical elements of successful AI teams. This alignment ensures that AI efforts are not given out as stand-alone technical experiments but rather encourage growth over the long term. According to Davenport and Kirby [13], suggest that multidisciplinary teams featuring an abundance of expertise, software engineering, including data science, machine learning and domain-relevant expertise are essential for projects involving AI.

**Organizational Culture: Fostering an Environment for Innovation**

Organizations need to promote a culture that facilitates experimentation, upholds innovation and supports data-driven decision-making if their goal is for AI to achieve its full potential. Based on Schein [32], an organization's beliefs, shared values, and practices determine its procedures for operation. In the opinion of Kotter [24], focusses on the fact that successful digital transformation including AI adoption, depends on maintaining a culture that is adaptable to change. The effective implementation and incorporation of AI in a business is significantly influenced by the organizational culture and a stimulating business environment is critical in the setting of AI. In the words of McAfee and Brynjolfsson [28], businesses that successfully integrate AI generally possess cultures that emphasize adjustments and continuous improvement. These organizations are adaptable and prepared to challenge established procedures for the purpose of opening up an environment for innovations enabled by AI. The authors suggest that even the most innovative technologies are unlikely to accomplish their full potential in the absence of an atmosphere that appreciates AI. In addition, organizational culture requires contending with unexpected challenges to the adoption of AI. Employees could perceive AI as a disruptive force or as a threat to their professions, as observed by Westerman et al. [34]. Therefore, leaders now have the responsibility to foster an environment of openness and acceptance, fostering a sense of security and inspiring employees to view AI as a technique that improves rather than replaces their skills.

**Training and Development: Equipping the Workforce for the Future**

Training and development are crucial in order to ensure that the workforce has the abilities and skills required to collaborate with AI. In the opinion of Brynjolfsson and McAfee [6], workforce abilities have to evolve in combination with AI and automation technological advancements. This involves improving critical thinking, problem solving and adaptability skills together with technical understanding of AI and data science. The capabilities of employees implementing AI technologies must evolve as well as with technological advancements. Continuous training must be provided for ensuring that employees are able to interact with AI systems in the most effective way Bughin et al. [9]. The authors illustrate that organizations that invest in training and development have a greater probability of effectively implementing AI, while their employees have a greater capacity to utilize AI tools for bettering business outcomes. Employees are required to engage in continuous training and development programs with the goal of developing the knowledge and abilities that are essential for effectively employing AI. In the AI era Huang and Rust [20], emphasize the intrinsic worth of both technical and soft skills. Employees possessing these soft capabilities could strategically and innovatively employ AI knowledge, resulting in long-term growth of the business. In addition, training and development programs must be continually adaptable, with an objective of keeping up with the rapidly increasing pace at which technology is progressing. In the investigation by Bessen [5], organizations that fail to continually instruct their employees in order to get better trained stand the risk of slipping behind as AI technologies evolve. Businesses can be confidence that their employees will remain at the forefront of AI innovation by boosting an environment of lifelong learning.

**Leadership: Guiding the Business Through Change**

Probably the most significant consideration to take into account while navigating the AI landscape for a long time in business development is leadership. Leadership is required in order to effectively navigate the business through the revolutionary modifications brought forward by AI. Effective leadership serves as essential for managing change, according to Kotter [24], this is particularly pertinent with regard to discussing the adoption of AI. A distinct understanding of how AI will assist with the company's goals in the long run must be expressed by leaders, who must also motivate and encourage employees to cooperate with AI-driven changes. In the opinion of Avolio et al. [1] transformational leadership, which prioritizes encouraging and facilitating employees particularly successful in encouraging the adoption of AI. Successful AI integration involves transformational leaders who additionally inspire innovation, support continuous training, and establish an environment that encourages collaboration. For the purpose of ensuring that AI is used responsibly and for the benefit of all stakeholders, these leaders are additionally proficient at analyzing the ethical and societal consequences of this technologies. On top of that in the AI era Westerman et al. [34] accentuate the relevance of digital leadership. Digital leaders are able to seamlessly integrate initiatives involving AI with the underlying business strategy due to their realization of the strategic potential of AI and they are also skilled at addressing the challenges that arise from data governance, privacy and security in initiatives related to AI. These individuals are able to assist businesses in navigating the difficulties of incorporating AI while additionally making sure that AI encourages long-term, sustainable business growth by employing strong leadership. Successful AI installations are more likely to succeed in businesses managed by employees whose services actively encourage a creative and data-driven culture [8]. With strong, visionary leadership capable of enabling AI initiatives are highly profitable right through future generations, businesses can efficiently tackle all the difficulties and obstacles associated with deploying AI.

## *1.2 Research Objectives*

This chapter provides recommendations and best practices concerning the creation of high performers and employee engagement based on the relationships between appropriate teams, organizational culture, training and development and leadership. The outcomes of this research are as follows: First, the methodology of AI initiatives depends significantly on four crucial standards and their impact on employee engagement to improve employee performance will be described in detail.

- To investigate the impact of appropriate teams on performance of employees, with employee engagement as a mediating component and AI as a moderating factor.
- To analyze the impact of leadership on performance of workers, with mediator as employee engagement and AI as moderator.
- To explore how training and development influence employee performance, involving the moderating role of AI and the mediating effect of employee engagement.
- To investigate the mediating impact of engagement and the moderating influence of AI between organizational culture and employee performance.
- To examine the mediating effect of employee dedication and the moderating effect of AI between the four crucial organizational factors like appropriate teams, organizational culture, training and development, leadership and employee performance.

## 2 Literature Review

### *2.1 Appropriate Teams*

Appropriate teams refer to groups of people who were extensively selected and categorized corresponding to their competencies, roles and respective fields of expertise with the aim of successfully completing a task or accomplishing a particular objective. These teams are established to make sure that an ideal combination of abilities and personality types strengthens other people with the objective to enhance productivity, collaboration and communication. An appropriate team's essential quality is its ability to operate harmoniously, integrating various members' strengths in order to guarantee successful outcomes. Hackman's Model of Team Effectiveness is a model that identifies five key components that improve team effectiveness is a real team, supportive context, compelling direction, expert coaching and enabling structure [17]. For teams to lay out the model for effective cooperation, Hackman emphasised the importance of making it clear when defining the purposes, roles and responsibilities of the team. Members of real teams have defined their tasks and rely on one another for support. Hackman's strategy focusses on providing teams with the resources they need to succeed and coordinating team structure with business goals.

In the study by Mathieu et al. [27], Hackman's Model of Team Effectiveness was implemented by reviewing team research conducted over the past decade. The authors emphasized the importance of understanding the nature of work teams and the substantive variations among various types of teams. They utilized an improved input framework and outcome process, aligning with Hackman's model that focusses on the conditions necessary for team effectiveness. They called for researchers to embrace the difficulty of current team-based organizational structures, which reflects Hackman's emphasis on the contextual factors influencing team performance. This comprehensive approach underscores the relevance of Hackman's model in contemporary team research.

### *2.2 Organizational Culture*

The shared principles, perspectives, behaviours and practices that determine an organization's both internal and external connections with stakeholders have been referred to as its organizational culture. It establishes the foundation for task completion and shapes the team's perspectives on accomplishing the organization's objectives.

Edgar Schein's Model of Organizational Culture by Edgar Schein [31], consists of three levels which includes artifacts, basic underlying assumptions and espoused values. Artifacts are distinct portion like rules on attire and company designs. Espoused values are the businesses declared norms and beliefs, such as mission statements. The deepest level, the basic underlying assumptions, are unconscious principles that impact behaviours. Schein's model highlights the challenging aspects of cultural change, as it requires addressing these deeply embedded assumptions.

In the study conducted by Srivastava [33], Edgar Schein's Model of Organizational Culture was utilized to examine the cultural frameworks of organizations rooted in American, Japanese and Indian contexts. The research focused on identifying shared values, which Schein posits as central to understanding organizational culture. In three multinational organizations like Maruti Udyog, Escorts Limited and Pepsi Foods Limited, the study correlated organizational culture through a set of values, which include self-realization and status enhancement, and examined their relationship with organizational effectiveness and employee competencies across different hierarchical levels such as supervisors, managers and executives. By employing Schein's three levels of culture, which consist of observable artifacts, values and basic underlying assumptions, the research aimed to highlight the differences and similarities in organizational cultures across these diverse cultural backgrounds.

## 2.3 *Training and Development*

Training and development is a process used by the businesses to strengthen the knowledge, capabilities and abilities of workers. Training is important on improving an employee's ability to perform their current job, while development is concerned with broader skill building for future roles. These efforts increase productivity, foster employee growth, and ensure that organizations stay competitive. In today's fast changing work environment, effective training and development programs are vital for cultivating a learning culture and driving innovation.

Kirkpatrick's Four-Level Training Evaluation Model [23], employs four levels for assessing the constructiveness of training, such as behaviour, reaction, results, and learning. Reaction gauges participants contentment and involvement, whereas learning assesses acquired knowledge or abilities. Behaviour looks at how well participants use these abilities at work, and results evaluate the overall effect on performance and productivity among the organization. In addition to providing immediate feedback, this inclusive model frequently examines the long-term efficacy of training.

Moroki [21] systematically applied Kirkpatrick's Four-Level Training Evaluation Model to assess the productiveness of the learning process and online teaching at the university. The reaction level determined the students' initial reactions to the online learning experience, marking the first phase of the evaluation. Considering pre- and post-test outcomes, the learning level subsequently focused on assessing enhancements in students' knowledge, attitudes and capabilities. Using questionnaires and interviews for obtaining data, the third-level behaviour change, evaluated how students employed their recently acquired competencies and skills in real-world circumstances. In the end, the results level evaluated how effectively the experience of learning online generally assisted students in accomplishing their educational objectives, highlighting out areas that require additional improvements in areas like technological and lecturer competencies.

## 2.4 Leadership

Leadership is a process of guiding and influencing individuals or teams toward accomplish a common objective by creating a perception, making strategic decisions and fostering collaboration. It builds trust, empowers individuals, and adapts to team needs. Effective leadership aligns personal and team efforts with organizational goals, which drives success and growth of an organization.

Expanding on the ideas of Max Weber [4] developed the Transactional Leadership Model, which focusses on structured tasks, rewards and penalties to manage performance. Leaders use exception management to intervene when performance deviates and to provide contingent rewards for achieving targets. Unlike passive management, which only intervenes when difficulties arise, active management conducts analysis in advance. While it might not possess the same level of resourcefulness and engagement as other leadership categories, it is effective for short-term goals and ensuring order.

In the study conducted by Oncer [2], Transactional Leadership Model was implemented through a two-dimensional framework consisting of contingent reward, and active management by exception. The focus of this study was to investigate the influence of these components on the entrepreneurial orientation of workers in multinational companies based in Istanbul. The investigation examined opinions on transactional leadership and its impact on entrepreneurial orientation, with a particular emphasis on proactiveness, innovative thinking, and risk taking. It encompassed 171 valid responses from non-executive employees. The outcomes revealed that transactional leadership positively impacted only the proactiveness measure, with no significant correlations found with innovativeness or taking risks. This demonstrates that, while transactional leadership can facilitate proactive behaviours, it may not encourage an extra comprehensive sense of entrepreneurship among employees.

## 2.5 Related Works

The study by Kadir and Amalia [22], investigates the relationships among job motivation, job satisfaction, organizational culture, compensation and performance of employees within the Ministry of Manpower. Adopting Structural Equation Modelling (SEM), this research reveals that compensation and organizational culture have no direct impact on job satisfaction, but job motivation has a considerable impact. On the other hand, job satisfaction has been confirmed for enhancing performance, and organizational culture has an encouraging impact on employee performance. The study's results indicate that an encouraging work environment and job satisfaction significantly influence employee performance, surpassing mere compensation or promotion. This underscores the significance of creating a assured work environment to amplify overall employee effectiveness. The research emphasises that by recognising those procedures, organizations can establish procedures into effect that promote employee productivity and engagement.

The enhancement of employee engagement and performance was modelled by integrating AI with talent management in a report by Rozman et al. [29]. A sample of 317 managers and owners of Slovenian businesses participated in the author's study. This study's key variables include leadership, organizational culture, workload reduction, employee training and development, and AI-supported talent acquisition. These organizational factors have a major influence on the outcomes since they show how well AI implementation affects employee engagement and the overall success of the businesses and the study highlights the need for enterprises to adopt AI strategically so that to bolster a competitive advantage in the advancing business landscape. For example, AI makes it easier to choose candidates and create customized training plans, both of which can increase employee performance as well as job satisfaction. Furthermore, through AI facilitation identifying and retaining talented employees and also optimizing the training techniques can enhance workplace efficiency and morale.

The study by the authors Rozman et al. [30] investigates the impact of AI on reducing employee workload and enhancing company performance in a uncertain, volatile, complex and ambiguous environment. It emphasises mainly three independent components, namely leadership bolster up by AI, organizational culture reinforced by AI, and training and development supported by AI. The results reveal that these elements play an immense part in employees opinions about a reduction in load, which in turn has an advantageous influence on employee engagement and the overall performance of the organization. In addition, the study proves that excellent leadership and a supportive AI culture generate engaged employees, which enhances the organizations performance and productivity. This study demonstrates the value of implementing AI into organizational practices to promote a more productive and creative workplace, as well as how important AI can be in conquering the challenges that accompany contemporary organizational issues.

The study "Satisfaction with the Work Done in University Employees" by Joselina Caridad et al. (2023) investigates the interconnections between job satisfaction and employee engagement among university staff in the Dominican Republic. In this research, job satisfaction is an important aspect that influences employee engagement, while employees feel satisfied with their jobs, they tend to become considerably more connected. The study emphasized a direct correlation between employee engagement and job satisfaction that could improve employee performance and dedication. It was also noted that variables such as leadership satisfaction, compensation or relationships have no impact on the level of engagement. This connection highlights the critical importance of creating supportive and encouraging workplace environments to foster employee commitment, a crucial factor for the success of educational institutions and the achievements of students.

## *2.6 Research Gap*

Based on various literature surveys, it was identified that the AI depends significantly on four crucial standard organizational factors and their impact on employee engagement to improve employee performance were not reported. It is also noticed that the standards of employee engagement impacting the performance of an employee are not analysed and reported. Hence, in this study we aimed to understand the synergy within the components and employee engagement leading to employee performance.

# 3 Research Methodology

The study evaluates how connections exist between the appropriate teams, organizational culture, training and development, and leadership as well as how these standards are impacted by AI on employee engagement and employee performance. The methodology employed in this chapter incorporates collecting and analysing of source data. A organized survey questionnaire was employed to gather the primary data. The questionnaire used in the survey is structured into many sections, each that emphasizes on one of the key variables of interest. This questionnaire was created to capture respondents opinions and personal experiences regarding the appropriate teams, organizational culture, training and development, and leadership. Employees from several companies comprise the target demographic of this study. The sample size of 177 respondents is determined to assure the reliability of the obtained results. A stratified sampling method was used to select employees from different departments, levels within the businesses and to generate a representative sample. The research methodology employed in this study is descriptive and the data collected is analysed using statistical techniques to examine the relationships between appropriate teams, organizational culture, training and development, and leadership to evaluate their impact on employee engagement and employee performance with AI. Cronbach's alpha is applied to evaluate the inmost stability of the survey items.

# 4 Research Model

This conceptual framework shows how AI impacts on four organizational factors; which influences employee engagement to increase employee performance that in turn drives business growth. Appropriate teams ensures a group of people who were selected and categorised based on their roles, while organizational culture focuses on sharing principles and practices to determine both internal and external connections with stakeholders whereas, training and development strengthen the capabilities, knowledge and abilities of employees. Leadership guides and encourages individuals or teams towards achieving a common goal. High employee engagement enhances their overall performance, productivity, minimise turnover and are the important components for the success of an organization. When these factors are employed with each anoother, it improves employee engagement which in turn reflects on employees performance and commitment to their work.

## 5 Hypothesis

The hypothesis that were developed in line with the model as illustrated in Fig. 1, and purpose of this study are as follows:

**H1**: How gender influences employee engagement.

**H2**: The effect of age on employee performance.

**H3**: The impact of predecessors of employee engagement on employee engagement.

**H3a**: The impact of appropriate teams on employee engagement.

**H3b**: The impact of organizational culture on employee engagement.

**H3c**: The impact of training and development on employee engagement.

**H3d**: The impact of leadership on employee engagement.

**H4**: The effect of mediating influence of employee engagement on employee performance.

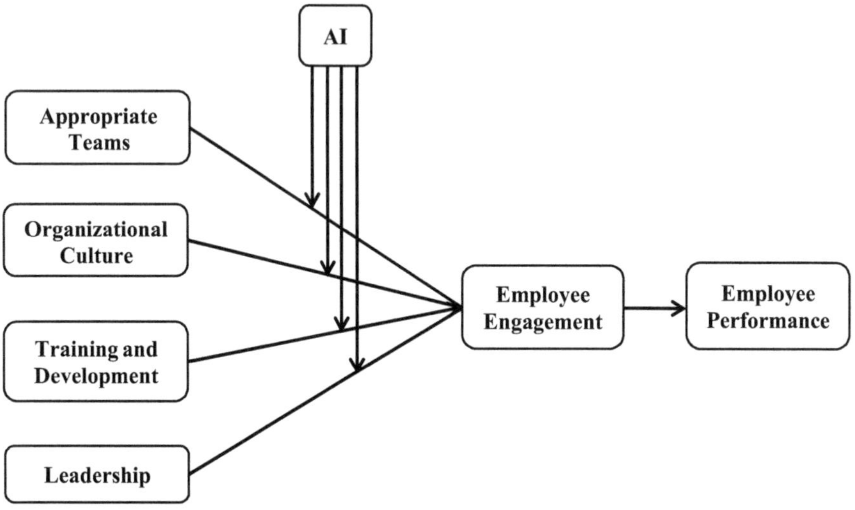


**Fig. 1** Conceptual model. *Source* Author's work

**H5**: The effect of moderating influence of AI on the predecessors of employee engagement.

**H5a**: The impact of AI on appropriate teams of employee engagement.

**H5b**: The impact of AI on organizational culture of employee engagement.

**H5c**: The impact of AI on training and development of employee engagement.

**H5d**: The impact of AI on leadership of employee engagement.

## 6 Data Analysis

Data analysis is an essential phase in the context of this research, since it guides researchers to gain an understanding of interconnections between the information collected from the data and to interpret the results. Data that has been collected went through the rigorous process of cleansing and transformation followed by reviewing and modelling with an intention to get clear information upon making hypothesis, drawing conclusions and decision making (Table 1).

AI will improve efficiency and sustainability as well as it is an essential part of successfully navigating for the purpose of achieving sustainable corporate growth. According to the analysis of the 177 participants, women made up 43.1% and men made up 56.9% of the group, while of the total number of participants, 38.0% fall into the age bracket of 30–39 years old, 29.2% fall into the age bracket of 18–29 years old, 20.8% fall into the age bracket of 40–49 years old, and 12.0% are aged 50 and above. On the other hand, 22.7% of individuals who participated in the study are single, while 77.3% are married. There are 39.8% of people who work in schools, 36.6% who work in software, and 23.6% who work in other sectors when split down by industry. 43.5% of the participants have more than 10 years of professional experience, 32.4% have between 5 and 10 years of experience, and 24.1% have less than 5 years of experience. This is a breakdown of the annual income, for your convenience: 46.3% of the population earns between 5 and 10 lakhs of Indian rupees, 23.1% earn between 3 and 5 lakhs of Indian rupees, 14.8% make less than 3 lakhs of Indian rupees, and 15.7% earn more than 10 lakhs of Indian rupees. With the help of this demographic split, we are able to conduct a more precise investigation into the role that AI plays in fostering the long-term growth of businesses across a wide variety of fields and degrees of skill (Table 2).

In the reliability test, we make use of Cronbach's Alpha, which reveals that the structures covered in the study are consistent with one another internally. Both Appropriate Teams and Employee Engagement have a Cronbach's Alpha value of 0.732, which indicates that they are reliable enough to be considered acceptable. Training and Development demonstrates the highest level of dependability with a Cronbach's Alpha score of 0.778, while Organizational Culture holds the highest reliability score with a score of 0.739. Leadership and Employee Performance both have Alpha scores of 0.767 and 0.764, respectively, indicating that they consistently demonstrate a high

**Table 1** Descriptive analysis of demographic variables taken up in the study

| Demographic variable | Frequency | Percentage |
|---|---|---|
| *Gender* | | |
| Male | 101 | 56.9 |
| Female | 76 | 43.1 |
| Total | 177 | 100.0 |
| *Age* | | |
| 18–29 years old | 52 | 29.2 |
| 30–39 years old | 67 | 38.0 |
| 40–49 years old | 37 | 20.8 |
| 50 years and above | 21 | 12.0 |
| Total | 177 | 100.0 |
| *Marital status* | | |
| Single | 40 | 22.7 |
| Married | 137 | 77.3 |
| Total | 177 | 100.0 |
| *Occupation* | | |
| Educational institutions | 70 | 39.8 |
| Software industries | 65 | 36.6 |
| Other industries | 42 | 23.6 |
| Total | 177 | 100.0 |
| *Professional experience* | | |
| Less than 5 years | 43 | 24.1 |
| 5–10 years | 57 | 32.4 |
| More than 10 years | 77 | 43.5 |
| Total | 177 | 100.0 |
| *Annual income* | | |
| Below 3,00,000 INR | 26 | 14.8 |
| 3,00,000–5,00,000 INR | 41 | 23.1 |
| 5,00,000 INR–10,00,000 INR | 82 | 46.3 |
| 10,00,000 INR and above | 28 | 15.7 |
| Total | 177 | 100.0 |

*Note* The table presents the demographic characteristics of the respondents, including gender, age, marital status, occupation, professional experience, and annual income
*Source* Author's calculation

**Table 2** Reliability test

| Dimensions | No of items | Cronbach's alpha |
|---|---|---|
| Appropriate teams | 5 | 0.732 |
| Organizational culture | 5 | 0.739 |
| Training and development | 5 | 0.778 |
| Leadership | 5 | 0.767 |
| Artificial intelligence | 5 | 0.747 |
| Employee engagement | 5 | 0.732 |
| Employee performance | 5 | 0.764 |

*Note.* The table presents the reliability analysis of the dimensions measured
*Source* Author's calculation

level of internal consistency. The reliability score for artificial intelligence is 0.747, which can be interpreted as a confirmation that all of the constructs that were utilized in the research are reliable and consistent for further investigation.

**H1: How gender influences employee engagement**.

Tables 3 and 4 provides one-sample t-test results and it examines the impact of gender on employee engagement. The average score for gender is 1.43, with a standard deviation of 0.496 and a standard error of 0.034. The t-value for gender is 42.363 with 176 degrees of freedom (df), and the p-value is 0.000, indicating that the mean difference of 1.431 is statistically significant. The mean score for employee engagement is 3.27, with a standard deviation of 1.344 and a standard error of 0.091. The t-value for employee engagement is 35.739 with 176 df, and the p-value is 0.000, indicating that the mean difference of 3.269 is statistically significant. Both gender and employee engagement exhibit substantial divergences from the test value of 0, indicating significant distinctions within the study's setting.

Banihani et al. [3] argue that work engagement itself can be "gendered", meaning it may inherently favour one gender's typical work traits or styles. Research by Hartman and Barber [18] explores how gender influences occupational self-efficacy and engagement, finding that women may face unique challenges that impact their engagement levels differently from men.

**Table 3** One-sample statistics in T test

| One-sample statistics | | | | |
|---|---|---|---|---|
| | N | Mean | Std. deviation | Std. error mean |
| Gender | 177 | 1.43 | 0.496 | 0.034 |
| Employee engagement | 177 | 3.27 | 1.344 | 0.091 |

*Source* Author's calculation

**Table 4** One-sample test in T test

| One-sample test | | | | | | |
|---|---|---|---|---|---|---|
| | Test value = 0 | | | | | |
| | t | df | Sig.(2-tailed) | Mean difference | 95% confidence interval of the difference | |
| | | | | | Lower | Upper |
| Gender | 42.363 | 176 | 0.000 | 1.431 | 1.36 | 1.50 |
| Employee engagement | 35.739 | 176 | 0.000 | 3.269 | 3.09 | 3.45 |

*Source* Author's calculation

**H2: The effect of age on employee performance.**

Table 5 provides analysis of variance (ANOVA) test results indicates the effect of age on employee performance results that there is a statistically significant difference between the age groups. The mean square number is 1.972, and the total of squares between groups is 5.915 with 3 df. A significance level of 0.029 is associated with 9.075 which is the F-value. This suggests that there is a substantial variation in employee performance across different age groups, since the p-value is less than 0.05, which implies that this is the case. Taking into account 173 df, the sum of squares for within-groups is 388.710, whereas the sum of squares for the entire set is 394.625. In light of these findings, it appears that age plays a vital role in determining the level of performance that employers expect from their employees.

A study on age diversity by Hoch et al. [19] found that age diversity within teams can affect overall team performance.

**H3: The impact of predecessors of employee engagement on employee engagement.**

Table 6 shows the model summary of the regression analysis of this table demonstrating that key factors such as appropriate teams, organizational culture, training and development, and leadership significantly impact employee engagement. With an R value of 0.843, the model shows a strong positive correlation, and an R Square of

**Table 5** ANOVA

| ANOVA | | | | | |
|---|---|---|---|---|---|
| Employee performance | | | | | |
| | Sum of squares | df | Mean square | F | Sig. |
| Between groups | 5.915 | 3 | 1.972 | 9.075 | 0.029 |
| Within groups | 388.710 | 173 | 1.834 | | |
| Total | 394.625 | 176 | | | |

*Source* Author's calculation

0.711 indicates that 71.1% of the variance in employee engagement can be explained by these factors, with an adjusted R Square of 0.704. Having an F-value of 105.830 and a p-value of 0.000, the results of the ANOVA provide evidence that this is the case. This highlight significance of leadership, organizational culture, appropriate training, and team cohesiveness in the process of encouraging employee engagement.

Table 7 illustrating the ANOVA analysis indicates that factors such as appropriate teams, organizational culture, training and development, and leadership significantly influence employee engagement. The regression sum of squares is 1553.382, and the F-value is 105.830, with a p-value of 0.000, demonstrating that the model is statistically significant. These outcomes show that these variables collectively account for a large portion of the variation in employee engagement. The relatively low residual sum of squares (631.160) supports the model's effectiveness in explaining employee engagement, emphasizing the importance of these factors in promoting an engaged workforce.

Table 8 showing the regression analysis results highlight the significant influence of key factors on employee engagement. The constant value (B = 2.222, p = 0.009) shows the baseline engagement level when all other variables are held constant. Among the variables, training and development has the most substantial positive effect on employee engagement, with a Beta value of 0.360 (p = 0.000). Organizational culture also plays an important role, with a Beta value of 0.252 (p = 0.001), indicating its strong influence. Leadership (Beta = 0.165, p = 0.034) and appropriate teams (Beta = 0.145, p = 0.048) also contribute positively to engagement, though

**Table 6** Regression summary with R values

| Model summary | | | | | | | | | |
|---|---|---|---|---|---|---|---|---|---|
| Model | R | R square | Adjusted R square | Std. error of the estimate | Change statistics | | | | |
| | | | | | R square change | F change | df1 | df2 | Sig. F change |
| 1 | 0.843[a] | 0.711 | 0.704 | 1.91560 | 0.711 | 105.830 | 4 | 172 | 0.000 |

a. Predictors: (Constant), Appropriate Teams, Organizational Culture, Training and Development, Leadership

*Source* Author's calculation

**Table 7** ANOVA-regression summary

| ANOVA[a] | | | | | | |
|---|---|---|---|---|---|---|
| Model | | Sum of squares | df | Mean square | F | Sig. |
| 1 | Regression | 1553.382 | 4 | 388.346 | 105.830 | 0.000[b] |
| | Residual | 631.160 | 172 | 3.670 | | |
| | Total | 2184.542 | 176 | | | |

a. Dependent Variable: Employee Engagement

b. Predictors: (Constant), Appropriate Teams, Organizational Culture, Training and Development, Leadership

*Source* Author's calculation

to a slightly lesser degree. The collinearity statistics, with tolerance values above 0.2 and VIF values below 10, confirm that there is no significant multicollinearity between the variables.

**H4: The effect of mediating influence of employee engagement on employee performance.**

Table 9 demonstrating regression results for Hypothesis 4 indicate that employee engagement significantly modifies the connections between employee performance and organizational outcomes. With an R-squared value of 0.580 and an R-value of 0.761, we can see that employee engagement explains 58% of the variation in employee performance. Relationship between two of them is significant having with an F-value of 241.442 and a p-value of 0.000, it can be inferred that the model has statistical significance. The importance of fostering employee engagement to improve performance outcomes is underscored by the fact that this is the case, which means that employee engagement significantly contributes to improved employee performance.

**Table 8** Regression summary coefficients

Coefficients[a]

| Model | | Unstandardized coefficients | | Standardized coefficients | T | Sig. | Collinearity statistics | |
|---|---|---|---|---|---|---|---|---|
| | | B | Std. error | Beta | | | Tolerance | VIF |
| 1 | (Constant) | 2.222 | 0.839 | | 2.648 | 0.009 | | |
| | Appropriate teams | 0.144 | 0.072 | 0.145 | 1.995 | 0.048 | 0.320 | 3.128 |
| | Organizational culture | 0.248 | 0.076 | 0.252 | 3.273 | 0.001 | 0.284 | 3.523 |
| | Training and development | 0.335 | 0.070 | 0.360 | 4.762 | 0.000 | 0.294 | 3.399 |
| | Leadership | 0.155 | 0.072 | 0.165 | 2.142 | 0.034 | 0.284 | 3.520 |

a. Dependent Variable: Employee Engagement
*Source* Author's calculation

**Table 9** Regression summary with R values

Model summary

| Model | R | R square | Adjusted R square | Std. error of the estimate | Change statistics | | | | |
|---|---|---|---|---|---|---|---|---|---|
| | | | | | R square change | F change | df1 | df2 | Sig. F change |
| 1 | 0.761[a] | 0.580 | 0.577 | 2.43913 | 0.580 | 241.442 | 1 | 175 | 0.000 |

a. Predictors: (Constant), Employee Engagement
*Source* Author's calculation

Table 10 showing employee engagement significantly improves performance, according to the regression analysis, having high positive correlation is indicated by the standardized coefficient (Beta) of 0.761. The effect is confirmed to be statistically significant with a t-value of 15.538 and a p-value of 0.000. The model's robustness is further validated by the collinearity statistics, which show no multicollinearity difficulties with a tolerance of 1.000 and a VIF of 1.000.

**H5: The effect of moderating influence of AI on the predecessors of employee engagement**.

Table 11 indicating the regression analysis using Process Macro shows that AI has a significant moderating effect on the factors influencing employee engagement. The model demonstrates a strong relationship with an R value of 0.8626, and the R-Square value of 0.7440 indicates that 74.4% of the variation in employee engagement can be explained by the interaction of AI with its predecessors. The F-value of 167.6171 and a p-value of 0.000 confirm the model's statistical significance. This highlights AI's important role in enhancing the influence of factors that contribute to employee engagement.

Table 12 showing the regression analysis using Process Macro states that AI amplifies the impact of leadership, training and development, corporate culture, and suitable teams on employee engagement. The coefficient for appropriate teams is 0.2186, with a p-value of 0.0007, indicating a meaningful impact. Organizational culture has a coefficient of 0.3150, and training and development has the highest coefficient at 0.3731, both with p-values of 0.0000. Leadership also has a notable effect, with a coefficient of 0.2545 and a p-value of 0.0000. These results highlight

**Table 10** Regression summary coefficients

Coefficients[a]

| Model | | Unstandardized coefficients | | Standardized coefficients | t | Sig. | Collinearity statistics | |
|---|---|---|---|---|---|---|---|---|
| | | B | Std. error | Beta | | | Tolerance | VIF |
| 1 | (Constant) | 3.459 | 1.006 | | 3.439 | 0.001 | | |
| | Employee Engagement | 0.811 | 0.052 | 0.761 | 15.538 | 0.000 | 1.000 | 1.000 |

a. Dependent Variable: Employee performance
*Source* Author's calculation

**Table 11** Model summary in process macro analysis

Model summary

| R | R-Sq | MSE | F | df1 | df2 | P |
|---|---|---|---|---|---|---|
| 0.8626 | 0.7440 | 3.2323 | 167.6171 | 3.0000 | 173.0000 | 0.0000 |

*Source* Author's calculation

AI's role in enhancing the effectiveness of these drivers of employee engagement (Table 13).

The correlation test results show strong effective relationships between all the constructs examined, with significance at the 0.01 level. Appropriate teams have a significant correlation with organizational culture (0.755**), training and development (0.756**), leadership (0.774**), AI (0.778**), employee engagement (0.734**), and employee performance (0.747**). Organizational culture is strongly interconnected with training and development (0.789**), leadership (0.785**), AI (0.827**), employee engagement (0.774**), and as well as employee performance (0.772**). Training and development shows strong correlations with leadership (0.772**), AI (0.792**), employee engagement (0.795**), and employee performance (0.765**). Leadership is associated with AI (0.771**), employee engagement (0.752**), and employee performance (0.726**). AI shows strong relationships with employee engagement (0.765**) and employee performance (0.779**). Employee engagement is also highly correlated with employee performance (0.761**), indicating the strong interdependence of these factors (Fig. 2).

The SEM diagram demonstrates the interrelationships between key organizational factors like Appropriate Teams (AT), Organizational Culture (OC), Training and Development (TD), and Leadership (LETO), and their impact on both Artificial Intelligence (AIET) and Employee Engagement (EE). These elements are shown to shape the use of AI and boost employee engagement. AI exerts a direct influence on both employee engagement (with a positive path coefficient of 0.23) and employee performance (with a negative path coefficient of –2.17). Furthermore, employee engagement significantly enhances employee performance, as indicated by a strong path coefficient of 0.82. By enhance these variable can further increase AI adoption and engagement levels, as the diagram also shows the mutual influence of LETO, TD, AT and OC. An explanation for the unexplained variation in engagement, performance, and AI can be found in the error terms (e1, e2, e3). In sum, the model demonstrates how engagement, organizational policies, and AI all play critical roles in boosting worker productivity (Fig. 3).

A model known as Confirmatory Factor Analysis (CFA) is used to explain the correlations that exist between a number of different constructs and the variables that correspond to them being observed. There are five indicators that are used to measure

**Table 12** Process macro analysis

| Model | Coeff | se | t | P | LLCI | ULCI |
|---|---|---|---|---|---|---|
| Appropriate teams | 0.2186 | 0.630 | 3.4701 | 0.0007 | 0.0943 | 0.3429 |
| Organizational culture | 0.3150 | 0.0719 | 4.3807 | 0.0000 | 0.1731 | 0.4570 |
| Training and development | 0.3731 | 0.0618 | 6.0360 | 0.0000 | 0.2511 | 0.4951 |
| Leadership | 0.2545 | 0.0608 | 4.1878 | 0.0000 | 0.1346 | 0.3745 |

*Source* Author's calculation

**Table 13** Pearson correlation coeffcient

| Construct | Appropriate teams | Organizational culture | Training and development | Leadership | Artificial intelligence | Employee engagement | Employee performance |
|---|---|---|---|---|---|---|---|
| Appropriate teams | 1 | – | – | – | – | – | – |
| Organizational culture | 0.755** | 1 | – | – | – | – | – |
| Training and development | 0.756** | 0.789** | 1 | – | – | – | – |
| Leadership | 0.774** | 0.785** | 0.772** | 1 | – | – | – |
| Artificial intelligence | 0.778** | 0.827** | 0.792** | 0.771** | 1 | – | – |
| Employee engagement | 0.734** | 0.774** | 0.795** | 0.752** | 0.765** | 1 | – |
| Employee performance | 0.747** | 0.772** | 0.765** | 0.726** | 0.779** | 0.761** | 1 |

**. Correlation is significant at the 0.01 level (2-tailed)
*Source* Author's calculation

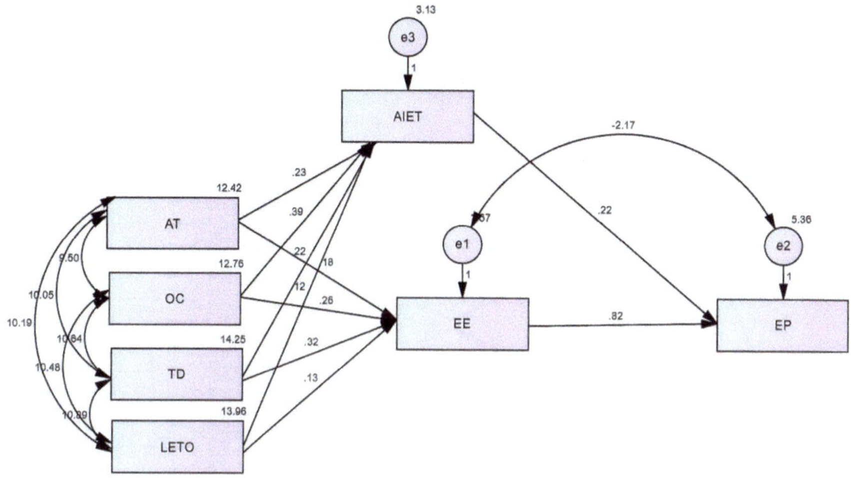


**Fig. 2** SEM model. *Source* Author's work

Appropriate Teams (Ap_te). These indicators are as follows: Organizational Culture (Or_Cu) by OC1-OC5, Training and Development (Tr_De) by TD1-TD5, Leadership (Le_Ip) by LE1-LE5, Artificial Intelligence (Ar_In) by AIE1-AIE5, Employee Engagement (Em_En) by EE1-EE5, and Employee Performance (Em_pe) by EP1-EP5. An emphasis is placed on the measuring structure of these organizational aspects by establishing a connection between each concept and the observed variables that correspond to it (Table 14).

## 7 Discussion

This study offers how AI and organizational factors contribute to the development of sustainable businesses. With participants from diverse industries, income levels, and professional backgrounds, the findings provide a well-rounded understanding of AI's role in business growth. The reliability test results show that training and development have the highest consistency, emphasizing the importance of employee development in maximizing AI's potential. There are notable gender disparities in employee engagement, as shown by the one-sample t-test. This suggests that measures tailored to individual genders might potentially increase engagement levels. Furthermore, the ANOVA underscores the impact of age on employee performance, underscoring the need to implement age-appropriate strategies to enhance performance. The outcomes of the regression test analysis suggest that cohesive teams, organizational culture, and leadership serve as key components for promoting employee engagement. There is a strong connection between AI, employee performance and employee engagement, which further emphasizes the necessity of an integrated approach. Based on

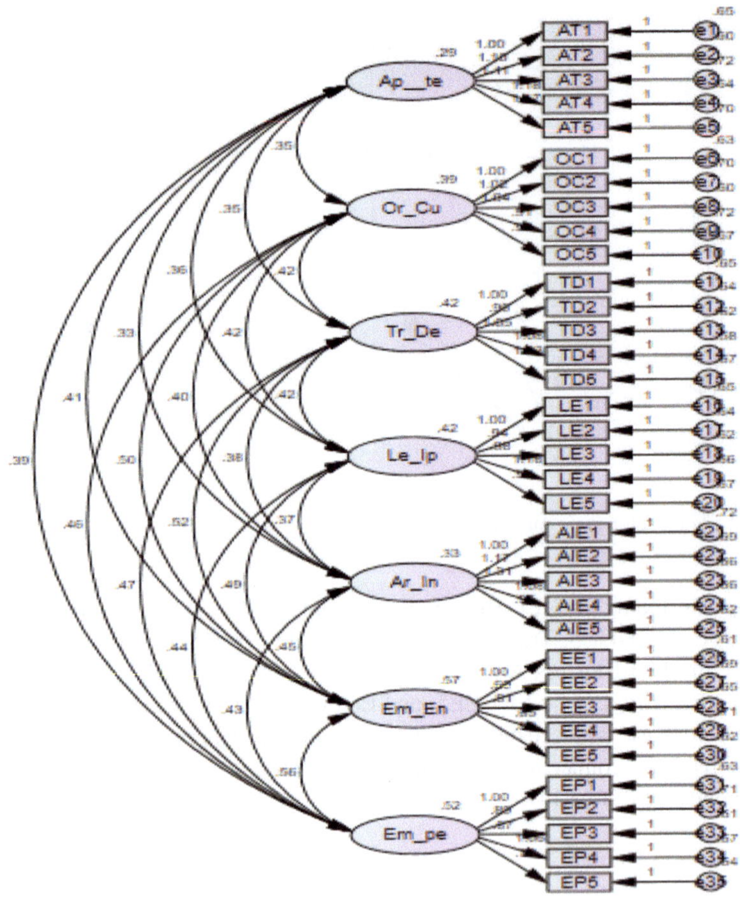

Fig. 3 CFA model. *Source* Author's work

**Table 14** Test summary

| Hypothesis | Statements | Tests | Accepted/ rejected |
|---|---|---|---|
| H1 | Gender influences employee engagement | T fest | Accepted |
| H2 | Age impacts employee performance | ANOVA | Accepted |
| H3 | Organizational factors influence employee engagement | Regression analysis | Accepted |
| H4 | Employee engagement impacts performance | Regression analysis | Accepted |
| H5 | AI moderates the influence of organizational factors on employee engagement | Process macro | Accepted |

*Source* Author's calculation

these findings, it appears that AI, when combined with supportive leadership, a robust organizational culture, and opportunities for development, has the potential to significantly boost employee engagement and drive sustainable growth in organizations.

## 8 Conclusion

According to this study, AI plays a significant part in fostering sustainable business growth by increasing the level of engagement and performance of employees. Additionally, not only in those organizational elements that are driven by employee engagement, but they have also improved employee performance, which ultimately adds to the success of the firm over the course of the long run. The fact that employee engagement acts as a mediator between organizational practices and performance is further evidence of the significance of cultivating an atmosphere that encourages engagement. Additionally, the moderating role of AI amplifies the effectiveness of these factors, demonstrating its potential as a catalyst for improving leadership, training, and organizational culture. The study accentuates the significance of considering demographic components such as gender and age when planning to increase employee engagement and performance. The effectiveness of leadership, teamwork, and company culture can be enhanced with the help of AI, which mediates and strengthens these linkages. Companies that invest in their employees education and careers, promote strong leadership, and use AI to improve their operations have a better chance of achieving long-term success and market dominance. With this all-encompassing plan, organizations have a better chance of long-term success and of adapting to new technology. Therefore, future studies should investigate the impact of employee performance interconnected with employee engagement as well as organizational factors in the direction of professional experiences and annual income of the individuals, which will provide much more comprehending knowledge and help an employee to achieve organizational long-term objectives.

## References


1. Avolio, B.J., Bass, B.M., Jung, D.I.: Re-examining the components of transformational and transactional leadership using the multifactor leadership questionnaire. J. Occup. Organ. Psychol. **72**(4), 441–462 (2004). https://doi.org/10.1348/096317999166789
2. Öncer, A.Z. (2013). Investigation of the effects of transactional and transformational leadership on entrepreneurial orientation. Int. J. Bus. Soc. Res., **3**(4), 153–166. https://doi.org/10.18533/IJBSR.V3I4.40
3. Banihani, M., Lewis, P., Syed, J.: Is work engagement gendered? Gend. Work. Organ. **20**(4), 528–542 (2013). https://doi.org/10.1111/j.1468-0432.2012.00591.x
4. Bass, B.M.: Stogdill's handbook of leadership: A survey of theory and research (Rev. ed.). Free Press, (1981)
5. Bessen, J.: AI and jobs: The role of demand. Econ. Innov. New Technol. **28**(2), 97–108 (2019). https://doi.org/10.1080/10438599.2018.1526352
6. Brynjolfsson, E., McAfee, A.: The second machine age: Work, progress, and prosperity in a time of brilliant technologies. W. W. Norton & Company, (2014)
7. Bughin, J., Chui, M., Manyika, J.: Notes from the AI frontier: Insights from hundreds of use cases. McKinsey Global Institute, (2019)
8. Bughin, J., Hazan, E., Ramaswamy, S., Chui, M., Allas, T., Dahlström, P., Henke, N., Trench, M.: Artificial intelligence: The next digital frontier? McKinsey Global Institute, (2017)
9. Bughin, J., Seong, J., Manyika, J., Chui, M., Joshi, R.: Skill shift: Automation and the future of the workforce. McKinsey Global Institute, (2018)
10. Chen, L., Patel, R.: Integrating artificial intelligence into sustainable business practices: Ethical and organizational considerations. Bus. Transform. J. **15**(2), 45–58 (2021). https://doi.org/10.1234/btj.2021.0215
11. Chui, M., Manyika, J., Miremadi, M.: Artificial intelligence in business: The future of work and innovation. McKinsey & Company, (2019)
12. Chui, M., Manyika, J., Miremadi, M.: The next normal in AI: How businesses are accelerating AI adoption. McKinsey Global Institute, (2020)
13. Davenport, T.H., Kirby, J.: Only humans need apply: Winners and losers in the age of smart machines. Harper Business, (2016)
14. Davenport, T.H., Ronanki, R.: Artificial intelligence for the real world. Harv. Bus. Rev. **96**(1), 108–116 (2018)
15. De Henríquez, J.C.T., Valerio, C.M.D.: Satisfaction with the work done in University employees: A study from a developing country. Administrative Sciences **13**(10), 221 (2023). https://doi.org/10.3390/admsci13100221
16. Gable, G.: AI in business: Overcoming the challenges of integration and sustainability. J. Bus. Innov. **34**(3), 44–59 (2021). https://doi.org/10.1016/j.jbi.2021.03.004
17. Hackman, J.R.: Leading teams: Setting the stage for great performances. Harvard Business School Press, (2002)
18. Hartman, R.O., Barber, L.K.: Gender differences in work engagement: The role of occupational self-efficacy and career aspirations. J. Vocat. Behav. **121**, 103470 (2020). https://doi.org/10.1016/j.jvb.2020.103470
19. Hoch, J.E., Pearce, C.L., Welzel, L.: Team performance and age heterogeneity: The benefits of age diversity in dynamic tasks. J. Appl. Psychol. **95**(6), 1076–1085 (2010). https://doi.org/10.1037/a0021180
20. Huang, M.H., Rust, R.T.: Artificial intelligence in service. J. Serv. Res. **21**(2), 155–172 (2018). https://doi.org/10.1177/1094670517752459
21. Moroki, I.: Evaluation of the effectiveness of the online teaching and learning process at the University: Based on Kirkpatrick's Model. https://doi.org/10.4108/EAI.11-12-2019.2302132
22. Kadir, M.H., Amalia, L.: The effect of job motivation, compensation, organizational culture towards job satisfaction and employee performance of the ministry of man power. Int. J. Bus. Manag. Inven. **6**(5), 73–80 (2017)

23. Kirkpatrick, D.L.: Techniques for evaluating training programs. J. Am. Soc. Train. Dir. **13**(3), 21–26 (1959)
24. Kotter, J.P.: Leading change. Harvard Business Review Press, (1996)
25. Kshetri, N.: 1 Artificial intelligence and the future of business: A practical guide for navigating change. Palgrave Macmillan, (2019)
26. Lee, J., Roberts, M.: AI for the long term: Balancing innovation and sustainability in business growth. J. Technol. Innov. **12**(4), 101–113 (2020). https://doi.org/10.1234/jti.2020.0408
27. Mathieu, J., Maynard, M.T., Rapp, T., Gilson, L.: Team effectiveness 1997–2007: A review of recent advancements and a glimpse into the future. J. Manag. **34**(3), 410–476 (2008). https://doi.org/10.1177/0149206308316061
28. McAfee, A., Brynjolfsson, E.: Machine, platform, crowd: Harnessing our digital future. W. W. Norton & Company, (2017)
29. Rožman, M., Oreški, D., Tominc, P.: Integrating artificial intelligence into a talent management model to increase the work engagement and performance of enterprises. Front. Psycho, **13**. https://doi.org/10.3389/fpsyg.2022.1014434
30. Rožman, M., Oreški, D., Tominc, P.: Artificial-intelligence-supported reduction of employees' workload to increase the company's performance in today's VUCA environment. Sustainability **15**(6), 5019 (2023). https://doi.org/10.3390/su15065019
31. Schein, E.H.: Organizational culture and leadership. Jossey-Bass (1985)
32. Schein, E.H.: Organizational culture and leadership, 4th ed. Jossey-Bass, (2010)
33. Srivastava, S.: Concepts of culture and organisational analysis: A perspective. Vis. J. Bus. Perspect., **4**(1_suppl), 32–42. https://doi.org/10.1177/09722629000040s105
34. Westerman, G., Bonnet, D., McAfee, A.: Leading digital: Turning technology into business transformation. Harvard Business Review Press, (2014)
35. Wilson, H.J., Daugherty, P.R.: Human + Machine: Reimagining work in the age of AI. Harvard Business Review Press, (2018)

23. Kirkpatrick, D.L.: Techniques for evaluating training programs. J. Am. Soc. Train. Dir. 13(3), 21–26 (1959)
24. Kotter, J.P.: Leading change. Harvard Business Review Press (1996)
25. [illegible]

# Exploring the Impact of Consumer Behaviour as a Mediator on the Payment Preferences Among Distinct Occupational Segments in West Bengal, India

**Camellia C. Burman and Swapan Sarkar**

**Abstract** A significant segment of the Indian populace faces difficulties in keeping abreast of technological advancements due to inadequate infrastructure, limited user awareness, concerns regarding cyber security, and issues related to technical literacy. The present study focuses on the intention of the people of West Bengal, India towards the acceptance of the digital modes of payment system. In order to understand the impact, 386 respondents have been surveyed from the state of West Bengal through a well-structured questionnaire thereby obtaining demographic details and their transaction behavioural information through a 34-item pool using Likert Scale. The study aims to examine the impact of five key factors—Consumer Behaviour, Consumer Awareness, Services, Infrastructure, and Promotion—on the adoption of cashless transactions within specific occupational domains. The research delves into the mediating roles of these factors in influencing individuals' payment preferences across diverse professional sectors. Remarkably, the study highlights that Consumer Behaviour emerges as a pivotal mediator in the relationship between causal and outcome variables.

**Keywords** Cashless economy · Payment preference · Consumer behaviour · Mediation analysis · Digital payment system

C. C. Burman (✉)
Department of Commerce, Maulana Azad College, Kolkata, West Bengal, India
e-mail: camelliaburman91@gmail.com

S. Sarkar
Department of Commerce, University of Calcutta, Kolkata, West Bengal, India



N. D. K. Nguyen et al. (eds.), *Digital Transformation and Sustainability: Methods and Applications*, Information Systems Engineering and Management 47,
https://doi.org/10.1007/978-3-031-91278-8_12

## 1 Introduction

In ancient times, barter systems were prevalent where goods were traded for goods. However, due to limitations like the absence of fixed exchange ratios and varying quality of goods, the concept of monetary exchange emerged, making money the medium of trade. This transition led to the evolution of money through different forms such as metallic currency, paper currency, plastic currency (like debit and credit cards), and the latest addition, digital currency. Barter systems, common not only in rural areas of underdeveloped countries but also in international trade due to economic policies and emerging international relations, have significantly influenced marketing strategies. Cashless digital marketing has transcended the boundaries of traditional barter systems, becoming a key tool in eliminating trade barriers and enhancing global business interactions, thereby reducing five categories of costs, namely, search, reproduction, transportation, tracking, and verification [13]. According to [15], marketing integrates the global market under the World Trade Organization (WTO), connecting developed countries with the rest of the world. Marketing has played a crucial role in the progress of underdeveloped, developing, and emerging economies, leading to the formation of trade groups such as the Central European Free Trade Association, the European Union (EU), the G-7, and others. Furthermore, the study conducted by [23], highlighted a positive and significant relationship between real GDP and cashless payment in the context of G-7 countries.

The history of barter systems dates back to around 6000 BC, initially introduced by Mesopotamian tribes and later adopted by the Phoenicians. [18], highlighted that Ancient India utilized a barter system as a primary mode of payment among individuals and communities, exchanging goods of equal value which even persisted until the time of the Indus Valley Civilization. Throughout history, the evolution of money signifies a shift in acceptable forms of exchange. It serves as a testament to society's ability to adapt and innovate. This section delves deep into the progression of money and offers a detailed overview of India's current Digital Payment System. From the barter system to modern digital currencies, money has consistently evolved to meet the changing demands of the global landscape. In ancient times, from the sixth century BCE to the sixth century CE, civilizations used various forms of exchange. Initially, bartering was common, with direct exchange of goods and services. However, as trade expanded, more efficient methods were sought. The earliest Indian coins were punch-marked, with symbols punched onto metal pieces. Different Mahajanapadas (ancient kingdoms) issued their distinct coins, showcasing regional diversity. Throughout this era, the evolution of exchange mediums mirrored the economic and social dynamics of each society. Advancing from barter to commodity money and then to standardized metal coins marked significant strides in monetary system development.

In a landmark decision, Indian Prime Minister Narendra Modi addressed the nation via live telecast at approximately 8:15 PM Indian Standard Time (IST) to announce the ban on the circulation of Rs. 500 and Rs. 1000 notes, effective from midnight on November 9, 2016. All other currency notes and coins would remain valid. Statistics indicated that a major portion of the transactions in India were conducted using the demonetized notes, causing significant disruption to the Indian economy as the Indian economy primarily relies on cash transactions. The connection between behavioural intention to utilize digital payment systems and the actual usage of these systems is influenced by the preference for cash payments [29]. Following the demonetization, citizens experienced serious cash shortages, as the new Rs. 500 and Rs. 2000 notes were not quickly produced to meet demand. This led to long queues at ATMs and compelled many individuals to shift towards online transactions, resulting in a notable increase in digital payments post-demonetization.

Prior to demonetization, approximately a major portion of transactions on the available e-commerce platforms like Amazon and Flipkart were made in cash. However, the landscape changed drastically after demonetization, with these online retailers experiencing a massive rise in unfulfilled Cash on Delivery (COD) orders, highlighting a concerning lack of digital readiness among the Indian populace. Cnaan et al. [10] pointed out that many people in India, particularly those in rural areas, still lag in adopting technology-based transactions due to various obstacles. While the younger, urban generation tends to be more adept and comfortable with digital transactions, many rural residents and senior citizens remain skeptical about the safety and usability of online platforms. Additionally, a significant portion of this demographic has little to no knowledge regarding how to use these services.

The payment system in the country has undergone significant evolution, particularly following the initiation of the Digital India campaign. This initiative has led to a substantial surge in the adoption of digital payment methods, notably the Unified Payments Interface (UPI), which was introduced in 2016. The uptick in UPI usage was notably catalyzed by the Demonetization of high-denomination banknotes in 2016 and the ensuing Covid-19-induced lockdown. In today's world, customers seek a straightforward, convenient, and efficient method to conduct transactions while keeping operational costs low. A key governmental objective is to integrate the entire nation into the banking framework. Recognizing the importance of strong e-payment instruments and channels is crucial for the success, sustainability and growth of cashless payment system network. The RBI and the Government of India are looking for a safe form of cashless economy for which different nations are in the process of developing electronic form of fiat money known as the Central Bank Digital Currency (CBDC) to stabilize the unregulated markets. With the users wanting fast and convenient payment methods the CBDC is subsequently garnering attention which facilitates the conversion of physical money into virtual money [4]. India has been entering rapidly in the cashless payment system network post demonetization and Covid-19 pandemic but much before that "another new electronic cash system," was conceived, being "a shared electronic money framework" [9] in the form of cryptocurrencies among which the most popular being the Bitcoin, which were designed to bypass the sovereign financial systems. Looking ahead, the CBDC emerges as a potential advancement in the digital payment system, with the Reserve Bank of India (RBI) currently exploring avenues to fulfill the economy's payment operation requisites [28]. To enhance the cashless economy in the country, the RBI has granted licenses to small finance banks and payments banks. This initiative aims to promote financial inclusion and introduce innovative banking solutions to benefit the users.

Additionally, the Covid-19 pandemic compelled many customers to alter their shopping behaviours. It also prompted a reassessment of the perceived risks associated with purchasing decisions. With stores closed for varying durations during the pandemic, many customers, particularly during periods of heightened illness rates, opted to steer clear of traditional shopping. During the Covid-19 outbreak, infected individuals could potentially transmit the virus via banknotes or coins through touch and respiratory droplets, raising concerns about physical currency as a potential vector for the virus. While there are no confirmed reports of virus transmission through cash, studies on virus survival on solid surfaces lend some support to this concern [27]. A conceptual model established by Amankwa et al. [1], by combining constructs from the health belief model and theory of planned behaviour to determine the usage intentions of the e-wallet payment options for business transactions among the Ghanaian workers in the formal and informal sectors which revealed that significant differences were observed in the usage. Consequently, a significant uptick in online shopping was noted in 2020 and 2021 globally. Since 2017, e-commerce revenues have surged dramatically, with current estimates indicating that a majority of individuals worldwide now utilize e-commerce platforms. China leads the market, followed by the United States, Japan, the United Kingdom, and Germany. The pandemic has hastened this transition. Mobility restrictions aimed at curbing the virus's spread resulted in a notable increase in online demand for various goods and services [30]. The corona virus has impacted global e-commerce, fundamentally altering the business landscape. According to research, 52% of consumers opted to avoid physical stores and crowded spaces, while 36% expressed a desire to stay away from brick-and-mortar shops until they received the corona virus vaccine [6]. E-commerce has experienced unprecedented growth, accompanied by a rapid expansion of cashless payment systems. The significance of these systems for businesses is manifold, including the advancement of card payment systems that have reduced the need for cash circulation, resulting in noteworthy economic benefits [12]. The growth of the cashless payments market is driven not only by the surge in online transactions but also by various factors that encourage customers to use card payments, NFC technology and even make payments on-site.

Currently, the payment system is shifting from a cash-centric model to a digital framework, a transition that presents challenges but is witnessing a significant increase in the adoption of electronic payment methods. Several factors contribute to the adoption and progression towards digital payment systems in India. The pivotal demonetization move in 2016 catalyzed a noteworthy shift towards digital payment methods among Indians, a trend further accelerated by the subsequent Covid-19 lockdown. The demonetization initiative spurred Indians, especially in rural areas, to embrace digital transactions. The rising acceptance of cashless payment methods in India has delivered numerous advantages to consumers. Nevertheless, this transition has also introduced certain challenges. The evolving information technology landscape has increased the risks associated with money laundering and cyber-crime. Moreover, reliance on technological systems exposes vulnerabilities in cashless payment infrastructures. In order to develop and validate a cashless transaction adoption model, the study conducted by [26], revealed that performance expectancy followed by perceptions of the economy's security and economic offence reduction, social influence, perceived trustworthiness, the expected level of effort and innovativeness form the strongest predictor of users' intention to use cashless payment systems while perceived risk and cost have been identified as the factors that adversely affect the users' behaviour to use cashless forms of payment. Despite these hurdles, the wave of digitization has significantly transformed transaction mechanisms in India. However, a considerable segment of the Indian population encounters obstacles in keeping pace with technological advancements due to insufficient infrastructure, limited user awareness, concerns about cyber-security, and challenges related to technical literacy.

## 2 Literature Review

The study conducted by Namahoot and Jantasri [21], in Thailand explores how the Unified Theory of Acceptance and Use of Technology (UTAUT) dimensions impact Behavioral Intentions (BIs) for adopting cashless payment systems. It integrates Theory of Reasoned Action (TRA), Theory of Planned Behavior (TPB), and Technology Acceptance Model (TAM), investigating indirect relationships through perceived risk and trust. Results show UTAUT, perceived risk, and trust significantly influence BIs. Effort expectancy can reduce risk and boost trust in cashless systems, potentially increasing customer engagement and understanding of customer intentions.

To investigate the factors influencing Cashless Payment Systems adoption in Chennai post demonetization, the research by Bhuvana and Vasantha [7], uses the Technology Acceptance Model to analyze variables like Perceived Usefulness and Perceived Ease of Use. The results reveal that demonetization acts as a mediator between attitudes and intentions to use cashless systems. Structural Equation Modeling confirms demonetization's partial mediation effect on behavioral intentions. Security concerns and user-friendly interfaces are highlighted for improved online transactions.

The paper by Nagdev et al. [20], aims to examine the immediate effects of demonetisation on the Indian economy, analyze the obstacles hindering the transition to a cashless economy, and explore the impact of demonetisation on moving towards a cashless system. A scale was developed to gauge the effectiveness and consequences of demonetisation on the Indian economy, specifically targeting individuals earning income from any source. The study involved Exploratory Factor Analysis (EFA) and Confirmatory Factor Analysis (CFA) on participants, followed by Structural Equation Modeling and path analysis. The findings indicated that demonetisation primarily aimed to pave the way for a cashless India. It was also noted that IT-enabled services and payment gateways saw significant growth during this period, while traditional banking suffered negative effects from demonetisation. Social impacts included a decrease in black money, increased income disclosure, more transparent business transactions, reduced cash hoarding, and other positive changes. Additionally, there was an indirect influence on the cashless economy due to economic concerns arising from demonetisation.

With the rise of digital technologies, customers have increasingly turned to various online platforms to meet their shopping needs. Following demonetisation, these digital shopping platforms have gained traction, enabling m-shoppers to shop conveniently from their desired locations. However, despite the numerous advantages of m-shopping, consumers still perceive certain risks when using their mobile devices for purchases. Thus, the study by Bhat and Gupta [5], investigates how various behavioural factors such as perceived risk, trust, commitment and purchase intention influence m-shopping. Data collected through standardized questionnaires were analyzed using various statistical methods. The results indicate that perceived risk negatively affects m-shopping behaviour. Additionally, the moderated mediation analysis shows that demonetisation has a positive influence on overall m-shopping behaviour. The study adds to the existing m-shopping literature and advises academics, policymakers and retailers to assess the extent of the risks perceived by customers in mobile shopping.

The growth of E-wallet services in Indonesia is promising, with 48 providers noted by the Bank of Indonesia. Understanding the intention of Indonesian adults to use e-wallets is crucial for grasping adoption factors, user preferences, effective marketing strategies, financial inclusion promotion, and guiding policy decisions. The study by Nawi et al. [22], addressed this gap by examining mediating factors affecting Indonesian users' e-wallet adoption intention, revealing how facilitating conditions, trust, usefulness, risk perception, compatibility, and perception of new technology positively influenced users' intention. However, social influence and ease of use did not have a significant impact on e-wallet adoption intention among Indonesian participants. The study's implications emphasize the importance of balancing ease of use and security in e-wallet systems to address safety concerns and unauthorized access, thus fostering user confidence. Future research should explore additional factors affecting e-wallet adoption and their long-term effects on user intention, providing valuable insights for e-wallet providers, policymakers, and researchers to enhance electronic wallet adoption in Indonesia.

Shoppers tend to spend more on unhealthy food when using cashless payments. The reason behind this behaviour could be the negative arousal triggered by non-cash transactions affecting their consumption choices. When paying with cash, this negative arousal is heightened, leading to increased attention towards decision risks. On the other hand, cashless payments lessen this negative arousal, subsequently reducing attention towards decision risks. This reduced focus on decision risks associated with cashless payments can result in riskier consumption behaviours. Two experiments were conducted by Park et al. [24], to test this hypothesis in the context of indulgent food items with health implications (vice food). The first study illustrated that cashless payments decrease negative arousal, as measured by skin conductance, and boost intentions to purchase vice food. The second study revealed that the impact of cashless payments is more pronounced for shoppers who are more health risk-sensitive.

The COVID-19 pandemic and associated measures, such as physical distancing, have led to a significant shift in consumer payment behaviours towards cashless options. As physical distancing is expected to continue for the foreseeable future and may be relevant in potential future pandemics, the study by Wang et al. [31], seeks to explore how consumers' perceived value of cashless payments influences their intention to use them within this context. It also examines the mediating role of psychological safety and the moderating effect of trust propensity. Data for this study was collected through a survey of 690 consumers in Vietnam, an emerging market in Asia. Various statistical methods, including structural equation modeling, were employed for analysis. The results indicate that the perceived value of cashless payments has a positive impact on usage intention, which is mediated by feelings of psychological safety. Additionally, trust propensity positively moderates the relationship between perceived value and psychological safety. The findings from this research offer valuable insights for retailers and stakeholders looking to implement and promote cashless payment systems, particularly in scenarios involving physical distancing. This study is one of the first to clarify the connections between consumer perceptions, psychological safety, and the intention to use cashless payments in the context of physical distancing, and its implications may extend to future pandemics.

With the rise of digitalization, mobile financial services have gained immense popularity, providing convenient, accessible, and secure options for managing finances. This trend is set to persist as more individuals globally access mobile technology and grow accustomed to digital financial services. Consequently, the study by Islam et al. [16], investigates the impact of consumption values, namely, functional, social and emotional on the acceptance of a cashless society, including the mediating role of cashless readiness. Employing a convenience sampling method, the researchers had gathered a total of 200 responses through a quantitative research approach. The findings indicate that functional, social and emotional values significantly affect cashless readiness, which in turn influences the adoption of a cashless society. Furthermore, the study highlights that cashless readiness mediates the relationship between consumption values and the acceptance of a cashless society. These insights can aid mobile financial service providers, banking institutions and government organizations in developing strategies to enhance the adoption of digital payment systems.

## 3 Objectives of the Study

The objectives of the current study can be enumerated as follows:

- To analyze the demographic distribution of the respondents under the study.
- To check whether the constructs Consumer Behaviour, Consumer Awareness, Service, Infrastructure, and Promotion mediate the role between consumers' choice of payment (transaction) and their type of work.

## 4 Research Methodology

The study's research methods elucidate the research design, variables along with their operational definitions, type of research, methodologies employed, sampling technique selection, target population for sampling, sample size determination, sampling procedure, observational design, questionnaire development, consumer questionnaire specifics, measurement items and scales for consumers, scoring and validation of tools, data collection methods, preparation and entry procedures, data coding, and the statistical techniques used for data analysis.

The study focuses on primary demand-side stakeholders of cashless payment systems, specifically consumers. The study is limited to the city of Kolkata and the South 24 Parganas district. These districts were selected due to the researcher's convenience and the extensive representation of diverse cultures and attitudes within the country. Hence, consumers utilizing cashless payment methods within these districts constitute the study's sample population.

The methodology employed involved referencing the Krejcie and Morgan table from the 1970s to determine the sample size for the consumer study. This table provides guidelines for sample sizes applicable to both finite and infinite populations [19]. As per Table 1, for populations exceeding one hundred thousand or considered infinite, the recommended sample size is 384. Given that the consumer study pertains to individuals utilizing cashless payment systems, which constitutes an infinite population, the sample size of 384 was selected based on the Krejcie and Morgan table from 1970.

**Table 1** Calculations showing the sample sizes for known and unknown populations

| Population size | Sample size | Population size | Sample size | Population size | Sample size |
|---|---|---|---|---|---|
| 10 | 10 | 220 | 140 | 1200 | 291 |
| 15 | 14 | 230 | 144 | 1300 | 297 |
| 20 | 19 | 240 | 148 | 1400 | 302 |
| 25 | 24 | 250 | 152 | 1500 | 306 |
| 30 | 28 | 260 | 155 | 1600 | 310 |
| 35 | 32 | 270 | 159 | 1700 | 313 |
| 40 | 36 | 280 | 162 | 1800 | 317 |
| 45 | 40 | 290 | 165 | 1900 | 320 |
| 50 | 44 | 300 | 169 | 2000 | 322 |
| 55 | 48 | 320 | 175 | 2200 | 327 |
| 60 | 52 | 340 | 181 | 2400 | 331 |
| 65 | 56 | 360 | 186 | 2600 | 335 |
| 70 | 59 | 380 | 191 | 2800 | 338 |
| 75 | 63 | 400 | 196 | 3000 | 341 |
| 80 | 66 | 420 | 201 | 3500 | 346 |
| 85 | 70 | 440 | 205 | 4000 | 351 |
| 90 | 73 | 460 | 210 | 4500 | 354 |
| 95 | 76 | 480 | 214 | 5000 | 357 |
| 100 | 80 | 500 | 217 | 6000 | 361 |
| 110 | 86 | 550 | 226 | 7000 | 364 |
| 120 | 92 | 600 | 234 | 8000 | 367 |
| 130 | 97 | 650 | 242 | 9000 | 368 |
| 140 | 103 | 700 | 248 | 10,000 | 370 |
| 150 | 108 | 750 | 254 | 15,000 | 375 |
| 160 | 113 | 800 | 260 | 20,000 | 377 |
| 170 | 118 | 850 | 265 | 30,000 | 379 |
| 180 | 123 | 900 | 269 | 40,000 | 380 |
| 190 | 127 | 950 | 274 | 50,000 | 381 |
| 200 | 132 | 1000 | 278 | 75,000 | 382 |
| 210 | 136 | 1100 | 285 | 1,000,000 | 384 |

The above table has been constructed using the following formula for determining the sample size.

- Formula for determining the sample size:

$$s = X^2NP(1 - P) + d^2(N-1) + X^2P(1 - P)$$

where,

$s$ Required sample size.
$X^2$ Table value of chi-square for 1 degree of freedom at the desired confidence level (3.841).
$N$ Population size.
$P$ Population proportion (assumed to be 0.50 since this would provide the maximum sample size).
$d$ Degree of accuracy expressed as a proportion (0.05).

(Source: [19])

A well-structured questionnaire was employed to gauge public receptiveness towards embracing cashless payment methods, with responses collected from 386 participants in the South 24 Parganas and Kolkata districts of West Bengal. Descriptive statistics were utilized to analyze the demographic distribution of the respondents. For the purpose of this study, factor analysis was utilized to categorize variables into pertinent factors, aiming to ascertain the key factors influencing consumers' adoption of cashless payment methods. Data collection involved a 34-item questionnaire comprising simple statements exploring various dimensions influencing consumers' inclination towards digital payment methods. Respondents rated these statements on a 5-point Likert Scale, ranging from "Strongly Disagree" to "Strongly Agree." Following the completion of the questionnaire comprising a 34-item pool using a five-point Likert scale, the data was processed for Factor Analysis. The results were obtained through Principal Component Analysis (PCA) with Varimax rotation. The 34-item pool was then grouped into 5 factors that represent the elements influencing consumer behaviour in the use of cashless payment methods. Subsequently, a reliability test using Cronbach's Alpha was conducted to assess the internal consistency among the factors. To investigate the relationship between demographic variables (such as age, gender, education level, current work area, monthly income and residential area) and the identified factors (Consumer Behaviour, Consumer Awareness, Service, Infrastructure and Promotion).This study aims to explore the impact of these five key factors on the adoption of cashless payments within distinct professional domains, shedding light on their mediating roles in shaping individuals' payment preferences across various occupational sectors.

## 5 Analysis and Interpretation

The data for this study was collected through a meticulously designed questionnaire from 386 respondents. The demographic composition of the respondents is detailed in Table 1. Analysis of the respondents' profile reveals that 59.6% were male, with the remaining 40.4% being female. The largest proportion of respondents, 35.5%, fell within the 18–24 age bracket. Regarding educational attainment, 43.3% were graduates, followed by 31.1% who held post-graduate qualifications. In terms of occupation, the majority were students (24.4%), followed by self-employed individuals (17.9%), government employees (18.1%), corporate professionals (17.1%), other professionals (14.5%), and unemployed individuals (8%). Concerning monthly income, the highest percentage of respondents (21.2%) fell within the INR 20,000 to INR 50,000 bracket, while 17.6% earned below INR 20,000. Lastly, in relation to residential settings, 60.9% resided in urban areas, with the remaining 39.1% residing in rural locales (Table 2).

The study is designed to evaluate respondents' perceptions regarding different facets of cashless transactions to determine their readiness to embrace digital payment systems. To achieve this, a meticulously crafted questionnaire comprising a 34-item pool is utilized, employing a five-point Likert scale. Subsequently, a Factor Analysis technique is employed on the gathered responses. In this study, varimax rotation is utilized to streamline the interpretation of factors by reducing the number of variables with substantial loadings on each factor. The five factors identified are Consumer Behaviour, Consumer Awareness, Service, Infrastructure, and Promotion. These factors are pivotal in influencing a consumer's inclination towards adopting cashless modes of transactions.

Mediation analysis was undertaken to investigate the mediating influence of Consumer Behaviour, Consumer Awareness, Service, Infrastructure, and Promotion on the relationship between consumers' choice of payment (dependent variable) and their type of work (independent variable).

**Analysis of the mediating role of consumer behaviour on the relationship between consumers' choice of payment (transaction) and their type of work**

A variable is a mediator, if the independent variable first has an effect on the mediator variable which in turn influences the dependent variable. A mediator therefore is also known as an intervening or process variable. The relationship between the independent variable, mediating variable and the dependent variable can be understood as under.

**Total Effect**: The total effect is the impact of independent variable on dependent variable without the involvement of the mediator.

**Indirect Effect**: The indirect effect is the impact of independent variable on the dependent variable through the mediating variable.

**Direct Effect**: The direct effect is the impact of independent variable on dependent variable in the presence of the mediator.

**Table 2** Analysis of the gender of the respondents

| Constructs | Characteristics | Frequency | Percentage |
|---|---|---|---|
| Gender | Male | 230 | 59.6 |
| | Female | 156 | 40.4 |
| Age | Below 18 years | 44 | 11.4 |
| | 18–24 years | 137 | 35.5 |
| | 25–44 years | 109 | 28.2 |
| | 45–65 years | 52 | 13.5 |
| | Above 65 years | 44 | 11.4 |
| Educational qualification | Secondary (10th standard) | 20 | 5.2 |
| | Higher secondary (12th standard) | 24 | 6.2 |
| | Graduate | 167 | 43.3 |
| | Post graduate | 120 | 31.1 |
| Current work area | Student | 94 | 24.4 |
| | Professional service | 56 | 14.5 |
| | Corporate service | 66 | 17.1 |
| | Government service | 70 | 18.1 |
| | Self employed/entrepreneur | 69 | 17.9 |
| | Currently not employed | 31 | 8.0 |
| Monthly income | Below INR 20,000 | 68 | 17.6 |
| | INR 20,000 to INR 50,000 | 82 | 21.2 |
| | INR 50,001 to INR 100,000 | 57 | 14.8 |
| | INR 100,001 to INR 250,000 | 52 | 13.5 |
| | Above INR 250,000 | 34 | 8.8 |
| | I don't earn yet | 51 | 13.2 |
| | I prefer not to answer | 42 | 10.9 |
| Living area | Urban | 235 | 60.9 |
| | Rural | 151 | 39.1 |

There are three types of mediation, namely, complete mediation, partial mediation and no mediation.

**Complete mediation**: Complete mediation exists if the independent variable exerts its total influence through the mediating variable.

**Partial mediation**: Partial mediation occurs when the independent variable exerts some influence through the mediating variable. It also exerts some of its influence directly on the dependent variable and not via the mediating variable. Therefore, the influence on the dependent variable is exerted both through the independent variable and the mediating variable.

**No mediation**: There exists no mediation when the indirect effect becomes insignificant.

## 6 Mediation Analysis Procedure

The mediation analysis procedure has been laid down by [14], which is illustrated as under.

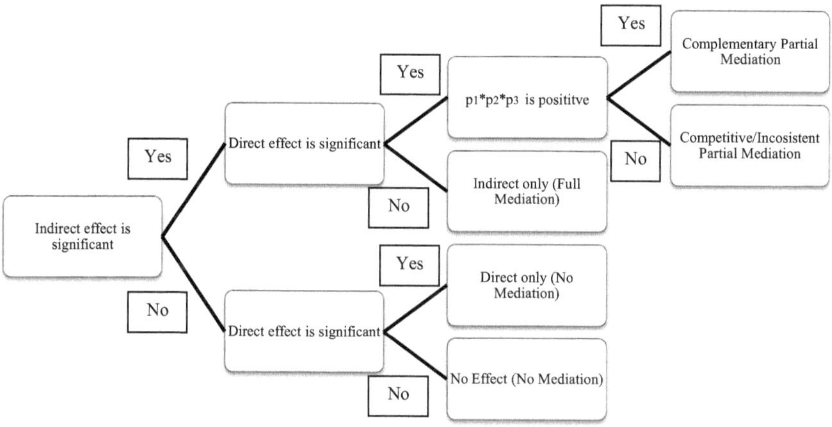


A simple mediation analysis was conducted using ordinary least square path analysis to test if Consumer Behaviour mediates the relationship between consumers' choice of payment (transaction) and their type of work (Table 3).

The results of the mediation analysis show the mediating role of Consumer Behaviour on the linkage between the choice of payment (transaction) and the type of work. The results reveal that the total effect of the choice of payment (transaction) on the type of work was significant (H1: $\beta = -0.03767$, $t = -2.751$, $p = 0.006$). With the inclusion of the mediating variable Consumer Behaviour, the impact of

**Table 3** Mediation estimates [type of work → consumer behaviour → choice of payment (transaction)]

| | | | | 95% Confidence interval | | | | |
|---|---|---|---|---|---|---|---|---|
| Effect | Label | Estimate | SE | Lower | Upper | Z | p | % Mediation |
| Indirect | a × b | −0.04217 | 0.01124 | −0.0637 | −0.0191 | −3.752 | <0.001 | 90.35 |
| Direct | c | 0.00450 | 0.00749 | −0.0105 | 0.0195 | 0.602 | 0.547 | 9.65 |
| Total | c + a × b | −0.03767 | 0.01369 | −0.0647 | −0.0107 | −2.751 | 0.006 | 100.00 |

the choice of payment (transaction) on the type of work was not significant (β = 0.00450, t = 0.602, p = 0.547). The indirect effect of the choice of payment (transaction) on the type of work through Consumer Behaviour was found significant (β = −0.04217, t = −3.752, p < 0.001). This shows that the relationship between the choice of payment (transaction) on the type of work is fully mediated by the construct Consumer Behaviour (Table 4).

See Chart 1.

Type of Work is negatively related with Consumer Behaviour (β = −0.09585, p < 0.001) and Consumer Behaviour is in turn positively related with the choice of payment (transaction) (β = 0.43996, p < 0 0.001). As evident from Table 3, there was a significant indirect effect on the choice of payment (transaction) by the type of work through increased Consumer Behaviour score (β = −0.04217, p < 0.001). The results suggest that after accounting for mediating role of Consumer Behaviour, type of work do not have a significant impact on the choice of payment (transaction) (β = 0.00450, p = 0.547) (Tables 5 and 6).

The results of the mediation analysis show the mediating role of Consumer Awareness on the linkage between the choice of payment (transaction) and the type of work. The results reveal that the total effect of the choice of payment (transaction) on the

**Table 4** Path estimates [type of work → consumer behaviour → choice of payment (transaction)]

| | | | | | | 95% Confidence interval | | | |
|---|---|---|---|---|---|---|---|---|---|
| | | | Label | Estimate | SE | Lower | Upper | Z | p |
| Work | → | CB | a | −0.09585 | 0.02557 | −0.1457 | −0.0440 | −3.749 | <0.001 |
| CB | → | Transaction | b | 0.43996 | 0.00788 | 0.4240 | 0.4556 | 55.842 | <0.001 |
| Work | → | Transaction | c | 0.00450 | 0.00749 | −0.0105 | 0.0195 | 0.602 | 0.547 |

CB: Consumer Behaviour

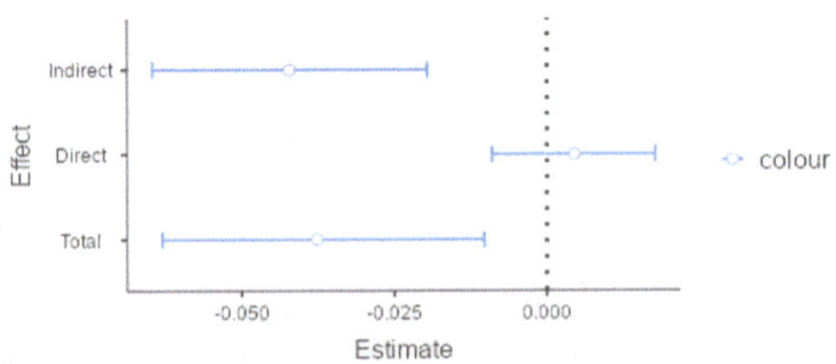


**Chart 1** Estimate plot

**Table 5** Mediation estimates [type of work → consumer awareness → choice of payment (transaction)]

| | | | | 95% Confidence interval | | | | |
|---|---|---|---|---|---|---|---|---|
| Effect | Label | Estimate | SE | Lower | Upper | Z | p | % Mediation |
| Indirect | a × b | 1.25e-4 | 0.00124 | −0.00270 | 0.00279 | 0.101 | 0.920 | 0.329 |
| Direct | c | −0.0378 | 0.01332 | −0.06354 | −0.01167 | −2.837 | 0.005 | 99.671 |
| Total | c + a × b | −0.0377 | 0.01331 | −0.06409 | −0.01178 | −2.829 | 0.005 | 100.000 |

**Table 6** Path estimates [type of work → consumer awareness → choice of payment (transaction)]

| | | | | | | 95% Confidence interval | | | |
|---|---|---|---|---|---|---|---|---|---|
| | | | Label | Estimate | SE | Lower | Upper | Z | p |
| Work | → | CA | a | −0.00339 | 0.0272 | −0.0566 | 0.0503 | −0.125 | 0.901 |
| CA | → | Transaction | b | −0.03679 | 0.0255 | −0.0870 | 0.0121 | −1.441 | 0.150 |
| Work | → | Transaction | c | −0.03779 | 0.0133 | −0.0635 | −0.0117 | −2.837 | 0.005 |

CA: Consumer Awareness

type of work was significant (H1: β = −0.0377, p = 0.005). With the inclusion of the mediating variable Consumer Awareness, the impact of the choice of payment (transaction) on the type of work was significant (β = −0.0378, p = 0.005). The indirect effect of the choice of payment (transaction) on the type of work through Consumer Awareness was not found significant (β = 1.25e-4, p = 0.920). This shows that the relationship between the choice of payment (transaction) on the type of work is not mediated by the construct Consumer Awareness (Tables 7 and 8).

The results of the mediation analysis show the mediating role of Service on the linkage between the choice of payment (transaction) and the type of work. The results reveal that the total effect of the choice of payment (transaction) on the type of work

**Table 7** Mediation estimates [type of work → service → choice of payment (transaction)]

| | | | | 95% Confidence interval | | | | |
|---|---|---|---|---|---|---|---|---|
| Effect | Label | Estimate | SE | Lower | Upper | Z | p | % Mediation |
| Indirect | a × b | −3.41e-4 | 0.00114 | −0.00304 | 0.00186 | −0.298 | 0.766 | 0.904 |
| Direct | c | −0.0373 | 0.01380 | −0.06484 | −0.01025 | −2.704 | 0.007 | 99.096 |
| Total | c + a × b | −0.0377 | 0.01377 | −0.06503 | −0.01094 | −2.736 | 0.006 | 100.000 |

**Table 8** Path estimates [type of work → service → choice of payment (transaction)]

| | | | | | | 95% Confidence interval | | | |
|---|---|---|---|---|---|---|---|---|---|
| | | | Label | Estimate | SE | Lower | Upper | Z | p |
| Work | → | Service | a | 0.0319 | 0.0288 | −0.0236 | 0.0907 | 1.109 | 0.267 |
| Service | → | Transaction | b | −0.0107 | 0.0251 | −0.0587 | 0.0387 | −0.425 | 0.671 |
| Work | → | Transaction | c | −0.0373 | 0.0138 | −0.0648 | −0.0102 | −2.704 | 0.007 |

was significant (H1: $\beta = -0.0377$, $p = 0.006$). With the inclusion of the mediating variable Service, the impact of the choice of payment (transaction) on the type of work was significant ($\beta = -0.0373$, $p = 0.007$). The indirect effect of the choice of payment (transaction) on the type of work through Service was not found significant ($\beta = -3.41\text{e-}4$, $p = 0.766$). This shows that the relationship between the choice of payment (transaction) on the type of work is not mediated by the construct Service (Tables 9 and 10).

**Table 9** Mediation estimates [type of work → infrastructure → choice of payment (transaction)]

| | | | | 95% Confidence interval | | | | |
|---|---|---|---|---|---|---|---|---|
| Effect | Label | Estimate | SE | Lower | Upper | Z | p | % Mediation |
| Indirect | a × b | −3.52e-4 | 0.00117 | −0.00301 | 0.00166 | −0.300 | 0.764 | 0.934 |
| Direct | c | −0.0373 | 0.01439 | −0.06962 | −0.00963 | −2.592 | 0.010 | 99.066 |
| Total | c + a × b | −0.0377 | 0.01436 | −0.06974 | −0.00997 | −2.623 | 0.009 | 100.000 |

**Table 10** Path estimates [type of work → infrastructure → choice of payment (transaction)]

| | | | | | | 95% Confidence interval | | | |
|---|---|---|---|---|---|---|---|---|---|
| | | | Label | Estimate | SE | Lower | Upper | Z | p |
| Work | → | Infrastructure | a | 0.0174 | 0.0294 | −0.0412 | 0.07518 | 0.591 | 0.555 |
| Infrastructure | → | Transaction | b | −0.0203 | 0.0264 | −0.0735 | 0.02966 | −0.766 | 0.444 |
| Work | → | Transaction | c | −0.0373 | 0.0144 | −0.0696 | −0.00963 | −2.592 | 0.010 |

The results of the mediation analysis show the mediating role of Infrastructure on the linkage between the choice of payment (transaction) and the type of work. The results reveal that the total effect of the choice of payment (transaction) on the type of work was significant (H1: $\beta = -0.0377$, $p = 0.009$). With the inclusion of the mediating variable Infrastructure, the impact of the choice of payment (transaction) on the type of work was significant ($\beta = -0.0373$, $p = 0.010$). The indirect effect of the choice of payment (transaction) on the type of work through infrastructure was not found significant ($\beta = -3.52\text{e-}4$, $p = 0.764$). This shows that the relationship between the choice of payment (transaction) on the type of work is not mediated by the construct Infrastructure (Tables 11 and 12).

The results of the mediation analysis show the mediating role of Service on the linkage between the choice of payment (transaction) and the type of work. The results reveal that the total effect of the choice of payment (transaction) on the type of work was significant (H1: $\beta = -0.0377$, $p = 0.006$). With the inclusion of the mediating variable Service, the impact of the choice of payment (transaction) on the type of work was significant ($\beta = -0.0373$, $p = 0.007$). The indirect effect of the choice of payment (transaction) on the type of work through Service was not found significant ($\beta = -3.41\text{e-}4$, $p = 0.766$). This shows that the relationship between the choice of payment (transaction) on the type of work is not mediated by the construct Service.

**Table 11** Mediation estimates[type of work → promotion→ choice of payment (transaction)]

| | | | | 95% Confidence Interval | | | | |
|---|---|---|---|---|---|---|---|---|
| Effect | Label | Estimate | SE | Lower | Upper | Z | p | % Mediation |
| Indirect | a × b | −2.13e-5 | 0.00107 | −0.00237 | 0.00262 | −0.0200 | 0.984 | 0.0567 |
| Direct | c | −0.0376 | 0.01392 | −0.06480 | −0.01057 | −2.7047 | 0.007 | 99.9433 |
| Total | c + a × b | −0.0377 | 0.01387 | −0.06486 | −0.01080 | −2.7161 | 0.007 | 100.0000 |

**Table 12** Path estimates [type of work → promotion→ choice of payment (transaction)]

| | | | | | | 95% Confidence interval | | | |
|---|---|---|---|---|---|---|---|---|---|
| | | | Label | Estimate | SE | Lower | Upper | Z | p |
| Work | → | Promotion | a | 0.0258 | 0.0274 | −0.0265 | 0.0786 | 0.9405 | 0.347 |
| Promotion | → | Transaction | b | −8.29e−4 | 0.0264 | −0.0542 | 0.0491 | −0.0314 | 0.975 |
| Work | → | Transaction | c | −0.0376 | 0.0139 | −0.0648 | −0.0106 | −2.7047 | 0.007 |

## 7 Theoretical Implications

Expanding on existing literature, the research delves into how consumers' payment preferences relate to their occupation. The study enhances current technological acceptance models by introducing new elements centered around users, such as Consumer Behaviour, Consumer Awareness, Service, Infrastructure, and Promotion as mediators. These aspects, previously overlooked in earlier studies, contribute to a more comprehensive understanding of the subject. The study's tests suggest that respondents' payment preferences may be influenced by their occupation type. Moreover, the presence of Consumer Behaviour plays a significant role in enhancing this influence, indicating that payment preferences are linked to respondents' work nature driven by their behaviours. The negative significant indirect relationship of the choice of payment (transaction) on the type of work through Consumer Behaviour implies that respondents will opt for cashless payment methods more frequently if the associated risks are mitigated. On the other hand, factors such as Consumer Awareness, Service, Infrastructure, and Promotion do not mediate between work nature and consumer payment preferences. The success of a cashless economy also relies on the availability and quality of telecom networks in both urban and rural areas of India. Even in metropolitan cities, online transaction issues are common, highlighting the need for serious attention as this is essential for a cashless economy. Other critical factors to consider include public awareness, financial literacy, advancements in technology and infrastructure, and strong government interventions, all of which are vital for the future of transactions.

## 8 Practical Implications

The research revealed that Consumer Behaviour has the potential to influence consumer payment preferences. The findings of this study could potentially affect how society embraces technological advancements in their everyday routines. This could also motivate users to appreciate the advantages of utilizing such services, subsequently influencing their inclination towards adopting cashless payment methods. The study recommends that banks need to customize their cashless payment systems to cater to the requirements of their customers. However, financial institutions face challenges in providing these services, including privacy concerns, trust issues, espionage and identity theft. Customers are increasingly worried about their privacy when making online transactions [2, 3, 32], a concern highlighted in various research studies. A customer's loyalty to an e-payment system from a financial institution hinges on trust. If customers feel satisfied and build trust, their perceived risk regarding digital payments will decrease [8, 25]. Transitioning from a cash-based system to a cashless one in the informal sector is a challenging endeavour. In India, many merchants tend to avoid keeping written records of their transactions in order to evade taxes. Additionally, the general public prefers cash payments, and business owners typically do not maintain records of their transactions or income, contributing to the prevalence of black money and tax evasion. The research revealed that students, individuals in professional services, and corporate sectors are more likely to prefer cashless payment methods. This preference is attributed to the convenience of making instant purchases from home using cashless options. Particularly, the younger generation of students favours cashless transactions as they spend significant time on smart phones and devices, granting them access to a wide range of products on websites. The adoption of cashless payment methods is on the rise. In such a landscape, companies involved in cashless payment systems and e-commerce needs to focus on building and enhancing trust. This can be achieved through transparent communication and privacy policies for their customers. Displaying security measures and quality certifications visually will assist clients in making informed decisions [11].

## 9 Conclusion and Scope for Future Research

The consumers' demand is taken into account progressively in almost all areas of service to the society and moreover, customers' demand for precaution has become a top priority for business owners. According to [17], this precaution is crucial for fostering trust in a cashless economy. The study aimed to investigate whether Consumer Behaviour, Consumer Awareness, Service, Infrastructure and Promotion could act as mediators between consumers' payment preferences and their occupation type. The tests performed in the study indicate that the occupation type could influence the choices of the respondents regarding their payment preferences. It further revealed that this influence is further enhanced in the presence of the construct consumer behaviour which suggests that the preferences regarding the payment type is influenced by the nature of the work of the respondents which is triggered by their behaviours. The other factors, namely, Consumer Awareness, Service, Infrastructure and Promotion do not play any mediating role between the nature of work and the payment preferences of the consumers. The connection between work nature and payment preferences is influenced by Consumer Behaviour, as indicated in this study. It suggests that consumers may be more inclined to choose cashless payment methods if the risks associated with them are minimized. This can be achieved through proactive steps by financial institutions to combat cyber fraud and enhance financial education. The study reveals that students and those working in corporate sectors and engaged in professional services predominantly opt for cashless payment methods. Conversely, individuals involved in government services, self-employment, and those currently unemployed tend to favour using cash. Among those in government services who opt for cash transactions or in-store purchases, the majority fall within the 45–65 age group, as they perceive cash transactions to be secure. Entrepreneurs, on the other hand, prefer cash due to the lack of official transaction records it entails. Those currently unemployed often rely on pocket money from family members as they may not have fully operational bank accounts.

The research reveals that the constructs Consumer Awareness, Service, Infrastructure, and Promotion do not act as mediators in the connection between work nature and payment preferences. This suggests that enhancing awareness, offering smooth and secure cashless services with strong infrastructure, and effective promotion can encourage people from all backgrounds to choose cashless payment methods.

It is noted that the current study focuses on the mediating role of the factors Consumer Behaviour, Consumer Awareness, Service, Infrastructure and Promotion between the nature of work and the payment preferences of the consumers only. The study can be further extended to find out the payment preferences of the respondents of varied age groups belonging to different generations and other factors like social media influence and income of the respondents can be taken into consideration.

## References

1. Amankwa, E., Amissah, G., Okoampa-Larbi, R.: Cashless economy–the nexus of COVID-19 and E-wallet usage intentions: a multi-group analysis between formal and informal sector workers in Ghana. J. Sci. Technol. Policy Manag. **14**(6), 1155–1183 (2023)
2. Akhter, S.H.: Privacy concern and online transactions: the impact of internet self-efficacy and internet involvement. J. Consum. Mark. **31**(2), 118–125 (2014)
3. Bansal, G., Zahedi, F.M., Gefen, D.: Do context and personality matter? Trust and privacy concerns in disclosing private information online. Inf. Manag. **53**(1), 1–21 (2016)
4. Barontini, C., Holden, H.: Proceeding with Caution—A Survey on Central Bank Digital Currency. BIS Paper, 101 (2019)
5. Bhat, I.H., Gupta, S.: Effect of consumers' perceived risk on relationship marketing for m-shopping in India: the moderated mediation role of demonetisation. Glob. Bus. Econ. Rev. **27**(3), 281–302 (2022)
6. Bhatti, A., Akram, H., Basit, H.M., Khan, A.U., Naqvi, S.M.R., Bilal, M.: E-commerce trends during COVID-19 Pandemic. Int. J. Future Gener. Commun. Netw. **13**(2), 1449–1452 (2020)
7. Bhuvana, M., Vasantha, S.: A mediating effect of demonetization of currency notes towards adopting cashless payment system. Int. J. Civ. Eng. Technol. **8**(6), 699–707 (2017)
8. Chang, K.C.: Examining the effect of tour guide performance, tourist trust, tourist satisfaction, and flow experience on tourists' shopping behavior. Asia Pac. J. Tour. Res. **19**(2), 219–247 (2014)
9. Chohan, U.W.: Cryptocurrencies and inequality. In: Cryptofinance: A New Currency for a New Economy, pp. 49–62 (2022)
10. Cnaan, R.A., Scott, M.L., Heist, H.D., Moodithaya, M.S.: Financial inclusion in the digital banking age: lessons from rural India. J. Soc. Policy **52**(3), 520–541 (2023)
11. Giovannini, A., Aubry, A., Panetto, H., El Haouzi, H., Pierrel, L., Dassisti, M.: Anti-logicist framework for design-knowledge representation. Annu. Rev. Control **39**(1), 144–157 (2015)
12. Goczek, Ł., Witkowski, B.: Determinants of non-cash payments. NBP Working Paper No. 196. Economic Institute, Warsaw (2015)
13. Goldfarb, A., Tucker, C.: Digital marketing. In: Handbook of the economics of marketing, vol. 1, pp. 259–290. North-Holland (2019)
14. Hair, J.F., Hult, G.T.M., Ringle, C.M., Sarstedt, M.: A Primer on Partial Least Squares Structural Equation Modeling (PLS-SEM), 2nd edn. Sage Publications Inc., Thousand Oaks, CA (2017)
15. Hoekman, B.M., Kostecki, M.M.: The Political Economy of the World Trading System: The WTO and Beyond. Oxford University Press (2009)
16. Islam, M.J., Hasan, S., Fahim, A.Y.: Influence of consumption values on adoption of cashless society in emerging economies: the mediating effect of cashless readiness. J. Financ. Serv. Mark., 1–13 (2024)

17. Jain, A., Jain, V.: Role of e-payment in cashless economy and its impact on consumer behavior. J. Stat. Manag. Syst. **25**(7), 1685–1696 (2022)
18. Khan, I.: Barter system, vedic rituals and their role in the introduction of metallic currency in Indo-Pakistan (2013
19. Krejcie, R.V., Morgan, D.W.: Determining sample size for research activities. Educ. Psychol. Measur. **30**(3), 607–610 (1970)
20. Nagdev, K., Kumar, P., Rajesh, A., Kumar, S.: Measuring demonetisation: a path towards the cashless India. Int. J. Public Sect. Perform. Manag. **4**(1), 114–132 (2018)
21. Namahoot, K.S., Jantasri, V.: Integration of UTAUT model in Thailand cashless payment system adoption: the mediating role of perceived risk and trust. J. Sci. Technol. Policy Manag. **14**(4), 634–658 (2023)
22. Nawi, N.C., Husin, H.S., Al-Jahwari, N.S., Zainuddin, S.A., Khan, N.U., Hassan, A.A., Ibrahim, W.S.A.A.W., Mohamed, A.F., Nasir, N.S.M., Hasan, M.Z.M.: The path to sustainability begins with going paperless: antecedents of intention to use electronic wallet using serial mediation approach. Heliyon **10**(2) (2024)
23. Noman, M., Maydybura, A., Channa, K.A., Wong, W.K., Chang, B.H.: Impact of cashless bank payments on economic growth: evidence from G7 countries. Adv. Decis. Sci. **27**(1), 1–20 (2023)
24. Park, J., Lee, C., Thomas, M.: Why do cashless payments increase unhealthy consumption? The decision-risk inattention hypothesis. J. Assoc. Consum. Res. **6**(1), 21–32 (2021)
25. Ponte, E.B., Carvajal-Trujillo, E., Escobar-Rodríguez, T.: Influence of trust and perceived value on the intention to purchase travel online: Integrating the effects of assurance on trust antecedents. Tour. Manage. **47**, 286–302 (2015)
26. Raj L, V., Amilan, S., Aparna, K.: Developing and validating a cashless transaction adoption model (CTAM). J. Sci. Technol. Policy Manag. (2023)
27. Ren, T., Tang, Y.: Accelerate the promotion of mobile payments during the covid-19 epidemic. The Innovation **1**(2) (2020)
28. Sharma, A., Rathod, N.S.: Central bank digital currency: assessment of digital financing approach towards safe cashless economy in India. In: Digital Currency Assets and Challenges to Financial System Stability, pp. 37–51. IGI Global (2024)
29. Sivathanu, B.: Adoption of digital payment systems in the era of demonetization in India: an empirical study. J. Sci. Technol. Policy Manag. **10**(1), 143–171 (2019)
30. Viviana Alfonso, C., Boar, C., Frost, J., Gambacorta, L., Liu, J.: E-commerce in the pandemic and beyond. BIS Bull. **36**, 1–7 (2021)
31. Wang, J., Nguyen, N., Jiang, X., Nguyen, H.V., Saleem, M.A.: Consumers' perceived value and use intention of cashless payment in the physical distancing context: evidence from an Asian emerging market. Asia Pac. J. Mark. Logist. **35**(6), 1513–1531 (2023)
32. Wu, L.: The antecedents of customer satisfaction and its link to complaint intentions in online shopping: an integration of justice, technology, and trust. Int. J. Inf. Manag. **33**(1), 166–176 (2013)

# Healing with Technology: The Convergence of AI and Robotics in Healthcare

**Ankita Pathak and Mohammad Shahidul Islam**

**Abstract** This chapter explores the transformative integration of artificial intelligence (AI) and robotics in healthcare, emphasizing their potential to revolutionize patient care, diagnostics, and operational workflows. To begin with, AI algorithms provide precise analysis of large datasets, enhancing diagnostic accuracy, minimizing errors, and enabling personalized treatments. In parallel, robotics complements these capabilities with unmatched efficiency in performing complex physical tasks, such as surgeries, rehabilitation, and patient monitoring. Furthermore, AI and robotics collectively align with the P5 medicine framework predictive, personalized, preventive, participatory, and precise shifting healthcare systems from reactive to proactive models. Notably, applications such as robot-assisted surgeries, which have seen a significant rise in adoption, AI-enabled telemedicine platforms extending healthcare access, and remote monitoring systems providing real-time insights for early interventions illustrate their growing impact. In addition to these advancements, the chapter delves into the theoretical implications, including contributions to systems theory, ethical AI frameworks, and precision medicine models. Equally important are the practical benefits, which range from streamlining operations and automating routine tasks to extending healthcare services to underserved regions. These innovations significantly enhance resource allocation and patient outcomes. However, it is essential to acknowledge the challenges, including high implementation costs, ethical concerns over patient privacy and algorithmic bias, and the need for extensive training and robust regulatory oversight. Despite these obstacles, integrating AI and robotics represents a paradigm shift in healthcare. Therefore, addressing these challenges through innovation, regulatory support, and ethical practices is critical to unlocking their full transformative potential and ensuring a more effective, efficient, and inclusive future in healthcare.

---

A. Pathak (✉)
GLA University, Institute of Business Management, Mathura, India
e-mail: ankitapthk76@gmail.com

M. S. Islam
BRAC Business School, BRAC University, Dhaka, Bangladesh
e-mail: mohd.sh.islam@bracu.ac.bd



N. D. K. Nguyen et al. (eds.), *Digital Transformation and Sustainability: Methods and Applications*, Information Systems Engineering and Management 47,
https://doi.org/10.1007/978-3-031-91278-8_13

**Keywords** Artificial intelligence · Robotics · Healthcare · Applications of AI robotics · Medical diagnosis · Patient monitoring · Ethical considerations

## 1 Introduction

The adage "Health is wealth" underscores the universal aspiration for a healthy life, yet modern living poses significant challenges to maintaining well-being. Factors like environmental pollution, sedentary lifestyles, lack of exercise, and unhealthy dietary habits are major barriers to achieving good health. Advances in healthcare science and technology, particularly through artificial intelligence (AI) and robotics, are transforming the medical landscape, offering innovative solutions to these challenges. The integration of these technologies has significantly enhanced diagnostic precision, treatment effectiveness, and patient care outcomes.

Robotics and AI have become pivotal in addressing healthcare needs, with the global market for medical robotics projected to grow from $7.24 billion in 2015 to $20 billion by 2023 [1]. These advancements highlight the increasing reliance on AI-powered systems for critical medical tasks. Early examples like the MYCIN expert system, developed in the 1970s, marked the beginning of AI's influence in healthcare, offering groundbreaking diagnostic capabilities [2]. Today, AI mimics doctors' decision-making processes, analyzing vast datasets to predict disease progression, improve diagnosis accuracy, and guide treatment plans. Simultaneously, robotics complements AI by executing complex physical tasks such as surgeries, patient monitoring, and rehabilitation.

The use of robotics in surgery exemplifies its transformative impact. The percentage of robot-assisted surgeries increased from 1.8% in 2012 to 15.1% in 2021, reflecting growing confidence in its precision and efficiency [3]. These systems offer unparalleled accuracy in minimally invasive procedures, leading to reduced trauma, shorter recovery times, and improved patient outcomes. Furthermore, AI algorithms integrated with robotic systems enhance their ability to interpret real-time data, providing critical insights during surgeries and ensuring better decision-making.

AI also plays a crucial role in reducing diagnostic errors, which account for approximately 80,000 deaths annually in the United States alone [4]. By analyzing medical records, imaging results, and genetic data, AI systems identify patterns and correlations that might elude human practitioners. This capability not only minimizes errors but also enables personalized treatment plans tailored to individual patients' needs.

Beyond surgery and diagnosis, the integration of AI and robotics is revolutionizing patient care through the P5 medicine framework—predictive, personalized, preventive, participatory, and precise healthcare [5]. AI-driven predictive analytics enable early detection of diseases, allowing for timely interventions that can save lives. For instance, wearable sensors powered by AI monitor vital signs and provide real-time alerts about potential health issues. These tools empower patients to actively participate in their care while reducing the burden on healthcare providers.

**Fig. 1** AI and Robotics in healthcare. *Sources* Adapted from [1, 6]

Robotics further enhances patient care by aiding in rehabilitation and daily assistance. Robotic systems are now equipped to assist patients with mobility challenges, enabling them to regain independence and improve their quality of life. In addition, robots are used for sanitation and medication delivery, crucial during pandemics like COVID-19, to minimize infection risks. While initial implementation costs of AI and robotics may be high, their long-term benefits are undeniable. These technologies streamline operations, reduce hospital stays, and enhance resource allocation, ultimately lowering healthcare costs and improving accessibility. As AI and robotics continue to evolve, they promise a future of more efficient, effective, and patient-centered healthcare (Fig. 1).

## 2 Significance and Potential Impact

AI-driven robotics are having a transformative effect on healthcare, offering numerous advantages that could redefine the industry. These robots can execute tasks with precision and dependability beyond human capabilities. Robots are capable of conducting surgeries with exceptional precision and minimal invasiveness, which promotes faster recovery and improved patient outcomes [7]. AI-driven robotics support telemedicine and deliver remote healthcare services, extending medical expertise to areas with limited healthcare access. Medical experts can control robotic systems globally, enabling them to perform surgeries and give consultations from a distance [8]. Numerous regions face a lack of specialized healthcare professionals, particularly in surgery and elderly care.

AI-powered robots can reduce the stress by helping healthcare workers with a range of tasks, including surgical assistance and daily care for the elderly or disabled. The automization of repetitive tasks would help in assisting with complex operations. These robots can decrease the likelihood of human error. Robots not only strengthens patient safety but also reduces the probability of adverse events during medical interventions. AI algorithms can process massive amounts of medical data to find patterns, predict outcomes, and aid healthcare providers in making well-informed choices [9]. The merger of robotics with AI, would directly help in patient care, providing personalized treatment strategies and improve diagnoses. Robots equipped with AI algorithms can support rehabilitation and physical therapy for patients healing from injuries or surgeries [10]. These robotic devices can provide customized exercises, observe progress, and fine-tune therapy regimens in real time based on how the patient responds, which enhance the success of rehabilitation programs. The initial cost of implementing AI-powered robotics may be high, but in long-term savings are substantial. AI based systems streamline operations, decrease hospital stays, and prevent complications, ultimately reduce healthcare expenses and boost the efficiency of care delivery [11] (Fig. 2).

## 3 Application of Robotics in Healthcare

In field of healthcare, robots are commonly classified into two categories: medical robots and surgical robots [12]. Medical robots assist doctors in examining patients in remote locations. These robots are known as "telepresence" robots [13]. Rehabilitation robots assist patients in carrying out specific movements during their recovery after surgery. Surgical robots are widely used in healthcare for various applications, such as organ transplants, spine surgery, urological procedures, and gastrointestinal operations [6]. It assist medical surgeons during operations and have contributed to the development of new techniques which minimally invasive surgeries and promote quicker patient recovery.

Medical transportation robots handle tasks such as delivering medications, dispensing prescribed drugs, which improve communication between patients, medical staff [14]. AI based robots are a blessing for healthcare workers, as it allow them to devote more attention to patients rather than administrative task. In response to pandemics such as Covid-19, the healthcare sector is focusing on sanitation and disinfection robots that can eliminate bacteria and viruses from surfaces in just minutes [15]. The speed and accuracy of robotic systems have been a boon to the healthcare industry.

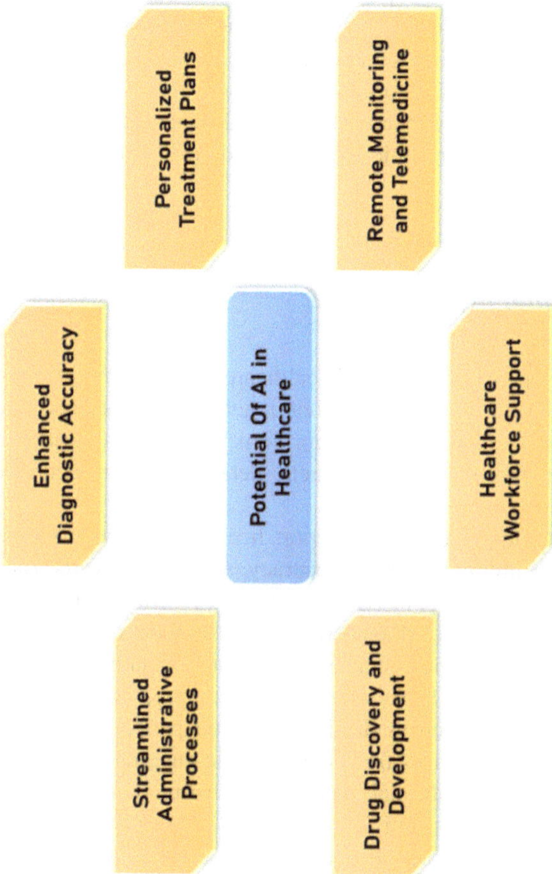


**Fig. 2** Potential of AI and Robotics in healthcare. *Sources* Adapted from [1, 10]

AI-integrated robotic surgical systems have significantly altered conventional surgical procedures. From minimally invasive procedures to intricate operations, Robots systems provide unmatched accuracy, flexibility, and control, in small and intricate operational procedure [16]. It lower surgical risks and enhance patient results. The integrating AI-powered robotics into telemedicine and remote monitoring, have now reached beyond traditional clinical settings. With the help of telepresence robots and wearable technology, healthcare professionals can remotely evaluate patients, track vital signs, and provide prompt interventions, particularly in underserved or distant regions [17]. AI-driven robotics can capture image and give early diagnosis. It can plan treatment, process the data and healthcare analytics. Utilizing image-guided techniques and automated radiology, AI can advance diagnostic accuracy, and facilitate workflow efficiency [18]. As AI-driven robotics become more common in healthcare, it is crucial to tackle ethical issues related to patient confidentiality, algorithmic prejudice, and the interaction between humans and AI.

## *3.1 Medical Diagnosis*

The application of AI and robotics in healthcare holds potential, particularly in medical diagnosis, where AI-enhanced robotics achieve extraordinary precision and accuracy [19]. The Machine learning algorithms can process patient records like medical history, imaging results, and genetic details, to offer accurate diagnostic predictions.

AI-enhanced robotic systems can support healthcare providers in interpreting diagnostic tests with improved speed, accuracy, and minimize the risk of human errors and diagnostic mistakes [11]. AI-driven robotics facilitate in providing personalized medicine on the basis of diagnostics and past records. Robotic systems can detect distinct biomarkers, genetic tendencies, and treatment responses, which leads to intervention of medicine [20]. This personalized approach boosts diagnostic accuracy and therapeutic effectiveness, leading to better patient outcomes and lower healthcare costs. AI-powered robotic systems enable ongoing monitoring of patient health metrics and real-time data analysis for predictive analytics [21]. These devices, fitted with sensors and AI algorithms, can track vital signs, physiological parameters, and disease symptoms in the early stage of health issues. AI based Robotics forecast how diseases might progress and identify key risk factors, so that healthcare providers can take preventive actions, and handle treatment of disease more effectively.

While AI-based robotics present great potential for medical diagnosis, still faces major challenges on data protection and security concerns, regulatory barriers, ethical considerations, and the continuous need for validating and refining AI algorithms. Furthermore, integrating AI and robotics into established healthcare frameworks requires funding for training, infrastructure to ensure compatibility standards.

### 3.2 Surgical Interventions

Integrating AI and robotics in surgical operations has greatly progressed medicine by minimizing human error, improving precision, and elevating patient results [22]. Surgical robots using AI technology have augmented the skills of medical practitioners, enabling them to handle intricate procedures with high accuracy and efficiency [23]. The combination of robotics and AI has drastically changed surgical approaches and presents significant potential for future medical advancements. AI-driven robotic systems provide submillimetre accuracy, surpassing human precision in delicate surgeries, like neurosurgery and eye operations, where minor errors could have serious impacts. Robotic assistance during surgery promotes less trauma, decreased blood loss, and faster patient recovery. Surgeons benefit from AI algorithms that enhance the precision of navigating complex anatomy, leading to smaller surgical openings and better visual outcomes [24].

AI-enhanced robotic platforms provide surgeons with real-time guidance, offering insights into tissue characteristics, vital statistics, and procedural specifics [25]. This personalized approach improves treatment effectiveness, reduces the risk of complications, and supports faster recovery after surgery. However, the high upfront cost of robotic systems limits their widespread adoption, particularly in resource-constrained environments [26]. Robust training programs are essential to prepare surgeons for proficient use of AI-based robotic platforms, with continuous professional development and certification ensuring safe and effective application in medical care [27, 28].

Well-defined regulatory guidelines and frameworks are essential for navigating the complex ethical and legal issues in this field. The future of surgical robotics is expected to be influenced by the integration of AI, robotics, and other advanced technologies like augmented reality, telemedicine, and nanorobotics [6]. New developments in AI-powered autonomous systems, surgical simulations, and remote surgery are anticipated to improve surgical outcomes and expand access to high-quality healthcare worldwide.

### 3.3 Drug Discovery

The contribution of assistive robotics is essential for progressing drug discovery, accelerating the screening and development stages, and improving personalized medicine and therapies. Robotics can automate the evaluation of thousands to millions of compounds for their biological effects on a specific target [29]. This significantly increases the efficiency of the screening process. Robotic platforms carry out functions such as liquid distribution, plate preparation, and assay implementation with precision and consistency [30]. This reduces the risk of human error and variability. Furthermore, the combination of robotic systems with microfluidic technology facilitates small-scale, high-throughput testing. This strategy allows for the swift evaluation of compounds by utilizing diminished volumes, thereby decreasing reagent costs and hastening the procedure. Integrating robotics with AI and machine learning algorithms permits the analysis of extensive data generated from screening assays [31]. AI algorithms can uncover patterns and relationships within complex biological data, supporting the discovery of potential drug candidates.

### *3.4 Personalized Medicine*

Robotics facilitates the examination of cells and tissues from patients to identify tailored treatment options. Automated systems can efficiently handle the requirements of personalized medicine by processing each patient's samples and testing them against a variety of pharmaceutical compounds or combinations [32].

### *3.5 Precision Drug Delivery*

Robotics plays a crucial role in drug delivery, especially for targeted treatments or specific drug formulations [33]. Automated systems are capable of controlling dosing and delivery conditions with great accuracy, optimizing therapeutic efficacy and reducing adverse effects.

### *3.6 Continuous Monitoring and Feedback Loop*

Robotics allows for the continual observation of cell cultures or biological systems, supplying real-time data that informs decision-making processes [34]. This continuous feedback loop enables the iterative refinement of both drug candidates and treatment protocols.

### *3.7 Rehabilitation—Assistive Robotics for Physical Therapy—Cognitive Rehabilitation Aid*

Rehabilitation involves a diverse array of strategies and tools aimed at restoring or improving physical and cognitive functions impacted by injuries or disabilities. Within this field, assistive robotics and cognitive rehabilitation present groundbreaking solutions to enhance therapeutic results and elevate overall quality of life [35].

### 3.7.1 Assistive Robotics for Physical Therapy

Wearable robotic systems are engineered to aid people with mobility limitations by facilitating limb movements through guidance and support [36]. These devices are especially useful for individuals recovering from strokes, spinal cord injuries, or musculoskeletal issues.

Robotic systems intended for gait training support individuals in regaining walking proficiency by providing controlled and repetitive movement patterns [37]. These systems can measure variables such as walking speed, step length, and degree of support according to the user's specific needs.

Those with upper limb impairments can find value in robotic devices that assist with rehabilitation exercises to improve arm function, strength, and coordination [38]. These devices frequently feature interactive games and exercises to captivate and encourage users during their therapy sessions.

### 3.7.2 Cognitive Rehabilitation Aids

VR systems can create immersive environments that simulate real-life scenarios and challenges. Brain-Computer Interfaces (BCIs) establish a direct link between the brain and external devices, allowing individuals with severe motor impairments to control computers or robotic systems through brain signals [27, 28]. BCIs hold great potential for cognitive therapy by offering avenues for neurofeedback training and brain stimulation methods [39].

Software applications and games created for cognitive rehabilitation present interactive challenges to boost memory, attention, language abilities, and other cognitive skills. These platforms often utilize adaptive algorithms to customize therapy tasks based on the user's strengths and track improvement over time.

## 4 Benefits of AI Robotics in Healthcare

The integration of AI robotics with healthcare systems has brought about groundbreaking innovations, providing various advantages that enhance patient care, streamline procedures, and support healthcare professionals. AI-powered robots are exceptionally proficient at performing precise and accurate tasks, including surgical procedures and medication delivery [22]. Robotic assistance has transformed the field of surgery, improving control, dexterity, and visual precision during operations [40]. AI-driven surgical robots enable minimally invasive surgeries with smaller incisions, reduced hemorrhage, and accelerated recovery times for patients [41]. This forward-thinking strategy allows for the early detection of potential health concerns, and results in timely interventions and improve patient outcomes. Utilizing AI algorithms, healthcare robots can process extensive patient data, such as medical records, genetic information, and real-time physiological metrics, to create individualized treatment plans meeting their specific requirements. This customized approach enhances treatment effectiveness and patient satisfaction.

AI robotics reduce the workload on healthcare professionals and optimize the use of available resources by automation of routine tasks and handling administrative responsibilities, and overseeing inventory [42]. The efficiency would allow employees to dedicate their efforts to more complex and high-value activities. Healthcare robots powered by AI operate continuously, and deliver persistent assistance and support to patients and healthcare professionals. They can be employed in medical facilities, clinics, and private residences, to care at any time of day or night. The use of AI robotics increases the reach of healthcare services, in distant locations where healthcare resources are limited.

Controlling infections is essential in healthcare, particularly in hospital settings. AI-enabled robots with cleaning functions can autonomously disinfect hospital rooms, devices, and surfaces [15]. It lowers the risk of infections and enhance the safety of patients. AI-enabled robotics provide significant assistance to elderly and disabled individuals with routine tasks. These robots promote independence and improve the living standards for patients AI-powered robots companies in healthcare continuously gather insights from their interactions with patients, healthcare providers, and their environment, for regular upgrades in their abilities and performance [43]. This cycle of learning and evolving ensures that AI robotics remain at the leading edge of innovation in the medical field.

## 5 Benefits of Implementing Robotics in Healthcare

The development of socially assistive robots (SARs) combines AI with physical assistance technologies. These emotionally intelligent robots provide specialized patient care by interacting with patients in a manner that triggers emotional responses. Responses from socially assistive robots encompass interaction, communication, companionship, and emotional connection [44]. Automated robots minimize the risk of human errors and negligence. The field of health and social care has been revolutionized by the development and continual evolution of SARs.

Robots are capable of effectively carrying out the roles of healthcare employees. Various types of robots, such as (i) receptionist robots, (ii) medical servers, and (iii) nurse robots, can perform these tasks with efficiency. These robots ensure the safety of healthcare workers by minimizing their exposure to pathogens and prevent infection transmission.

Automated mobile robots (AMRs) facilitate in regularization of standard tasks, decrease the physical workload for healthcare personnel, and ensure more precise execution of procedures [45]. These robots are effective in addressing staffing gaps, maintaining records, and placing orders punctually. It ensure the timely availability of drugs and medical supplies, and automated robots facilitate in efficient room cleaning and sanitization. This allows healthcare providers to engage in other vital patient-related responsibilities. Furthermore, AI-powered robots are proficient in diagnosing a range of diseases [46]. Robots with radiological imaging capabilities apply AI and deep learning to assist in diagnostics [20]. These robots are employed for procedures like MRIs and X-rays, as it prevent the medical staff from the detrimental radiation typically used in these tests [47].

# 6 Integration Technologies with Robotics in Healthcare

The fusion of robotics with integration technologies, such as the Internet of Things (IoT) and blockchain, has unlocked innovative applications in healthcare. These integrations enhance the capabilities of robotic systems, ensuring better accuracy, security, and efficiency in diverse medical tasks. This section explores the significant role of these technologies in reshaping healthcare delivery while maintaining alignment with the overarching theme of AI and robotics.

1. **IoT in Robotic Healthcare Systems**

IoT technology augments robotic systems by enabling them to collect, analyze, and act on data in real time [48]. In healthcare, this capability has profound implications for patient monitoring, surgical precision, and operational efficiency.

- **Remote Patient Monitoring**: IoT sensors integrated into robotic systems facilitate real-time tracking of vital signs, such as blood pressure, heart rate, and oxygen levels [49]. These robots provide healthcare professionals with continuous updates, improving the management of chronic diseases and enabling early detection of complications [50]. For instance, IoT-enabled robots have proven effective in managing elderly care by detecting falls and monitoring activity levels remotely.
- **Enhanced Surgical Precision**: IoT empowers robotic surgical systems with superior navigation and control by connecting them to real-time data sources [51]. This minimizes the risk of human error and ensures safe and precise operations, particularly in complex procedures like neurosurgery or cardiac interventions [52].
- **Hospital Workflow Optimization**: In hospital settings, IoT-enabled robots automate logistical tasks such as medication delivery, sanitation, and patient transport [53]. This reduces the workload on healthcare staff and improves resource allocation.

2. **Blockchain for Data Security in Robotic Systems**

The integration of blockchain technology with robotic systems addresses critical concerns about data security and patient privacy, which are central to healthcare operations. Blockchain's decentralized and immutable ledger ensures that sensitive data collected and used by robots is protected against breaches or unauthorized access [54].

- **Data Integrity**: Blockchain ensures the accuracy and consistency of patient records, even as they are shared across multiple platforms [55]. This is crucial for robotic systems that rely on these records for diagnostics, treatment planning, or monitoring.
- **Secure Collaboration**: By enabling secure and decentralized access to medical data, blockchain fosters seamless collaboration between healthcare providers [56]. For instance, surgical robots can access encrypted patient histories during procedures without risking privacy violations [57].

3. **Combined Applications in Rehabilitation and Elderly Care**

Robotics, integrated with IoT and blockchain, offers groundbreaking solutions in rehabilitation and elderly care:

- **Rehabilitation Robots**: IoT sensors in robotic exoskeletons gather real-time feedback from patients recovering from surgeries or injuries [58]. This data is analyzed to adjust therapy programs dynamically, enhancing recovery outcomes.
- **Elderly Assistance**: IoT-powered service robots in eldercare settings detect emergencies, such as falls, and alert caregivers immediately. Additionally, these robots promote independence by assisting with mobility and daily tasks while maintaining health records on secure blockchain systems.

4. **Pandemic Management and Emergency Response**

During global crises like the COVID-19 pandemic, IoT-enabled robots demonstrated their value in ensuring healthcare continuity while minimizing risks:

- **Disinfection and Sanitization**: Robots equipped with IoT sensors autonomously disinfected hospital rooms and public spaces, reducing pathogen transmission risks [59].
- **Contactless Delivery**: IoT and robotics facilitated the safe delivery of medications and medical supplies, especially in quarantined zones [60].
- **Remote Monitoring**: IoT-linked robots monitored patients' vitals from a distance, ensuring timely interventions without exposing healthcare workers to infection [61].

The integration of IoT and blockchain with robotics addresses several critical challenges in healthcare, including accessibility, security, and efficiency [62]. IoT enhances the functionality of robotic systems through real-time data collection and dynamic responsiveness. Meanwhile, blockchain ensures the confidentiality and integrity of patient data, fostering trust in robotic-assisted care. This synergy expands the potential of AI-powered robotics to provide remote care, personalized treatment, and operational optimization, revolutionizing healthcare delivery.

# 7 Future Perspective

The 2020 study by Sheetz, Clafin, and Dimick showed that the share of robot-assisted surgeries grew from 1.8% in 2012 to 15.1% in 2021 [3]. This significant increase suggests a rise in confidence in robotic systems and an expansion in the availability of robotic surgeries. A report from Credence Research highlights that the medical robotics market, which was valued at $7.24 billion in 2015, is projected to $20 billion globally by 2023 [63].

The healthcare sector is substantial global industries, which aims to speed up progress through modernization and inventive strategies. In the past, healthcare relied on manual processes that were time-consuming and prone to human errors. Recent advancements in machine learning have initiated a revolution in the healthcare sector, focusing on creating intelligent machines that replicate human actions and responses. While AI and robotics are still in their infancy in healthcare, the future appears promising for their adoption and practicality. The area's most likely to rapidly adopt AI and robotics in healthcare are as follows.

## *7.1 Growth in Robotic-Assisted Surgeries*

Robot-assisted surgeries have experienced a steady rise in adoption, with their share growing from 1.8% in 2012 to 15.1% in 2021 [3]. This trend reflects growing confidence in the precision, safety, and efficiency offered by surgical robots. Future developments aim to incorporate AI-driven decision support systems that assist surgeons in real-time, enhancing outcomes in complex procedures like neurosurgery and orthopedic interventions.

## *7.2 Expanding Applications in Non-surgical Domains*

Beyond surgery, robots are expected to revolutionize areas such as:

- **Diagnostics**: AI-enhanced robotic systems will enable faster, more accurate diagnosis by integrating data from imaging, genetic analyses, and patient histories [64].
- **Drug Delivery**: Precision robotics will optimize drug administration by targeting specific tissues, reducing side effects, and improving treatment efficacy [65].
- **Rehabilitation**: Advanced robotic exoskeletons equipped with IoT sensors will provide tailored recovery programs, monitoring progress and adjusting therapies dynamically [66].

## 7.3 AI-Driven Predictive Healthcare

AI's role in predictive analytics will expand significantly, enabling early detection of diseases and personalized treatment plans. By 2030, predictive models are expected to integrate real-time patient data, environmental factors, and genomics to forecast health risks, empowering preventative care strategies [67].

## 7.4 Enhanced Remote Monitoring and Telemedicine

The rise of wearable devices and IoT-enabled robotics will redefine remote healthcare. These systems will monitor patient vitals continuously, alert providers to anomalies, and facilitate virtual consultations [68]. Projections suggest that by 2025, over 50% of patients in developed nations will rely on AI-driven remote monitoring solutions.

## 7.5 Ethical AI in Autonomous Systems

The development of autonomous robotic systems for surgery, rehabilitation, and elder care raises ethical questions. Future advancements must address issues such as accountability, bias, and informed consent to ensure equitable and trustworthy AI systems.

## 7.6 Integrating Nanorobotics in Healthcare

Nanorobotics represents a frontier for internal treatments, such as targeted drug delivery and microsurgery. Research into biohybrid microbots capable of navigating the human body autonomously promises groundbreaking applications in oncology and regenerative medicine [69].

## 7.7 Market Growth and Adoption

The global medical robotics market, valued at $7.24 billion in 2015, is projected to surpass $20 billion by 2023 [63]. This growth is driven by increasing investments in AI-powered healthcare solutions and widespread adoption of robotic systems in both developed and developing countries.

# 8 Ethical Consideration and Privacy

The integration of AI and robotics in healthcare offers unprecedented opportunities but also raises significant ethical and privacy challenges. Addressing these concerns is critical to building trust, ensuring equitable access, and safeguarding patient welfare. This section highlights key issues and actionable solutions.

## *8.1 Patient Autonomy*

AI-driven robotics often assist in making complex medical decisions, such as treatment recommendations and diagnostic interpretations. To uphold patient autonomy, transparency in how these systems operate is essential. AI algorithms must provide clear explanations for their outputs, enabling patients and healthcare providers to make informed decisions collaboratively. For example, explainable AI (XAI) systems offer visual interpretations of decisions, fostering trust in AI-assisted care [70].

## *8.2 Accountability and Liability*

Determining responsibility for errors or adverse outcomes involving AI and robotics is a pressing concern. Legal frameworks must address questions such as:

- Who is liable if an AI-powered robot malfunctions during surgery?
- How can accountability be established for algorithmic biases leading to diagnostic errors? Best practices suggest adopting shared accountability models where developers, healthcare providers, and regulators collaboratively ensure system safety [71].

## *8.3 Bias and Fairness*

AI algorithms may perpetuate biases embedded in training datasets, leading to unequal treatment across demographic groups. For instance, AI diagnostic tools have shown discrepancies in accuracy for different racial and ethnic groups due to insufficient representation in datasets [72]. To mitigate this:

- Datasets must be diverse and inclusive.
- Regular audits should ensure fairness and eliminate algorithmic biases.

### 8.4 Data Security and Privacy

Robotic systems equipped with AI rely on vast amounts of sensitive patient data, raising concerns about breaches and misuse. Blockchain technology offers a potential solution by providing immutable and secure data storage, ensuring that only authorized personnel can access patient records [73]. Furthermore, adopting stringent cybersecurity measures, such as encryption and two-factor authentication, can reduce risks [74].

### 8.5 Informed Consent and Data Transparency

Patients must be fully informed about how their data is collected, analyzed, and stored. Clear communication of AI capabilities and limitations is essential to gaining consent. For example, during clinical trials involving robotic systems, participants should understand how AI predictions will influence medical decisions [75].

### 8.6 Ethical Use of Autonomous Systems

As autonomous robotic systems advance, ethical considerations around human oversight become paramount. Autonomous surgical robots, for instance, should be designed to alert human operators in ambiguous situations, ensuring a fail-safe mechanism [76]. This hybrid model of autonomy and oversight reduces risks while maintaining human control.

## 9 Conclusion

The integration of artificial intelligence (AI) and robotics into healthcare represents a transformative advancement, redefining the way medical services are delivered. These technologies have enabled precision in diagnostics, minimally invasive surgical interventions, and the customization of treatments through predictive analytics. For instance, AI-driven algorithms analyze extensive patient data to detect patterns that may elude human observation, while robotics performs tasks with unparalleled accuracy. Together, these innovations align with the principles of predictive, personalized, preventive, participatory, and precise (P5) medicine, shifting healthcare from a reactive to a proactive model. This paradigm shift not only improves patient outcomes but also optimizes the allocation of resources, streamlining workflows and reducing operational inefficiencies.

Building on these advancements, the theoretical implications of AI and robotics in healthcare are substantial. Firstly, these technologies expand human-technology interaction models, emphasizing collaborative decision-making between medical professionals and intelligent systems. Furthermore, they enrich precision medicine frameworks by demonstrating how AI and robotics enable tailored healthcare interventions. The integration of these technologies also advances systems theory, showcasing the seamless collaboration between human and artificial agents in a socio-technical healthcare environment. Importantly, the ethical considerations highlighted in this chapter contribute to developing fair and accountable AI frameworks, ensuring that the benefits of these technologies are equitably distributed across diverse populations.

From a practical perspective, the applications of AI and robotics have far-reaching implications. By automating routine tasks, such as administrative work and medication delivery, these technologies reduce the workload on healthcare professionals, allowing them to focus on more complex responsibilities. Additionally, telepresence robots and AI-enabled telemedicine platforms extend healthcare services to underserved regions, bridging accessibility gaps. Robotic rehabilitation systems further enhance patient recovery by providing personalized therapy regimens, addressing shortages in specialized care. Despite the high initial costs, these technologies offer long-term economic benefits, including reduced hospital stays, efficient resource utilization, and streamlined operations, which justify the investment in their implementation.

Nevertheless, the adoption of AI and robotics in healthcare is not without limitations. High implementation costs remain a significant barrier, particularly for low-resource healthcare systems. Moreover, challenges related to data security, patient privacy, and algorithmic bias raise critical ethical and legal concerns. The successful integration of these technologies also depends on extensive training and skill development for healthcare professionals, which may delay widespread adoption. Furthermore, the effectiveness of AI algorithms hinges on access to accurate and diverse datasets, and interoperability issues between AI systems and existing healthcare infrastructure present additional technical challenges. Addressing these limitations is essential to fully realize the transformative potential of AI and robotics in healthcare. With continued advancements in research, regulatory frameworks, and training initiatives, these technologies can revolutionize healthcare delivery, enhancing patient outcomes and global well-being.

## References


1. Poushneh, A., Vasquez-Parraga, A., Gearhart, R.S.: The effect of empathetic response and consumers' narcissism in voice-based artificial intelligence. J. Retail. Consum. Serv. **79**, 103871 (2024)
2. Cohen, T.A., Patel, V.L., Shortliffe, E.H.: Intelligent systems in medicine and health. In: The role of AI 2022 (2021)

3. Sheetz, K.H., Claflin, J., Dimick, J.B.: Trends in the adoption of robotic surgery for common surgical procedures. JAMA Netw. Open **3**(1), e1918911–e1918911 (2020)
4. Newman-Toker, D.E., Wang, Z., Zhu, Y., Nassery, N., Tehrani, A.S.S., Schaffer, A.C., Siegal, D., et al.: Rate of diagnostic errors and serious misdiagnosis-related harms for major vascular events, infections, and cancers: toward a national incidence estimate using the "Big Three." Diagnosis **8**(1), 67–84 (2021)
5. Longo, U.G., Carnevale, A., Massaroni, C., Lo Presti, D., Berton, A., Candela, V., Denaro, V., et al.: Personalized, predictive, participatory, precision, and preventive (P5) medicine in rotator cuff tears. J. Personal. Med. **11**(4), 255 (2021)
6. Chatterjee, S., Das, S., Ganguly, K., Mandal, D.: Advancements in robotic surgery: innovations, challenges and future prospects. J. Robot. Surg. **18**(1), 28 (2024)
7. Johansson, B., Eriksson, E., Berglund, N., Lindgren, I.: Robotic surgery: review on minimally invasive techniques. Fusion Multidiscip. Res. Int. J. **2**(2), 201–210 (2021)
8. Haleem, A., Javaid, M., Singh, R.P., Suman, R.: Medical 4.0 technologies for healthcare: features, capabilities, and applications. Internet Things Cyber Phys. Syst. **2**, 12–30 (2022)
9. Zewail, A., Saber, S.: AI-powered analytics in healthcare: enhancing decision-making and efficiency. Int. J. Appl. Health Care Anal. **8**(5), 1–16 (2023)
10. Sumner, J., Lim, H.W., Chong, L.S., Bundele, A., Mukhopadhyay, A., Kayambu, G.: Artificial intelligence in physical rehabilitation: a systematic review. Artif. Intell. Med., 102693 (2023)
11. Prabhod, K.J.: The role of artificial intelligence in reducing healthcare costs and improving operational efficiency. Q. J. Emerg. Technol. Innov. **9**(2), 47–59 (2024)
12. Ragno, L., Borboni, A., Vannetti, F., Amici, C., Cusano, N.: Application of social robots in healthcare: review on characteristics, requirements, technical solutions. Sensors **23**(15), 6820 (2023)
13. Naseer, F., Khan, M.N., Nawaz, Z., Awais, Q.: Telepresence robots and controlling techniques in healthcare system. Comput. Mater. Continua **74**(3) (2023)
14. Holland, J., Kingston, L., McCarthy, C., Armstrong, E., O'Dwyer, P., Merz, F., McConnell, M.: Service robots in the healthcare sector. Robotics **10**(1), 47 (2021)
15. Sarker, S., Jamal, L., Ahmed, S.F., Irtisam, N.: Robotics and artificial intelligence in healthcare during COVID-19 pandemic: a systematic review. Robot. Auton. Syst. **146**, 103902 (2021)
16. Lee, H.C., Pacheco, N.E., Fichera, L., Russo, S.: When the end effector is a laser: a review of robotics in laser surgery. Adv. Intell. Syst. **4**(10), 2200130 (2022)
17. Li, Y., Wang, M., Wang, L., Cao, Y., Liu, Y., Zhao, Y., Kang, H., et al.: Advances in the application of ai robots in critical care: scoping review. J. Med. Internet Res. **26**, e54095 (2024)
18. Mun, S.K., Wong, K.H., Lo, S.C.B., Li, Y., Bayarsaikhan, S.: Artificial intelligence for the future radiology diagnostic service. Front. Mol. Biosci. **7**, 614258 (2021)
19. Farooq Mohi-U-din, S., Tariq, M.: Advancing healthcare: the power of AI in robotics, diagnostics, and precision medicine. Revista de Inteligencia Artificial enMedicina **15**(1):87–112 (2024)
20. Habuza, T., Navaz, A.N., Hashim, F., Alnajjar, F., Zaki, N., Serhani, M.A., Statsenko, Y.: AI applications in robotics, diagnostic image analysis and precision medicine: current limitations, future trends, guidelines on CAD systems for medicine. Inf. Med. Unlocked **24**, 100596 (2021)
21. Rath, K.C., Khang, A., Rath, S.K., Satapathy, N., Satapathy, S.K., Kar, S.: Artificial intelligence (AI)-enabled technology in medicine-advancing holistic healthcare monitoring and control systems. In: Computer Vision and AI-Integrated IoT Technologies in the Medical Ecosystem, pp. 87–108. CRC Press (2024)
22. Iftikhar, M., Saqib, M., Zareen, M., Mumtaz, H.: Artificial intelligence: revolutionizing robotic surgery. Ann. Med. Surg. **86**(9), 5401–5409 (2024)
23. Moglia, A., Georgiou, K., Georgiou, E., Satava, R.M., Cuschieri, A.: A systematic review on artificial intelligence in robot-assisted surgery. Int. J. Surg. **95**, 106151 (2021)
24. Rokhshad, R., Keyhan, S.O., Yousefi, P.: Artificial intelligence applications and ethical challenges in oral and maxillo-facial cosmetic surgery: a narrative review. Maxillofac. Plast. Reconstr. Surg. **45**(1), 14 (2023)

25. Varghese, C., Harrison, E.M., O'Grady, G., Topol, E.J.: Artificial intelligence in surgery. Nat. Med., 1–12 (2024)
26. Afzal, A., Katz, D.S., Goues, C.L., Timperley, C.S.: A study on the challenges of using robotics simulators for testing (2020). Preprint at arXiv:2004.07368
27. Zhang, H., Jiao, L., Yang, S., Li, H., Jiang, X., Feng, J., Wei, B., et al.: Brain-computer interfaces: the innovative key to unlocking neurological conditions. Int. J. Surg., 10–1097 (2024)
28. Zhang, M., Wu, S., Ibrahim, M.I., Noor, S.S.M., Mohammad, W.M.Z.W.: Significance of ongoing training and professional development in optimizing healthcare-associated infection prevention and control. J. Med. Signals Sens. **14**(5), 14 (2024)
29. Singh, A., Verma, A., Sharma, A., Malviya, R., Sekar, M.: Use of artificial intelligence and robotics: making the drug development process easier. In: Pharmaceutical Industry 4.0: Future, Challenges and Application, pp. 145–185. River Publishers (2023)
30. Medina, D.A.V., Maciel, E.V.S., Lanças, F.M.: Modern automated sample preparation for the determination of organic compounds: a review on robotic and on-flow systems. TrAC Trends Anal. Chem., 117171 (2023)
31. Iqbal, M.J., Javed, Z., Sadia, H., Qureshi, I.A., Irshad, A., Ahmed, R., Sharifi-Rad, J., et al.: Clinical applications of artificial intelligence and machine learning in cancer diagnosis: looking into the future. Cancer Cell Int. **21**(1), 270 (2021)
32. Hassan, M., Awan, F.M., Naz, A., deAndrés-Galiana, E.J., Alvarez, O., Cernea, A., Kloczkowski, A., et al.: Innovations in genomics and big data analytics for personalized medicine and health care: a review. Int. J. Mol. Sci. **23**(9), 4645 (2022)
33. Sanadgol, N., Wackerlig, J.: Developments of smart drug-delivery systems based on magnetic molecularly imprinted polymers for targeted cancer therapy: a short review. Pharmaceutics **12**(9), 831 (2020)
34. Banerjee, A., Chakraborty, C., Kumar, A., Biswas, D.: Emerging trends in IoT and big data analytics for biomedical and health care technologies. In: Handbook of Data Science Approaches for Biomedical Engineering, pp. 121–152 (2020)
35. Taravati, S., Capaci, K., Uzumcugil, H., Tanigor, G.: Evaluation of an upper limb robotic rehabilitation program on motor functions, quality of life, cognition, and emotional status in patients with stroke: a randomized controlled study. Neurol. Sci., 1–12 (2022)
36. Bardi, E., Gandolla, M., Braghin, F., Resta, F., Pedrocchi, A.L., Ambrosini, E.: Upper limb soft robotic wearable devices: a systematic review. J. Neuroeng. Rehabil. **19**(1), 87 (2022)
37. Louie, D.R.: Walking once again: use of a robotic exoskeleton during subacute stroke rehabilitation to promote functional recovery. Doctoral dissertation, University of British Columbia (2021)
38. Khalid, S., Alnajjar, F., Gochoo, M., Renawi, A., Shimoda, S.: Robotic assistive and rehabilitation devices leading to motor recovery in upper limb: a systematic review. Disabil. Rehabil. Assist. Technol. **18**(5), 658–672 (2023)
39. McFarland, D.J., Daly, J., Boulay, C., Parvaz, M.A.: Therapeutic applications of BCI technologies. Brain-Comput. Interfaces **4**(1–2), 37–52 (2017)
40. Dehghani, H., Kim, P.C.: Robotic automation for surgery. In: Digital surgery, pp. 203–213 (2021)
41. Ness, S., Xuan, T.R., Oguntibeju, O.O.: Influence of AI: robotics in healthcare. Asian J. Res. Comput. Sci. **17**(5), 222–237 (2024)
42. Javaid, M., Haleem, A., Singh, R.P., Suman, R.: Artificial intelligence applications for industry 4.0: a literature-based study. J. Ind. Integr. Manag. **7**(1), 83–111 (2022)
43. Bekbolatova, M., Mayer, J., Ong, C.W., Toma, M.: Transformative potential of AI in Healthcare: definitions, applications, and navigating the ethical Landscape and Public perspectives. Healthcare **12**(2), 125 (2024). MDPI
44. Robinson, F., Nejat, G.: An analysis of design recommendations for socially assistive robot helpers for effective human-robot interactions in senior care. J. Rehabil. Assist. Technol. Eng. **9**, 20556683221101388 (2022)
45. Zachariae, A., Plahl, F., Tang, Y., Mamaev, I., Hein, B., Wurll, C.: Human-robot interactions in autonomous hospital transports. Robot. Auton. Syst. **179**, 104755 (2024)

46. Mohanty, K., Subiksha, S., Kirthikka, S., Sujal, B. H., Sokkanarayanan, S., Bose, P., & Sathiyanarayanan, M. (2021). AI-Powered Robotics and COVID-19: Challenges and Opportunities. In: Robotic Technologies in Biomedical and Healthcare Engineering, 117–128.
47. Reardon, S.: Rise of robot radiologists. Nature **576**(7787), S54–S58 (2019)
48. Kaur, J.: FutureCare: AI robots revolutionizing health and healing. In: Revolutionizing the Healthcare Sector with AI, pp. 311–340. IGI Global (2024)
49. Shajari, S., Kuruvinashetti, K., Komeili, A., Sundararaj, U.: The emergence of AI-based wearable sensors for digital health technology: a review. Sensors **23**(23), 9498 (2023)
50. Igwama, G.T., Olaboye, J.A., Cosmos, C., Maha, M.D.A., Abdul, S.: AI-powered predictive analytics in chronic disease management: regulatory and ethical considerations (2024)
51. Ali, O., Abdelbaki, W., Shrestha, A., Elbasi, E., Alryalat, M.A.A., Dwivedi, Y.K.: A systematic literature review of artificial intelligence in the healthcare sector: benefits, challenges, methodologies, and functionalities. J. Innov. Knowl. **8**(1), 100333 (2023)
52. Morris, M.X., Fiocco, D., Caneva, T., Yiapanis, P., Orgill, D.P.: Current and future applications of artificial intelligence in surgery: implications for clinical practice and research. Front. Surg. **11**, 1393898 (2024)
53. Rjab, A.B., Mellouli, S.: Smart cities in the era of artificial intelligence and internet of things: promises and challenges. In: Smart Cities and Smart Governance: Towards the 22nd Century Sustainable City, pp. 259–288
54. Zuo, Y.: Exploring the synergy: AI enhancing blockchain, blockchain empowering AI, and their convergence across IoT applications and beyond. IEEE Internet Things J. (2024)
55. Tamilarasi, M., Hemalatha, S., Jothimani, K., Ashokkumar, L.: Existing tools and technologies in biomedical and healthcare system. In: Blockchain for Biomedical Research and Healthcare: Concept, Trends, and Future Implications, pp. 17–49. Springer Nature, Singapore (2024)
56. Merhej, J., Harb, H., Abouaissa, A., Idoumghar, L.: Toward a new era of smart and secure healthcare information exchange systems: combining blockchain and artificial intelligence. Appl. Sci. **14**(19), 8808 (2024)
57. Singh, J., Sillerud, B., Singh, A.: Artificial intelligence, chatbots and ChatGPT in healthcare—narrative review of historical evolution, current application, and change management approach to increase adoption. J. Med. Artif. Intell., 6 (2023)
58. Muheidat, F., Tawalbeh, L.A.: AIoMT artificial intelligence (AI) and internet of medical things (IoMT): applications, challenges, and future trends. In: Computational Intelligence for Medical Internet of Things (MIoT) Applications, pp. 33–54. Academic Press (2023)
59. Thangam, D., Malali, A.B., Subramaniyan, G., Mariappan, S., Mohan, S., Park, J.Y.: Relevance of artificial intelligence in modern healthcare. In: Integrating AI in IoT Analytics on the Cloud for Healthcare Applications, pp. 67–88. IGI Global (2022)
60. Matthew, U.O., Kazaure, J.S., Amaonwu, O., Adamu, U.A., Hassan, I.M., Kazaure, A.A., Ubochi, C.N.: Role of internet of health things (IoHTs) and innovative internet of 5G medical robotic things (IIo-5GMRTs) in COVID-19 global health risk management and logistics planning. In: Intelligent Data Analysis for COVID-19 Pandemic, pp. 27–53. Springer, Singapore (2021)
61. Virman, T., Sharma, A., Sharma, A., Kumar, G., Bhati, M.: Artificial intelligence-based IoT applications in future pandemics. In: Machine Intelligence for Internet of Medical Things: Applications and Future Trends, pp. 83–106. Bentham Science Publishers (2023)
62. Firouzi, F., Farahani, B., Daneshmand, M., Grise, K., Song, J., Saracco, R., Luo, A., et al.: Harnessing the power of smart and connected health to tackle COVID-19: IoT, AI, robotics, and blockchain for a better world. IEEE Internet Things J. **8**(16), 12826–12846 (2021)
63. Siddiqi, A., Horan, T., Molloy, R.M., Bloomfield, M.R., Patel, P.D., Piuzzi, N.S.: A clinical review of robotic navigation in total knee arthroplasty: historical systems to modern design. EFORT Open Rev. **6**(4), 252–269 (2021)
64. Zeb, S., Nizamullah, F.N.U., Abbasi, N., Fahad, M.: AI in healthcare: revolutionizing diagnosis and therapy. Int. J. Multidiscip. Sci. Arts **3**(3), 118–128 (2024)
65. Abbasi, N., Hussain, H.K.: Integration of artificial intelligence and smart technology: AI-driven robotics in surgery: precision and efficiency. J. Artif. Intell. Gen. Sci. (JAIGS) **5**(1), 381–390 (2024). ISSN: 3006-4023

66. Anmar Albarwary, S., Mohammed, R.B.: Design development of robotic exoskeletons in rehabilitation. J. China Univ. Pet. **48**(3) (2024)
67. Singhania, K., Reddy, A.: Improving preventative care and health outcomes for patients with chronic diseases using big data-driven insights and predictive modeling. Int. J. Appl. Health Care Anal. **9**(2), 1–14 (2024)
68. Pramanik, P.K.D., Upadhyaya, B.K., Pal, S., Pal, T.: Internet of things, smart sensors, and pervasive systems: enabling connected and pervasive healthcare. In: Healthcare Data Analytics and Management, pp. 1–58. Academic Press (2019).
69. Zarepour, A., Khosravi, A., Iravani, S., Zarrabi, A.: Biohybrid micro/nanorobots: pioneering the next generation of medical technology. Adv. Healthc. Mater., 2402102 (2024)
70. Adeniran, A.A., Onebunne, A.P., William, P.: Explainable AI (XAI) in healthcare: enhancing trust and transparency in critical decision-making (2024)
71. Sittig, D.F., Belmont, E., Singh, H.: Improving the safety of health information technology requires shared responsibility: it is time we all step up. Healthcare **6**(1), 7–12 (2018). Elsevier
72. Daneshjou, R., Smith, M.P., Sun, M.D., Rotemberg, V., Zou, J.: Lack of transparency and potential bias in artificial intelligence data sets and algorithms: a scoping review. JAMA Dermatol. **157**(11), 1362–1369 (2021)
73. Attaran, M.: Blockchain technology in healthcare: challenges and opportunities. Int. J. Healthc. Manage. **15**(1), 70–83 (2022)
74. Gadde, S., Rao, G.S., Veesam, V.S., Yarlagadda, M., Patibandla, R.S.M.: Secure data sharing in cloud computing: a comprehensive survey of two-factor authentication and cryptographic solutions. Ingénierie des Systèmes d'Information **28**(6) (2023)
75. Asan, O., Bayrak, A.E., Choudhury, A.: Artificial intelligence and human trust in healthcare: focus on clinicians. J. Med. Internet Res. **22**(6), e15154 (2020)
76. Liu, T., Wang, J., Wong, S., Razjigaev, A., Beier, S., Peng, S., Wu, L., et al.: A review on the form and complexity of human-robot interaction in the evolution of autonomous surgery. Adv. Intell. Syst. **6**(11), 2400197 (2024)

# Exploring the Relationship Between Financial Inclusion and Broader Sustainable Development Within Low-Income Sectors

**K. Shriya and T. Velmurugan**

**Abstract** This study examines the complex relationship between digital financial inclusion and broader sustainable development objectives, particularly within low-income sectors. Through a comprehensive methodology incorporating systematic literature review, theoretical frameworks, and both quantitative and qualitative analyses, the research investigates how financial inclusion correlates with poverty alleviation, economic growth, and social development. The study employs a structured questionnaire methodology with 134 respondents and utilizes statistical tools including correlation analysis, regression analysis, and Partial Least Square Structural Equation Modeling (PLS-SEM). Findings reveal significant positive correlations between financial inclusion initiatives and sustainable development outcomes, with behavioral intention, social influence, and usability emerging as key determinants. The research demonstrates that participation in the financial system stimulates business creation, employment opportunities, and economic growth while reducing income disparities. Additionally, it examines financial inclusion's influence on enhancing investment knowledge and other critical Sustainable Development Goals (SDGs) components. These insights prove valuable for policymakers, financial institutions, and development professionals working to create more inclusive and sustainable financial systems.

K. Shriya (✉) · T. Velmurugan
Faculty of Management, SRM Institute of Science and Technology, Kattankulathur, Tamil Nadu, India
e-mail: shriyakrishanan18@gmail.com

T. Velmurugan
e-mail: velmurut@srmist.edu.in

T. Velmurugan
Assistant Professor, Faculty of Management, SRM Institute of Science and Technology, Kattankulathur, Chennai, India

K. Shriya
Research Scholar, Faculty of Management, SRM Institute of Science and Technology, Kattankulathur, Chennai, India



N. D. K. Nguyen et al. (eds.), *Digital Transformation and Sustainability: Methods and Applications*, Information Systems Engineering and Management 47,
https://doi.org/10.1007/978-3-031-91278-8_14

**Keywords** Digital financial inclusion · Economic empowerment · Financial literacy · Decision-making strategies · Financial technology

## 1 Introduction

Recent developments have highlighted the critical role of financial inclusion in achieving sustainable development, particularly for economically disadvantaged populations. Financial inclusion extends beyond basic banking services to encompass broader socio-economic factors, including poverty eradication, gender equality, and economic empowerment. This study addresses a key research question: What is the relationship between financial services utilization and sustainable development in low-income areas? The research aims to provide insights into the mechanisms shaping socio-economic development from both top-down and bottom-up perspectives by examining how financial access initiatives impact sustainable development goals. The intersection of financial inclusion and sustainable development has garnered increasing attention in academic and policy circles. Financial inclusion—defined as the access to and utilization of affordable and regulated financial services—serves as a cornerstone for achieving several United Nations Sustainable Development Goals (SDGs), including poverty alleviation, reduced inequalities, and gender equity [1, 2]. By integrating underserved populations into formal financial systems, financial inclusion not only enhances individual and collective economic resilience but also facilitates investments in health, education, and entrepreneurship, which are pivotal for sustainable economic growth [2, 3]. In low-income sectors, the lack of access to formal financial services exacerbates socio-economic disparities and stymies community-level growth. For instance, unbanked populations—estimated at 1.7 billion globally—often rely on informal systems, which are less secure and limit economic empowerment (World Bank 2017). Digital financial technologies, such as mobile banking and peer-to-peer lending platforms, have emerged as transformative tools in bridging this gap, enabling inclusive growth while supporting environmental sustainability through reduced carbon footprints from traditional banking operations [4, 5].

Despite these advancements, challenges such as technological literacy gaps, infrastructural deficiencies, and regulatory hurdles persist, particularly in underdeveloped regions. Addressing these challenges requires a multidimensional approach that integrates financial literacy, inclusive policy frameworks, and innovative fintech solutions tailored to local needs. This paper explores the multifaceted relationship between financial inclusion and sustainable development, analyzing its role in fostering equitable socio-economic transformation while addressing associated risks and opportunities. The findings provide actionable insights for policymakers and stakeholders committed to fostering inclusive and sustainable financial ecosystems. Through an interdisciplinary approach combining economics, finance, sociology, and development studies, this research seeks to understand the complex interactions between financial inclusion and sustainable development. This paper highlights the multifaceted challenges and opportunities in expanding economic access for underserved populations. The study considers various aspects including cultural factors, institutional frameworks, and technological advances in policy formulation.

## 2 Background and Context of Financial Inclusion

There has been a notable emphasis on financial inclusion worldwide in recent times due to the recognition that having access to regulated public utility financial services is not only a fundamental human right but also crucial for economic development and reducing poverty. Financial inclusion refers to the provision as well as the accessibility of credit, savings, insurance and other forms of payments to the people as well as the business entities regardless their standard of living. As evidenced in the case of several developing nations and particularly in the low-income groups, a large part of the population still remains outside the what can be termed as the formal financial sector. According to the World Bank's Global Findex database, there are more than 1 billion individuals who lack access to any banking or mobile money services. The global population of still unbanked individuals amounts to 1.7 billion people who cannot use basic financial services. Most of the world's unbanked population resides in the financially inaccessible areas and low-income areas with few or no access to physical branch banks.

The relationship between financial inclusion and sustainable development is multifaceted. Empirical evidence suggests that access to financial services facilitates poverty alleviation, fosters entrepreneurship, and promotes human capital investment, all of which are vital components of economic growth and social development [2, 5]. For instance, financial services enable individuals to invest in education and health, secure against income shocks, and engage in productive economic activities, contributing to improved household welfare and broader developmental outcomes [1].

## 3 Importance of Sustainable Development

Sustainable development forms a critical perspective when analyzing the complexity between financial access and micro foundations of sustainable development in the low-income context. It also stresses engagements for current and necessary consumption and production without destructing future generations' capacity to replicate and use monetary facilities and products. However, as the principles of sustainable development imply that the social, economic, and environmental aspects are linked, there is the need for intervention that works to eradicate poverty, bring social justice, and preserve the environment. Thus, in low-income sectors, where people's sensitivity to adverse economic fluctuations and natural disasters is high, the principles of sustainable development establish the financial environments appropriate for developing the communities supported by natural assets but non-harming them. Also, sustainable development top-down view, which encompasses economic, social, and environmental aspects, prompts the researchers to explore not only short-term economic effects of financial initiatives but also their long-term social and environmental implications. Hence, this ambitious aim is well achieved by this systematic approach as it fronts an ideal guide that provides insights of the comprehensive link between monetary attachment and maintainable progress goals hence aid the formulation of evidence-based interventional policies and strategies geared towards supporting inclusive growth without compromising the environment. Thus, it is crucial to incorporate the sustainable development principles regarding the financial access and sustainable development in low-income segments of economy into the further investigation of the connection between Fintech solutions.

Sustainable development emphasizes the integration of economic, social, and environmental goals to ensure the well-being of current and future generations. It serves as a guiding framework for addressing global challenges such as poverty, inequality, and climate change. Financial inclusion plays a critical role in this context by providing marginalized populations with the tools to participate in economic systems and build financial resilience. This, in turn, supports the attainment of SDGs by fostering inclusive growth, reducing inequalities, and enhancing social cohesion [2, 6]. In low-income sectors, where vulnerabilities to economic fluctuations and environmental risks are pronounced, financial inclusion offers a pathway to sustainable development. By enabling access to credit, insurance, and savings, it empowers communities to invest in sustainable livelihoods, adapt to economic challenges, and contribute to environmental preservation. Additionally, financial inclusion facilitates the transition towards a digital economy, reducing reliance on cash and promoting transparency and efficiency in financial transactions [7, 8].

## 4 Research Objectives

1. Examine the behaviour intention on financial inclusion among the low-income sectors.
2. Analyse the correlation between the accessibility of financial services and the degree of integrating the low-income population.
3. Evaluate on how perceptions of risk affect the take-up and utilization of financial access in the low-income segments.
4. Understand how social influence factors can either support or challenge attempts to advance purposeful financial inclusion of disenfranchised groups of people.
5. Assess how effort expectancy influences use and availability of financial services to the low-income earners.
6. Review the effectiveness of financial solutions addressing the problems of low-income population.

## 5 Review of Literature

This research conducted a systematic review of published articles from Embase, PubMed, and Web of Science databases spanning 2000 to February 2020 to understand the relationship between demographic and socio-economic factors influencing credit accessibility. The study focused on evaluating empirical evidence of credit scoring discrimination based on variables including education, income, gender, race, and age. The findings emphasized the importance of financial education for formal credit inclusion while highlighting challenges faced by disadvantaged groups such as women, minorities, and low-income earners. Additionally, the research underscored the significance of addressing these gaps in credit access to achieve UN Sustainable Development Goals (SDGs), particularly in poverty eradication and inequality reduction [1].

Hussain et al. [2] examined the complexities of Financial Inclusion (FI), Information and Communication Technology (ICT), and Sustainable Economic Development (SED) across 110 developing countries from 2004–2020. Using Principal Component Analysis (PCA) and Robust Panel Estimations (RPE), their research established a sustained positive association between FI, ICT, and SED regardless of country income levels. By incorporating innovation diffusion theory and environmental considerations, their study addressed a significant gap in empirical literature.

Research by Chibba [9] investigated the relationships between Financial Inclusion and poverty eradication within the context of Millennium Development Goals (MDGs). The study determined that conventional approaches were insufficient and proposed a comprehensive framework incorporating key FI constituents. Their findings highlighted the urgent need for increased attention to FI demand and accessibility in the context of global financial changes.

Ratnawati [10] analyzed the linkages between financial access and development parameters in ten Asian nations. The study assessed multi-bank competition's impact and usage rates on banking service availability and economic well-being through banking penetration rates. The results demonstrated that improving financial inclusion quality positively affected these factors, though opportunities for enhancement remained.

Omar et al. [5] evaluated the importance of outreach in affecting poverty and income distribution across 116 developing nations from 2004 to 2016. By examining factors such as income, internet usage, and purchasing power, they found that increased access to banking and state-affiliated sources correlated with decreased poverty indices and reduced income disparity. Their findings suggested the need for optimizing user experiences in banking and financial services to benefit marginalized populations.

## 6 Impact of Financial Inclusion on Poverty Alleviation and Economic Growth

Financial inclusion therefore means the ability of people and firms to access and use financial products and services especially for people in poverty, lower-income groups, and other excluded groups. It has a humongous impact on the annotations to poverty eradication as well as on economic growth and prosperity.

Poverty Alleviation:

1. Capital Accessibility: Specifically, on the usage aspect of the financial system, FI assists people to acquire credit and manage their savings to fund education, health, and business opportunities. This accessibility also entails that people are capable of coming out of the poverty wagon and embrace productive investments.
2. Risk Mitigation: By way of bringing about an improved economic state, insurance forms part of financial liberalization to allow for insurance services to be of help in risk management in the occurrence of something like sickness or disasters. This ensures that the vulnerable groups do not fall even more into extreme poverty as a result of unfavourable events.
3. Income Stability: Some of the advantages include the provision of more effective ways and means of handling financial resources for useful consumption plans and predictable events, resulting to less chances of people falling into the bracket of poverties during economic sensitive periods.

Economic Growth:

1. Entrepreneurship Promotion: Financial inclusion promotes self-employment because it enables those who desire to venture into business by providing them with the essential monetary needs for business, create employment, increase productivity and invention that spur economic growth.
2. Human Capital Investment: This is one of the ways through which, by acquiring credit, financial services facilitate investment on education and skill development which at large improve human capital. Well trained employees mean more efficiency and thus will lead to enhancement of the growth of any economy.
3. Market Expansion: It widens financial service's outreach, increases customer base of financial institutions, supports process of financial intermediation and accumulation of financial resources which is important for development of an economy.
4. Economic Stability: An economically inclusive economy is usually more secure helping to eliminate extreme income disparity and increase the budgetary security of individuals, groups and communities hence adding value to investors, creating demand for goods and services and thus triggering economic development.

Social Development:

1. Empowerment and Inclusion: The process of providing the underprivileged sections of society with the same opportunities as others in accessing financial

services and tools helps advance social inclusion by allowing the formation of economically independent groups/individuals and inclusion in related processes.
2. Inequality Reduction: Since FI enables wealth creation and the access and use of production resources by the poor, and empowers poor people to access vital financial services, it shall contribute positively to poverty alleviation by closing the wealth and asset ownership gap within communities.

Policy Implications:

1. Regulatory Support: People's governments, at national level as well as subnational entities, and other policy makers have a central part in ensuring FSPs by enacting such important policy instruments as supporting regulation and own interventions in the financial markets and encouraging investment into the related infrastructure.
2. Collaborative Approach: Breaking the barriers to financial inclusion requires cooperation between governments, financial institution, CSOs and private sector for results to be sustainable.
3. Fintech Integration: The integration of the technological advancement in the financial sector (fintech) can boost the financial inclusion process through the following; the reduction of transaction costs, diversity of the channels of accessing the financial services and effectively covering the hard-to-reach populations.

**Challenges and opportunities for achieving financial inclusion in low-income sectors**

Establishing financial inclusion within the limited income sectors of the society is not an easy task, and it requires an integration of the various programs. Still, major challenges remain in place to provide the marginalized population with adequate access to financial services. This discourse looks at the complex nature of issues and possibilities of extending the financial inclusion to the economically weakened parts.

Challenges:

1. Infrastructural Deficiency: Due to the lack of traditional banking facilities like branch and ATM the biggest challenge to receive basic formal financial services in the LMI situation persists.
2. Financial Illiteracy: The limited understanding and awareness of the many financial goods and services offered in the market greatly impede individuals' comprehension of formal banking services, leading to apprehension and disengagement from the system.
3. Dominance of Informal Systems: As for the background, the cultures of borrowing from a HS/ML or participating in a ROSCA are rather profound and widely used by low-income people due to the generally informal organization of their financial systems.
4. Income Volatility: Uncertain and volatile income sources common with low-income earner's sectors, pose a challenge to both the individual and the financial

institutions, meaning conventional credit scoring techniques are not effective in evaluating credit worthiness of the irregular earners.
5. Technological Barriers: Technology and digital literacy within the low-income brackets restrain their utilization of the digital financial services which are rapidly going mainstream.
6. Regulatory Constraints: Banking regulation that makes provisions like identification necessary actually limits low-income groups of people who cannot afford any recognized identification.

Opportunities:

1. Mobile Technology Integration: The use of mobile phones is prevalent in poor areas hence a great opportunity in delivering mobile banking and digital wallet services evident by the success of mobile money.
2. Agent Banking: The utilisation of agent banking where the agents offer simple banking services for the banks relieves the problem of lack of physical structures such as retail outlets in the low-income areas by using the existing business establishments to perform the role of business banks.
3. Financial Literacy Initiatives: Structured financial literacy programmes targeted at the low-income populations equip people to act responsibly financially as they avail themselves to those emerging structures of formal financial services.
4. Innovative Product Development: Designing financial products that are appealing to the low-income population which includes micro financial services such as microloans and microinsurance increases their usage among the low-income earners.
5. Collaborative Partnerships: The partnership models of cooperation between various governmental agencies, financial organizations, NGOs, and high-tech companies mean the enhancement of pooled resources and knowledge when addressing the complex issues of financial access more efficiently.
6. Regulatory Innovations: Regulatory environments that can be put in place by governments include; Integrated regulatory measures that ease the KYC process on financial service providers or having graduated regulation for such innovations as the fintech startups (Fig. 1).

## 7 Research Methodology

The kind of analysis used here is descriptive statistics analysis. It is a questionnaire that is being conducted on based millennials and only after rigorous checks on whether he/she has basic investment knowledge and also the financial literacy level. Recalling the methods used in the study above, one can identify the specific technique used—The Judgmental sampling. The researcher has excluded those samples, which were marked with incorrect responses, and, therefore, the remaining number of samples reached 134. The research techniques used in the study were highly recommended by several researchers with high usage rates in several studies as well as

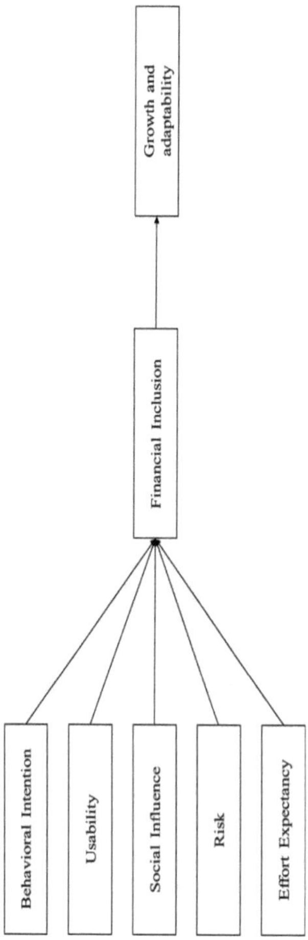


**Fig. 1** Proposed model. *Source* Authors Work

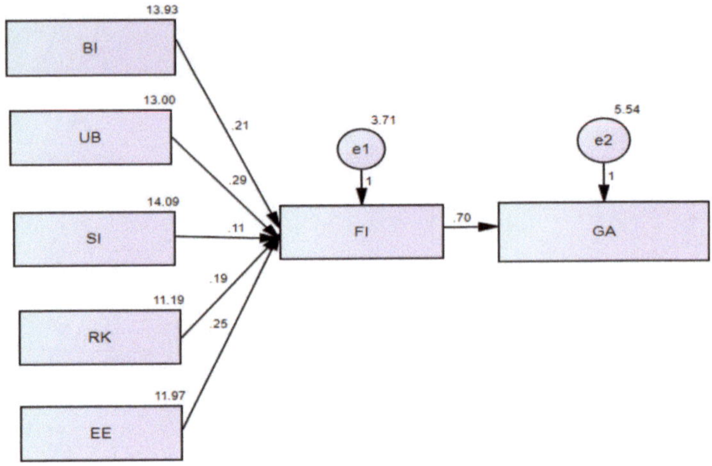


**Fig. 2** SEM analysis. *Source* Authors work

pertinent to the research aims and objectives, reliability, validity, as well as internal consistency. Using a pilot study approach was made. For the reliability coefficient identified in this study it is seen that the researchers used ranges between 0. 7 to 0. 8 While eight of the proposed criteria are appropriate for descriptive research, Smith et al., used these results to establish that the research was dependable (2002). In the study, the statistical tools used are the percentage analysis, correlation analysis, regression analysis, and Partial Least Square Structural Equation Modeling (PLS-SEM). For each part of the work, there are statistical software programmes used as follows: For data analysis—MS Excel; For SEM analysis—PL SEM; and for power analysis—SPSS (Table 1).

The demographic data of the respondents is shown in the analysis generated using the Statistical Package for the Social Sciences (SPSS). gender distribution, with 51.49% males and 48.51% females. The age group is diverse, with the highest representation from individuals aged 35–42 years (21.64%), followed by 18–26 years (20.89%) and 43–50 years (20.14%). Most respondents are married (51.49%). Regarding occupation, "Others" comprise the largest group (41.81%), followed by households (32.08%) and entrepreneurs (26.11%). Income levels vary, with the majority earning between ₹21,001 to ₹30,000 (23.13%), while a notable proportion earns less than ₹10,000 (17.19%) or more than ₹50,000 (15.67%), reflecting a mix of financial backgrounds (Table 2).

The reliability analysis conducted as part of the research assessed the internal consistency of the measured variables using SPSS. The results indicated favourable reliability levels across the constructs under investigation. Specifically, Cronbach's alpha coefficients ranged from 0.687–0.782, demonstrating satisfactory to excellent reliability. Notably, variables such as usability, social influence, risk, effort expectancy, financial inclusion, and growth and adaptability exhibited particularly

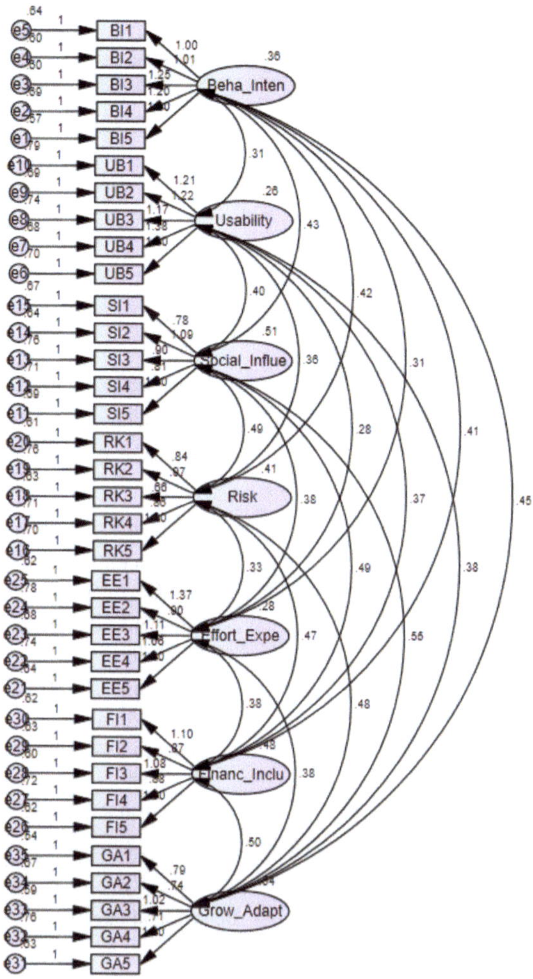


**Fig. 3** Confirmatory factor analysis. *Source* Authors work

strong internal consistency, with coefficients ranging from 0.731–0.782. These findings are crucial for affirming the reliability of the data collected, thereby bolstering the validity of subsequent statistical analyses and interpretations within the research study (Table 3).

| ANOVA[a] | | | | | | |
|---|---|---|---|---|---|---|
| Model | | Sum of squares | df | Mean square | F | Sig |
| 1 | Regression | 1261.94 | 5 | 252.389 | 68.976 | 0.000b |
| | Residual | 468.363 | 128 | 3.659 | | |
| | Total | 1730.31 | 133 | | | |

[a] Dependent variable: GA

[b] Predictors: (Constant), EE, RK, UB, BI, SI

**Table 1** Demographic profile

| S. No | Demographic variable | Category | Frequency | Percentage (%) |
|---|---|---|---|---|
| 1 | Gender | Male | 69 | 51.49 |
| | | Female | 65 | 48.51 |
| 2 | Age | 18–26 | 28 | 20.89 |
| | | 27–34 | 27 | 20.14 |
| | | 35–42 | 29 | 21.64 |
| | | 43–50 | 27 | 20.14 |
| | | above 50 | 23 | 17.19 |
| 3 | Marital status | Married | 69 | 51.49 |
| | | Unmarried | 65 | 48.51 |
| 4 | Occupation | Household | 43 | 32.08 |
| | | Entrepreneur | 35 | 26.11 |
| | | Others | 54 | 41.81 |
| 5 | Income | Less than Rs.10,000 | 23 | 17.19 |
| | | Between Rs.11,000 to Rs. 20,000 | 25 | 18.65 |
| | | Between Rs.21,001 to Rs.30,000 | 31 | 23.13 |
| | | Between Rs.31,001 to Rs.40,000 | 19 | 14.17 |
| | | Between Rs.41,000 to Rs.50,000 | 15 | 11.19 |
| | | More than Rs.50,000 | 21 | 15.67 |

**Table 2** Reliability

| | |
|---|---|
| Behavioural intention | 0.776 |
| Usability | 0.722 |
| Social influence | 0.750 |
| Risk | 0.692 |
| Effort expectancy | 0.710 |
| Financial inclusion | 0.782 |
| Growth and adaptability | 0.747 |

| Coefficients[a] | | | | | | | | |
|---|---|---|---|---|---|---|---|---|
| Model | | Unstandardized coefficients | | Standardized coefficients | t | Sig | Collinearity statistics | |
| | | B | Std. Error | Beta | | | Tolerance | VIF |
| 1 | (Constant) | 1.235 | 1.009 | | 1.224 | 0.223 | | |

(continued)

(continued)

Coefficients[a]

| Model | | Unstandardized coefficients | | Standardized coefficients | t | Sig | Collinearity statistics | |
|---|---|---|---|---|---|---|---|---|
| | | B | Std. Error | Beta | | | Tolerance | VIF |
| | BI | 0.266 | 0.084 | 0.276 | 3.167 | 0.002 | 0.279 | 3.585 |
| | UB | 0.125 | 0.088 | 0.126 | 1.427 | 0.156 | 0.273 | 3.661 |
| | SI | 0.304 | 0.087 | 0.318 | 3.513 | 0.001 | 0.258 | 3.875 |
| | RK | 0.091 | 0.095 | 0.085 | 0.955 | 0.341 | 0.269 | 3.711 |
| | EE | 0.145 | 0.08 | 0.14 | 1.827 | 0.07 | 0.36 | 2.775 |

[a] Dependent variable GA

The correlation table displays the Pearson correlation coefficients among seven variables: There is a series of factors such as Behavioural Interactions which is BI, Usability which is US, Social Influence is SI, Risk is RI, Effort Expectancy which is EE, Financial Inclusion is FI while the last one is Growth and Adaptability which is GA. Each of the correlation coefficients reported is presented together with the associated probability level or p-value, which refers to the 2-tailed test and the sample size of the study (N). It is worth noting the all the variables are strongly positively correlated; the array goes from 0. 734–808. Particularly, BI has significant positive correlations with US with the coefficient of correlation equal to 0. 784 for SI with the coefficient of correlation equal to 0. 758 for RI with the coefficient of correlation equal to 0. 802 for EE with the coefficient of correlation equal to 0. 752 and for FI with the coefficient of correlation equal to 0. 808 Finally, the coefficient of correlation between BI and GA is The same analysis applies to such other variables too; they practically have high, positive coefficients of inter-correlation. These conclusions promote the hypothesis that there is a strong positive correlation between various spheres of business processes, innovation, satisfaction, integration, and engagement with financial issues and administrative activities, which explain the existence of a broad integration within the organization. The statistical values of the correlation coefficients are high, which suggests that changes in one facet are likely to be linked with similar changes in another aspect of the school organization; thus, these relationships should be taken into consideration when adopting the organizational strategies\interpolated intervention (Table 4).

Dependent variable: Growth and adaptability.

Independent variable: Behavioural intention, usability, social influence, risk, effort expectancy.

Mediating variable: Financial inclusion.

Regression analysis examines the causal link between numerous variables. Regression analysis examines the relationship between the dependent variable and independent variable. The study's table value demonstrates how growth and adaptability have a big impact on Behavioural intention, usability, social influence, risk, effort expectancy and financial inclusion. Financial mindset is unaffected by the privacy dimension (Table 5).

**Table 3** Correlation

| Correlations | | | | | | | | |
|---|---|---|---|---|---|---|---|---|
| | | BI | UB | SI | RK | EE | FI | GA |
| BI | Pearson Correlation | 1 | 0.740** | 0.766** | 0.802** | 0.723** | 0.776** | 0.782** |
| | Sig. (2-tailed) | | 0.000 | 0.000 | 0.000 | 0.000 | 0.000 | 0.000 |
| | N | 134 | 134 | 134 | 134 | 134 | 134 | 134 |
| UB | Pearson Correlation | 0.740** | 1 | 0.801** | 0.776** | 0.735** | 0.797** | 0.753** |
| | Sig. (2-tailed) | 0.000 | | 0.000 | 0.000 | 0.000 | 0.000 | 0.000 |
| | N | 134 | 134 | 134 | 134 | 134 | 134 | 134 |
| SI | Pearson Correlation | 0.766** | 0.801** | 1 | 0.773** | 0.748** | 0.771** | 0.800** |
| | Sig. (2-tailed) | 0.000 | 0.000 | | 0.000 | 0.000 | 0.000 | 0.000 |
| | N | 134 | 134 | 134 | 134 | 134 | 134 | 134 |
| RK | Pearson Correlation | 0.802** | 0.776** | 0.773** | 1 | 0.680** | 0.770** | 0.744** |
| | Sig. (2-tailed) | 0.000 | 0.000 | 0.000 | | 0.000 | 0.000 | 0.000 |
| | N | 134 | 134 | 134 | 134 | 134 | 134 | 134 |
| EE | Pearson Correlation | 0.723** | 0.735** | 0.748** | 0.680** | 1 | 0.761** | 0.727** |
| | Sig. (2-tailed) | 0.000 | 0.000 | 0.000 | 0.000 | | 0.000 | 0.000 |
| | N | 134 | 134 | 134 | 134 | 134 | 134 | 134 |
| FI | Pearson Correlation | 0.776** | 0.797** | 0.771** | 0.770** | 0.761** | 1 | 0.755** |
| | Sig. (2-tailed) | 0.000 | 0.000 | 0.000 | 0.000 | 0.000 | | 0.000 |
| | N | 134 | 134 | 134 | 134 | 134 | 134 | 134 |
| GA | Pearson Correlation | 0.782** | 0.753** | 0.800** | 0.744** | 0.727** | 0.755** | 1 |
| | Sig. (2-tailed) | 0.000 | 0.000 | 0.000 | 0.000 | 0.000 | 0.000 | |
| | N | 134 | 134 | 134 | 134 | 134 | 134 | 134 |

** Correlation is significant at the 0.01 level (2-tailed)

**Table 4** Regression

| Model summary | | | | | | | | | |
|---|---|---|---|---|---|---|---|---|---|
| Model | R | R Square | Adjusted R Square | Std. Error of the Estimate | Change Statistics | | | | |
| | | | | | R Square Change | F Change | df1 | df2 | Sig. F Change |
| 1 | 0.854[a] | 0.729 | 0.719 | 1.91287 | 0.729 | 68.976 | 5 | 128 | 0.000 |

[a] Predictors: (Constant), EE, RK, UB, BI, SI

The mediating analysis utilizing Hayes procedure in SPSS reveals compelling insights into the relationships between variables. Each predictor, including Behavioral Intention, Usability, Social Influence, Risk, and Effort Expectancy, demonstrates a significant impact on the outcome variable. The high R-squared values indicate substantial variance explained by the predictors, suggesting a robust model

**Table 5** Mediation analysis

| | R | R square | Standard error of estimate | F change | DF1 | DF2 | SIG.F.CHANGE |
|---|---|---|---|---|---|---|---|
| Behavioural intention | 0.8423 | 0.7095 | 3.8761 | 105.8128 | 3.0000 | 130.0000 | 0.0000 |
| Usablity | 0.8247 | 0.6801 | 4.2584 | 92.1090 | 3.0000 | 130.0000 | 0.0000 |
| Social influence | 0.8512 | 0.7245 | 3.6667 | 113.9651 | 3.0000 | 130.0000 | 0.0000 |
| Risk | 0.8248 | 0.6804 | 4.2541 | 92.2363 | 3.0000 | 130.0000 | 0.0000 |
| Effort expetancy | 0.6936 | 0.6936 | 4.0778 | 98.1076 | 3.0000 | 130.0000 | 0.0000 |

**Table 6** Discriminant validity

| | CR | AVE | MSV | ASV |
|---|---|---|---|---|
| Financ_Inclu | 0.785 | 0.423 | 1.096 | 1.029 |
| Beha_Inten | 0.776 | 0.410 | 1.192 | 1.022 |
| Usability | 0.722 | 0.343 | 1.203 | 1.091 |
| Social_Influe | 0.752 | 0.379 | 1.184 | 1.077 |
| Risk | 0.691 | 0.310 | 1.203 | 1.105 |
| Effort_Expe | 0.705 | 0.326 | 1.080 | 1.008 |
| Grow_Adapt | 0.751 | 0.380 | 1.128 | 1.026 |

fit. Moreover, the low standard errors of estimate reflect the precision of the model's predictions. The significant F-changes further affirm the importance of the predictors in influencing the outcome variable, emphasizing their mediating roles in the relationship under investigation (Table 6).

# 8 Findings

1. The study found a strong positive correlation between financial inclusion initiatives and broader sustainable development within low-income sectors.
2. The findings underscored the importance of supportive infrastructure and policy frameworks in facilitating financial inclusion and sustainable development.
3. Gender distribution: Slight majority of males (50.38%) compared to females (49.61%).
4. Internal consistency assessed using Cronbach's alpha coefficients (ranging from 0.687–0.782).

5. Variables such as usability, social influence, risk, effort expectancy, financial inclusion, and growth and adaptability exhibited particularly strong internal consistency (coefficients ranging from 0.731–0.782).
6. Strong positive correlations observed among all variables, ranging from 0.734–0.808.
7. BI exhibits strong correlations with US ($r = 0.784$), SI ($r = 0.758$), RI ($r = 0.802$), EE ($r = 0.752$), FI ($r = 0.808$), and GA ($r = 0.769$), indicating significant positive relationships.

## 9 Future Research Directions

The paper provided both the primary and secondary thematic areas for future study to advance knowledge and enhance the understanding of the interconnection between financial access and sustainable development. Expansion research possibilities regather tendencies that innovative FTs have on global financing, a function of social capital inside the monetary incorporation procedure, and efficiency and durability of manifold inclusive finance methodologies.

## 10 Recommendations and Suggestions

It is essential to include the low-income sectors in the financial market to promote overall development, eliminate poverty, and accomplish sustainable development objectives. Financial inclusion encompasses the integration of all persons, particularly those who have been excluded, by providing them with convenient and affordable access to financial services that cater to their personal and business needs. Based on the research, the connection between financial inclusion and other elements of sustainable development presents many potentials for collaboration across different sectors and for mutual support. Below are recommendations and suggestions for realizing financial inclusion within low-income sectors while concurrently advancing sustainable development: Below are recommendations and suggestions for realizing financial inclusion within low-income sectors while concurrently advancing sustainable development:

1. Localized Needs Assessment: The most important thing about servicing the low-income groups is that the financial services have to be customized to their needs and wants. When market research is detailed, and when entrepreneurs interact with the target consumer base, they can gather an enormous amount of information regarding the target market's financial propensities, special challenges, and dreams.
2. Technological Advancements and Innovation: They present possibilities of bridging geographical barriers through uses of mobile banking, alternative

means of paying bills online among the highly developed platforms and biometric authentication to ease the expenses that come with other online associated services as embraced by conventional banks. Multi-stakeholder initiatives involving governments, financial institutions, and fintech firms are critical in arriving at effective ideas and executing them to make monetary facilities more reasonable and effective for the minorities.

3. Support for Microfinance and Community-Based Endeavours: Supporting micro finance institutions and grassroots organisations may be useful in extending credit to the micro entrepreneurs /informal economy players. Many of such organizations may have within them detailed information of the functioning of a specific region and are capable of providing the necessary financial services that are singularly appropriate for low-income customers.
4. Gender-sensitive Approaches: Thus, the identification and analysis of the gender differences in the usage of financial services are important in observing the gender equality and development. Measures that would be useful to reduce the level of disparity between male and female persons for the purpose of financial accommodation includes increasing access to education for women, property rights, and employment opportunities.
5. Resilience-building Strategies: To build up the financial capital among poor groups, therefore it not only requires the use of conventional monetary facilities but also the insurance and saving mechanisms. Coordinating the efforts of financial inclusion with other resilience interventions can strengthen the resilience of vulnerable groups when it comes to shocks and adversities.
6. Policy Reforms and Regulatory Frameworks: Most policies and regulations are usually initiated by governments making it easy for them to set an enabling environment for financial inclusion. These may include promotion of competition, protection of consumers, and regulation of FSPs in the Financial sector. Through its curtain, a regulatory sandbox provides an environment for the implementation of new financial instruments, together with their risks.

## 11 Conclusion

In conclusion, this research has positively and firmly examined FI's multifaceted relationship with sustainable development with low-income sector, using a composite model including the factors including the behavioural intention, the usability, perceived risks, social influence and the expectancy of the undertaking and has clearly defined growth and adaptability as the dependent variable controlled by FI. In the presence of the results of the study, one has to spotlight the general applicability of financial inclusion as the pathway to building the capacity to cope with adversities and chart development at the boarder communities' level. Knowledge and perceived behavioural control also emerge as critical constructs, demonstrating that clients' intentions to utilise inclusive financial services affect their socioeconomic outcomes. Moreover, usability turns out to be an effective enabler while eventually postulating effort expectancy; usability is a key enabler that highlights the proposition that for fin-tech asset to be useful and adopted among people with the weaker and vulnerable section convenience is paramount. Furthermore, the research contributes to the understanding of the level of visibility for the segment and the challenges related to the perceived risk and social influence as well as the importance of perceived risks and using the social network to enhance the consequences of economic insertion. Ch Depending on the concept of change, the role of mediation that can be depicted by financial inclusion is the role that helps to draw the periphery groups closer to the sphere of sustainable development. Thus, through the elimination of access and participation barriers, the programmes of monetary attachment enhance the ability of the people as well as the firms or companies in a country to manage the economic risk and to optimally realise the economic opportunities and effectively manage changes wherever and whenever they may happen at individual and country level. Therefore, this paper offered a valuable discussion to the area of financial inclusion and sustainable development by presenting the policy and strategy type of intervention on the low-income segment of society.

## References


1. Kara, A., Zhou, H., Zhou, Y.: Achieving the United Nations sustainable development goals through financial inclusion: a systematic literature review of access to finance across the globe. Int. Rev. Financ. Anal.Financ. Anal. **77**, 101833 (2021)
2. Hussain, S., Gul, R., Ullah, S.: Role of financial inclusion and ICT for sustainable economic development in developing countries. Technol. Forecast. Soc. Chang. **194**, 122725 (2023)
3. Kim, D.W., Yu, J.S., Hassan, M.K.: Financial inclusion and economic growth in OIC countries. Res. Int. Bus. Financ.Financ. **43**, 1–14 (2018)
4. Ahamed, M.M., Mallick, S.K.: Is financial inclusion good for bank stability? International evidence. J. Econ. Behav. Organ.Behav. Organ. **157**, 403–427 (2019)
5. Omar, M.A., Inaba, K.: Does financial inclusion reduce poverty and income inequality in developing countries? A panel data analysis. J. Econ. Struct. **9**(1), 37 (2020)
6. Banerjee, R., Donato, R., Maruta, A.A.: The effects of financial inclusion on development outcomes: New insights from ASEAN and East Asian countries. In Financial inclusion in Asia and Beyond (pp. 54–97) Routledge. (2021)
7. Ediagbonya, V., Tioluwani, C.: The role of fintech in driving financial inclusion in developing and emerging markets: issues, challenges and prospects. Technol. Sustain. **2**(1), 100–119 (2023)
8. Khan, I., Khan I.: Financial inclusion matter for poverty, income inequality and financial stability in developing countries: new evidence from public good theory. Int. J. Emerg. Mark. (2023)
9. Chibba, M.: Financial inclusion, poverty reduction and the millennium development goals. Eur. J. Dev. Res. **21**, 213–230 (2009)
10. Ratnawati, K.: The impact of financial inclusion on economic growth, poverty, income inequality, and financial stability in Asia. J. Asian Financ. Econ. Bus. **7**(10), 73–85 (2020)

# A Study on Stress Management Techniques and Workforce Sustainability Among IndianOil Refinery Employees in Assam

**Sneha Hazarika and Joyjit Sanyal**

**Abstract** One of the major challenges that modern organizations face is relieving and mitigating employee stress. The labour environment is changing rapidly due to the shifting global order, which puts an unsustainable amount of strain on the workforce. As technology advances, workers must keep up with new developments to maintain their positions in a competitive market. They also need to provide high-quality work and possess relevant skills. In order to manage the stress and anxiety that employees experience, it is crucial for both organizations and individuals to adopt suitable stress management practices. According to Nekzada and Tekeste [27], it is important to stress causes and its management at the work place-A qualitative study on the causes of stress and management mechanisms at Volvo Trucks AB, Umea. Employees have a significant amount of self-control when it comes to managing their stress levels. They use different strategies and approaches to handle their emotions effectively. This study aims to analyse the different coping mechanisms that employees at Indian Oil refineries in Assam use to manage their stress. Personnel from executive and non-executive categories of three Indian Oil refineries in Assam were surveyed using stratified random sampling to collect relevant data for the study. The study highlights the pertinence of considering the demographic characteristics of the respondents while choosing a stress-reduction technique. If not managed properly, employee stress can have negative effects on their physical and mental health, job satisfaction, and job performance.

**Keywords** Occupational stress · Stress management techniques · Demographic profile of the respondents · Workforce sustainability · Individual stress management strategies

S. Hazarika (✉)
Assistant Professor, Golaghat Commerce College, Golaghat, Assam, India
e-mail: snehahazarika123@gmail.com

J. Sanyal
Assistant Professor, Department of Commerce, Assam University Diphu Campus (A Central University), Diphu, Assam, India



N. D. K. Nguyen et al. (eds.), *Digital Transformation and Sustainability: Methods and Applications*, Information Systems Engineering and Management 47,
https://doi.org/10.1007/978-3-031-91278-8_15

## 1 Introduction

Stress can be defined as "a non-specific response of the body to any demand made upon it" [1]. The vicious effect of job stress is a fact of modern life and is known to affect employees and their employers negatively. Stress has become a part of our lives, but it needs to be managed in order to reduce its harmful effect on a person's physical and mental health [2]. Stress is a condition arising from the interaction of people and their jobs and characterised by changes within people that force them to deviate from their normal functioning [3]. Stress occurs when the demands on an individual exceed that person's psychological resources [4]. Stress can be defined as a psychological response to demands for which there is something at stake and where coping with those demands taxes or exceeds a person's capacity or resources [5]. Stress can be defined as a state of well-being where people experience difficulties, opportunities, or losses of something they desire, and their outcome is both unexpected and significant [6]. Thus, stress is anything that puts excess pressure on a person and affects the normal functioning of the stressed person. Stress is a demand of a situation, often vital for the person; thus, the person experiences pressure, anxiety and restlessness.

Stress is a factor that carries positive and negative symptoms. In a good way, stress is required to motivate a person to do a task or to undertake an assignment. Nevertheless, in its vicious form, excess stress affects a person's performance, whether in their job or personal life. People may face stress for various reasons, such as life situations, job responsibilities, relationship issues, etc. When stress exceeds an optimum level, it brings drastic consequences like depression, anger, sleep deprivation, more sleep, poor appetite, overeating, poor job performance, job dissatisfaction and job burnout. Poor working conditions, conflicts with supervisors, traumatic events or internal harassment of employees, sometimes result in anguish, neurotic disorders or even suicide of the employees [7]. Stress-related health problems, such as depression, anxiety, exhaustion, sleep disorders, diabetes, etc., have increased drastically in recent years, leading to adverse situations like the death of a person [8]. With increased job pressure and demands at the workplace, employees are often found to be under stress, affecting their careers and lives. Job stress affects a person's professional and family life, thus making the person depressed and sad.

To avoid such consequences, people should find ways to mitigate their stress. The stress management strategies could be organisational or strategies undertaken by the employees at a personal level. In the case of job stress, organisations need to develop strategies and policies to keep an eye on and manage the job stress of their employees. Some common organisational stress management strategies include job redesign, preparing a working schedule, mentoring, improved communication and decision-making, good relationships with superiors and peers, timely and adequate conflict management and proper career planning strategies. Moreover, strategies like mindfulness training, exercise, relaxation, meditation, employee assistance programs, counselling, resilience training, and post-traumatic stress training are practical measures that employees can resort, to relieve their stress [9]. Individual stress intervention strategies focus on developing employees' skills to manage, cope, and reduce stress

[10]. Every person should try to understand their stress reasons and find ways and means to cope with it. Strategies for managing stress are significant as they assure the well-being of employees in an organisation and thus prevent stressful situations [11]. There may be various ways like relaxing, exercising, punctuality, discipline, etc., with the help of which a person can alleviate their stress. Some ways to cope with stress include proper sleep, sports, conversations with a close one, relaxation through various activities and quitting addictive products [3]. Some other commonly used individual approaches to stress management include physical activities, yoga, meditation, cognitive methods, biofeedback relaxation methods, emotional freedom techniques, massage and aromatherapy, music, humour, adequate sleep, good eating habits and time management [11]. Miller (1982), suggested specific ways of coping with stress, such as getting rid of illogical or irrational ideas, breathing, time management, engaging in some recreational activities, good nutrition, and changing usual routines, which could help people manage stressful situations in life and at work [12]. Stress management is a continuing process that requires the willingness, commitment, and self-awareness of the person involved. Small steps of attempting to reduce one's stress become routine over time, turning into a habit and making the person build a strong resilience to the stressor [2]. It is necessary to adopt stress management strategies by the top management of enterprises in order to ensure efficient, smooth and sustainable business running [13]. Stress management strategies must be a part of the regular management activity of an organisation [8]. The HR department should formulate contemporary and practical policies and practices to reduce the stress of the organisation's employees and conduct regular programs and exercises for stress reduction.

One of the significant aspects of job stress that affects an organisation is employee job performance. Performance is defined as the attained outcomes of actions with skills of employees who perform in some situation [14]. However, stressed employees are unable to meet the expectations of their organisation as they experience constant physical, psychological and organisational burnout [15]. Stressed and dissatisfied employees often think of leaving the organisation or losing focus and attention in the workplace. The performance of satisfied and stress-free workers is always better as they are motivated towards achieving the firm's goals [7].

Individual differences in the workers, such as personality, age, education, experience, and coping style, are significant in predicting whether certain job specifications will result in stress [9]. Accordingly, the demographic factors of the respondents, like age, gender, marital status, salary, educational level, and family structure, should be considered when framing and undertaking stress management strategies. Moreover, studies also suggest that demographic factors influence the stress management strategies adopted by the employees to reduce their stress.

IndianOil is committed to emphasising occupational health, safety, and environmental protection by enriching the quality of life of its employees, customers, and the community. The organisation is committed to the well-being and prosperity of its employees, and all its occupational health programs are designed to improve their health, well-being, and productivity. One of IndianOil's main objectives is to maintain and promote the health and well-being of its employees. IndianOil has published six health manuals to create awareness in areas like work stress, nutrition, physical health, the well-being of employees, and home remedies.

## 2 Rationale of the Study

The oil refining industry is inherently stressful due to safety concerns, long hours, and demanding work conditions. Investigating coping strategies is vital for employee well-being and operational efficiency. Demographic characteristics, such as age, gender, education, and job role, can significantly influence how individuals respond to stress. Understanding these differences can help tailor stress management programs to meet diverse employee needs. Effective coping strategies can mitigate stress-related health issues and enhance productivity. This study aims to identify the most effective stress management methods for IndianOil employees and contribute to a healthier workforce. Assam has unique socio-cultural dynamics that may affect stress perceptions and coping mechanisms. This research will provide insights specific to the region, facilitating culturally relevant interventions. While there is extensive literature on stress coping strategies, particular studies focusing on India's oil and gas sector, particularly in Assam, are limited. This research seeks to fill that gap, providing valuable insights for academia and industry. Findings from this study can help IndianOil's HR policies and wellness programs, promoting a supportive work environment that enhances employee satisfaction and retention. The study can serve as a foundation for future research exploring stress management and coping strategies in similar industrial contexts, broadening the understanding of workforce sustainability.

## 3 Literature Review

Lian and Tam [16] observed correlation between work stressors and the adopted coping strategies, which may vary depending on the type of problems being dealt with and the interplay between the employee and the situation's demands. Programmes that teach resilience building are more likely to reduce the health effects of stress in employees. The study indicates a negative relationship between job stressors and stress management strategies [13]. The study suggests that to retain the talented employees of an organisation, the organisation needs to reduce its job stressors and concentrate on employee effectiveness. A study conducted on 1016 students between 12–19 years of age in Maharashtra, India, found an intercorrelation among stress coping strategies used interchangeably based on the stressful situation [4]. The study also highlights that the annual family income of the respondents is not a significant variable in determining the type of stress-coping strategy used by the respondents. However, gender, age, and parents' education are variables having influence on stress coping strategies used by respondents. In a study conducted on employees of fourteen private universities, it was observed that employees use two coping methods, i.e., approach and avoidance [17]. The study reveals that as the employees' stress increases, their performance also increases, thus implying that stress is necessary for more productivity. A study of 296 residents of Szczecin over 60 years old showed that demographic factors such as gender, education, financial situation, and living conditions need to be considered while providing them with care for their stressful life situations [18]. The study demonstrated that respondents with low economic status and a low level of education have problems with actively handling stress and taking the initiative. The study revealed that respondents with a higher level of education and an excellent financial position prefer active stress-coping strategies. A study conducted on 204 employees employed in the service industry reveals that the perceived workplace stress and stress coping mechanisms differ significantly with age [19]. At the same time, other demographic factors were found to have insignificant differences. The perceived stress of older employees was found to be higher than younger employees, and the use of coping mechanisms was also found to be common among older employees. The study attributes high-stress perceptions in older employees to the new stressors at work. Tran et al. [9] found that when stress exceeds a certain limit, it adversely affects the respondents' bodies, behaviour and psychology. The study found a relationship between tensions from stress and ill effects on a person's body. The study mentions that stress cannot be completely avoided; therefore, appropriate strategies must be developed to tolerate and resist stress. Catherine and Fonceca [6] mentioned in their study that effective stress management strategies minimise stress and develop resilience in a person to tolerate stress and handle obstacles. The study indicated a negative relationship between job stress and job performance; however, stress-coping strategies like seeking social support and problem-solving strategies mediate the relationship between job stress and performance [20]. These strategies enable employees to improve their performance in their workplace. A study on 280 non-academic staff of Bowen University in Nigeria showed a substantial correlation between employee job performance and personal stress management strategies used by the employees [21]. The study suggests that effective stress management strategies improve employees' productivity at the organisational or personal level. A study conducted on 394 employees of a bank in Nigeria found that a positive relationship exists between stress-coping strategies and job performance [22]. The study has emphasised providing counselling and flexible office timing as effective solutions to stress-related issues. Flexible work hours, like employees themselves scheduling their work hours and teleworking, are great ways of reducing employee stress and tension. A study conducted on 301 employees of a manufacturing firm in Edo State in Nigeria found stress management strategies to be significantly correlated with job performance, leading to stress alleviation of employees in the workplace [14]. The significant stress coping strategies included in the study were workplace social support, coping competence, time management and mind–body relaxation techniques. Sharma and Goswami [23] conducted a study on 35 traffic police personnel, finding that high stress exists among police personnel and that no effective stress management strategies exist to relieve employees' stress. The study emphasised that sustainability should not always be from an environmental perspective but should be included within an organisational framework. Adopting sustainable practices, such as adopting stress management policies to reduce the stress of employees, ensures a business's sustainability in the long run. A study conducted on 270 respondents revealed that stress management strategies positively affect employee job performance [24]. Specifically, meditation and relaxation techniques help employees be more focused, less stressed, and peaceful. These techniques give people time to reflect on stressful situations, develop focus and concentration on planning their work, and reduce their stress. Less stress will also lead to less absenteeism, turnover, job discontentment and low productivity.

## 4 Objectives

1. To study the various stress-coping strategies used by the employees of IndianOil refineries in Assam.
2. To determine if there is a significant difference in the stress coping strategies used by the employees based on their demographic characteristics.
3. To determine the relationship between stress management strategies and employees' job performance.

## 5 Hypothesis

H1: There is a significant difference in the stress-coping strategies used by the employees based on their grades in the refinery.

H2: There is a significant difference in the stress coping strategies used by the employees based on their gender.

H3: There is a significant difference in the stress coping strategies used by the employees based on their age.

H4: There is a significant difference in the stress-coping strategies used by the employees based on their tenure in the oil sector.

H5: There is a significant difference in the stress coping strategies used by the employees based on their marital status.

H6: There is a significant difference in the stress coping strategies used by the employees based on the nature of their families.

H7: There is a significant association between stress management strategies personally adopted by employees and their job performance.

## 6 Research Methodology

**Participants**: Using the sample survey method, primary data is collected by surveying employees working in the three IndianOil refineries of Assam, i.e. Bongaigaon Refinery, Digboi Refinery and Guwahati Refinery. The total population of the study is 2695 permanent employees working in the executive and non-executive grades at the three refineries of the state. The probability sampling method used in the study is stratified random sampling method. Structured questionnaires were distributed to 410 employees, and 309 usable questionnaires were returned for further analysis. The sample size for the current study is 309 employees who work in the three selected IndianOil refineries in Assam. The demographic characteristics of the respondents considered in the study are the grade of the employees in the refinery, their gender, age, tenure in the refinery sector, marital status and structure of the family to which they belong. The meaning of the terms considered for the study is mentioned below:

- Grade: Grade indicates the two broad categories of employees in the refineries, i.e., executives and non-executives.
- Gender: Gender specifies the actual or perceived sex or gender identity and expression associated with an individual. The three genders recognised by Indian law are included in the study: man, woman, and transgender.
- Age: The term indicates the age of the respondents, which is divided into four groups with equal distribution ranging from 21–30 years, 31–40 years, 41–50 years and 51–60 years.
- Tenure of service in the refinery sector: Tenure is the period a respondent has worked in the refinery sector. The tenure is divided into five categories in the study, ranging from 0–5 years, 6–10 years, 11–15 years, 16–20 years and 20 years and above.
- Marital status: The term specifies the marriage status of a person, whether the person is married or unmarried. Married indicates a person who is currently married, and unmarried suggests a person who is a bachelor. Also, a third category is included in the study, 'others', 'indicating widowed, separated or divorced people'.
- Structure of family: The term indicates the two broad types of families, i.e., joint and nuclear. A joint family is a family consisting of various generations, and a nuclear family is made up of two parents and children.

Of these 309 employees, 160 are from the non-executive category, and 149 are executives. For the gender of the respondents, 83.5% of the respondents are male, and 16.5% are female. In terms of age group, the majority of the respondents, 33%, are from 31–40 years of age, 32% are from 21–30 years of age, 20.7% are between 51–60 years, and 14.2% are between 41–50 years. The analysis reveals that 33% of the respondents have served in the sector for five or less than 5 years, 22.7% of the respondents have worked between 16–20 years in the refinery sector, 18.8% of the respondents are employed for more than 20 years in the oil refinery sector, 15.5%

**Table 1** Shows the characteristics of the respondents (n = 309)

| Variable | Categories | Frequency | (%) |
|---|---|---|---|
| Grade | Executive | 149 | 48.2 |
| | Non-Executive | 160 | 51.8 |
| Gender | Male | 258 | 83.5 |
| | Female | 51 | 16.5 |
| Age | 21–30 years | 99 | 32.0 |
| | 31–40 years | 102 | 33.0 |
| | 41–50 years | 44 | 14.2 |
| | 51–60 years | 64 | 20.7 |
| Years of experience | 0–5 years | 102 | 33.0 |
| | 6–10 years | 48 | 15.5 |
| | 11–15 years | 31 | 10.0 |
| | 16–20 years | 70 | 22.7 |
| | 21 years and above | 58 | 18.8 |
| Marital status | Married | 213 | 68.9 |
| | Unmarried | 96 | 31.1 |
| Structure of family | Nuclear | 223 | 72.2 |
| | Joint | 86 | 27.8 |

of the respondents are employed for 6–10 years. Only 10% of the respondents are employed between 11–15 years. Only 31.1% of the respondents are unmarried, and 68.9% are married. Most respondents, i.e., 72.2%, are from nuclear families, and 27.8% are from joint families (Table 1).

**Data Collection**: The research investigation was carried out using both primary and secondary data. Primary data was gathered using a questionnaire with closed-ended questions in a descriptive research design. Secondary data is collected from already-published books and research papers.

## 7 Measurement

The various stress coping strategies considered in the present study include seeking counselling from experts, good sleep, regular physical exercise, meditation, spending time with family, following a properly balanced diet, self-appreciation, seeking help from employers or supervisors, reducing working hours while at the workplace, taking frequent breaks at work, scheduling the work to be done, vacation, developing hobbies, prioritising the tasks in hand, spending time with friends and engaging oneself in recreational activities. The reliability of the 16 scale items is measured

**Table 2** Shows the mean of the sixteen stress-coping strategies used by the respondents

| | N | Mean | Std. Deviation | Skewness | Kurtosis |
|---|---|---|---|---|---|
| Mean of the stress coping strategies | 309 | 52.8058 | 8.72494 | 0.096 | 1.390 |

with Cronbach's Alpha, which is 0.825, implying that the scale is within reliability limits and can be used for the study.

## 8 Data Analysis

Statistical tests are conducted using the SPSS software package version 26, to analyse the data. Likert's Five Point Scale is used to rate the primary data. Descriptive statistics like mean, standard deviation, percentages, independent sample t-test, ANOVA and Pearson Correlation are used to analyse the data.

Table 2 shows that the mean of the stress coping strategies, which the employees use to cope with their stress is 52.80.

## 9 Demographic Particulars of the Respondents, Along with Their Mean and Significance of the Stress Management Strategies

**Category of employees**: Table 2 indicates no significant difference in employees' stress coping strategies amongst the two categories, i.e., executives and non-executives ($p = 0.180 > 0.05$). Therefore, we accept the null hypothesis that no statistically significant mean difference exists in the stress-coping strategies used by the two categories of respondents, executives and non-executives. The typical three strategies executives use to manage stress include good sleep, prioritising tasks, and spending time with family. The three primary strategies used by the non-executive respondents to relieve their stress involve spending quality time with family members, getting proper sleep, and scheduling the work to be done (Table 3).

**Gender**: There is no noticeable difference in stress coping strategies used by male and female employees ($p = 0.846 > 0.05$). Therefore, we accept the null hypothesis that there is no statistically significant mean difference in the stress-coping strategies used by the two genders of respondents, male and female. The study reveals that female respondents' most commonly used stress management methods include good sleep, spending time with family, and prioritising work. The most widely used stress coping strategies by male employees include proper sleep, prioritising tasks, and spending some good time with the family.

**Table 3** Shows significant differences in stress management strategies used by the respondents based on their demographic characteristics

| Variable | | Mean of stress management strategies | St dev | Significance | |
|---|---|---|---|---|---|
| Grade | Executive | 53.49 | 8.33 | 0.180 | Not significant |
| | Non-Executive | 52.16 | 9.04 | | |
| Gender | Male | 52.84 | 8.86 | 0.846 | Not significant |
| | Female | 52.58 | 8.06 | | |
| Age | 21–30 years | 51.64 | 8.66 | | |
| | 31–40 years | 51.86 | 8.09 | 0.006 | Significant |
| | 41–50 years | 52.70 | 8.34 | | |
| | 51–60 years | 56.17 | 9.36 | | |
| Years of experience | 0–5 years | 51.20 | 8.28 | | |
| | 6–10 years | 53.25 | 8.91 | 0.006 | Significant |
| | 11–15 years | 51.87 | 7.78 | | |
| | 16–20 years | 51.45 | 8.17 | | |
| | 21 years and above | 56.00 | 9.36 | | |
| Marital status | Married | 53.56 | 8.36 | 0.022 | Significant |
| | Unmarried | 51.11 | 9.30 | | |
| Structure of family | Nuclear | 52.68 | 8.81 | 0.698 | Not significant |
| | Joint | 53.11 | 8.54 | | |

**Age**: The ANOVA results indicate a significant difference in the stress management strategies used by the respondents in the various age groups ($p = 0.006 < 0.05$). Therefore, we reject the null hypothesis that there is no statistically significant mean difference in the stress-coping strategies used by the respondents of different age groups. The individual stress management strategies are used more by respondents in the age group of 51–60 years, followed by the age group of 41–50. In the age group 21–30 years, the principal stress management strategies used by the respondents include good sleep, prioritising the tasks at hand, and regular physical exercise. For the age group 31–40 years, the respondents' most widely used stress management strategy includes good sleep followed by prioritising work and spending time with family. In the age group 41–50 years, the primary stress management strategies used by the respondents include spending time with family, prioritising work, and consuming a balanced diet. For the age group 51–60 years, the respondents' most widely used stress management strategy includes good sleep followed by prioritising work, scheduling the work to be accomplished and spending time with family.

**Years of Experience**: The ANOVA results indicate a significant difference in stress-coping strategies used by the respondents based on the years of work experience in the oil refinery sector ($p = 0.006 < 0.05$). Therefore, we reject the null hypothesis that there is no statistically significant mean difference in the stress-coping strategies used by the respondents with varying years of work experience in the refinery sector. The most commonly used stress management strategy by employees with work experience in the refinery sector below 5 years includes good sleep, prioritising tasks, and exercising regularly. Respondents with experience between 6 to 10 years mostly use strategies like proper sleep, followed by spending time with family and prioritising the work at hand. The most widely used strategy among respondents with work experience between 11 and 15 years in the refinery sector includes spending time with family, getting good sleep and scheduling the work to be done. Respondents with experience between 16 and 20 years mostly use prioritising work, proper sleep strategies, spending time with family and scheduling the job to be accomplished as measures to cope with their stress. The everyday stress management strategies with senior employees above 20 years of experience include good sleep, spending quality time with family and prioritising the work.

**Marital Status**: The study revealed a significant difference in the stress management strategies used by married and single employees ($p = 0.022 < 0.05$), with married respondents making more use of individual stress management strategies than the unmarried ones. Therefore, we reject the null hypothesis that no statistically significant mean difference exists in the stress-coping strategies used by the married and unmarried respondents. The study reveals that for the married respondents, the most commonly used methods include good sleep followed by spending time with family and prioritising the work at hand. For the unmarried respondents, the strategies mainly included prioritising the job at hand, followed by getting good sleep and regular physical exercise.

**Structure of Family**: The study identifies no statistically significant mean difference in stress-coping strategies used by employees from nuclear and joint families ($p = 0.698 > 0.05$). Therefore, we accept the null hypothesis that no statistically significant mean difference exists in the stress-coping strategies used among nuclear and joint families' employees. The study reveals that for the respondents from joint families, the most commonly used stress coping methods include good sleep followed by self-appreciation and scheduling the work to be done. The strategies for the respondents from nuclear families mainly include getting good sleep, prioritising tasks, and spending time with family.

## 10 Correlation Between Stress Management Strategies Used by Employees and Their Job Performance

From Table 4, it is seen that the personal stress management strategy of good sleep is maximum used by the respondents as a measure to cope with their stress, followed by the strategies of prioritising the tasks at hand, spending time with family, scheduling the work to be done, following a balanced diet, engaging in regular physical exercise and spending time with friends and self-appreciation. The strategies used least by the respondents include engaging in recreational activities, developing hobbies, going on vacation, seeking help from employers, taking frequent breaks at work, and seeking counselling from experts.

The strategies that correlate with job performance include good sleep, regular physical exercise, meditation, spending time with family, following a properly balanced diet, self-appreciation, scheduling the work to be done, developing hobbies, prioritising the tasks at hand, spending time with friends and engaging oneself in recreational activities. The association between the strategies and job performance is positive for the correlated factors, indicating that using such strategies helps respondents improve their job performance. Thus, the null hypothesis is rejected, stating that there is no significant association between stress management strategies personally adopted by the employees and the job performance of the employees.

## 11 Conclusion

The study found that for the demographic factors of age, years of experience, and marital status, there is a significant difference in the stress-coping strategies used by the employees. No significant difference in stress coping strategies is observed for the factors of the grade of the employees, gender of employees, and family structure of the employees. Demographic factors, thus, play a vital role for employees when choosing stress management strategies to manage their stress. Understanding this relationship between demographic characteristics and employee stress management strategies enables organisations to develop stress-coping techniques that are more employee-individual-oriented. Adequate and proper stress management strategies would allow employees to check their stress levels, leading to the organisation's and the economy's overall sustainability. Oil refineries, one of the country's critical sectors, need to function efficiently, which requires its employees to perform well and be stress-free. Timely and appropriate stress management strategies will help employees keep their job performance intact and improve, leading to organisational progress and long-term sustainability. Stress cannot be eliminated as it is associated with various challenges in running an organisation. However, it should be managed to ensure a good working environment with properly designed jobs and flexible policies [15]. Knowing the causes of stress, its signs, and the strategies for eliminating it, will enable employees and organisations to take appropriate steps to alleviate the

**Table 4** Shows the association between the stress management strategies and the job performance of the respondents

| Sl no | Stress management strategies | Mean of stress management strategies | Standard deviation | Correlation of stress management strategies with job performance | Sig (2-tailed) |
|---|---|---|---|---|---|
| 1 | Seeking counselling from experts | 2.37 | 1.111 | 0.007 | 0.901 |
| 2 | Good sleep | 4.02 | 0.901 | 0.171** | 0.003 |
| 3 | Regular physical exercise | 3.62 | 0.962 | 0.144* | 0.011 |
| 4 | Meditation | 2.88 | 1.196 | 0.141* | 0.013 |
| 5 | Spending time with family | 3.91 | 0.944 | 0.133* | 0.019 |
| 6 | Following a proper balanced diet | 3.68 | 0.970 | 0.220** | 0.000 |
| 7 | Self-appreciation | 3.55 | 1.010 | 0.247** | 0.000 |
| 8 | Seeking help from employers or supervisors | 2.96 | 1.012 | 0.095 | 0.094 |
| 9 | Reducing working hours while at the workplace | 2.30 | 1.059 | -0.090 | 0.114 |
| 10 | Taking frequent breaks at work | 2.43 | 0.967 | -0.022 | 0.694 |
| 11 | Scheduling the work to be done | 3.72 | 1.030 | 0.169** | 0.003 |
| 12 | Vacation | 3.10 | 0.924 | 0.107 | 0.059 |
| 13 | Developing hobbies | 3.31 | 0.975 | 0.192** | 0.001 |
| 14 | Prioritising the tasks at hand | 3.94 | 0.901 | 0.238** | 0.000 |
| 15 | Spending time with friends | 3.59 | 0.938 | 0.231** | 0.000 |
| 16 | Engaging oneself in recreational activities | 3.43 | 1.032 | 0.209** | 0.000 |

**Correlation is significant at the 0.01 level (2-tailed)
*Correlation is significant at the 0.05 level (2-tailed)

stress experienced. By doing this, organisations will have satisfied, committed, and productive workers and reduce the costs related to stress [25].

## 12 Suggestions and Future Implications

Thus, it can be said that stress is present in all our lives. However, it depends on how much effort a person puts into managing the stressor to relieve it. Moreover, organisations also need to make efforts to initiate stress management programmes and policies to keep their employees' stress at a minimum level.

The study is confined to the oil refinery sector and the state of Assam. Studies can be conducted in other sectors to analyse the strategies used by employees for managing work-related stress. Studies can also be undertaken by considering other demographic factors and analysing any significant difference in stress-coping strategy used by the respondents based on their demographic characteristics. Further studies could be conducted to determine the factors that cause organisational stress and the effect of work-related stress.

## References

1. Biriowu, C.S., Chikwe, J.E.: Stress management strategies and organizational performance: insight from selected private and public hospitals in Port Harcourt, Nigeria. Int. J. Bus. Manag. Res. 27–36 (2019)
2. Brunner, H.: Stress Management: strategies for a healthier and happier life. J. Metab. Syndr. 1–2 (2023)
3. Panigrahi, A.: Managing stress at workplace. J. Manag. Res. Anal. 154–162 (2017)
4. Pardeshi, P., Jha, P.C.: A study of demographic factors affecting adolescent students in their coping strategies. JABER **I**, 6665–6681 (2015)
5. Kokutensa, A.T.: Stress and coping strategies among distance working employees during covid-19. Malardalen University Sweden, Swede (2021)
6. Catherine, A.V., Fonceca, C.M.: Employee stress and its impact on their job performance. J. Acad. Ind. Res. (JAIR), 34–38 (2022)
7. Mathur, G., Vigg, S., Sandhar, S., Holani, U.: Stress as a correlate of job performance: a study of manufacturing organisations. J. Adv. Manag. Res. 79–85 (2007)
8. Toufaili, B.E.: Consequences of organizational stress on workforce productivity. Research on strategies–promote a sustainable workforce. Bus. Excel. Manag. 42–52 (2018)
9. Tran, C.T., Tran, H.T., Nguyen, H.T., Mach, D.N., Phan, H.S., Mujtaba, B.G.: Stress management in the modern workplace and the role of human resource professionals. Bus. Ethics Lead. 26–40 (2020)
10. Holman, D., Johnson, S., O'Connor, E.: Stress management interventions: improving subjective psychological well-being in the workplace. In *Handbook of well-being* (pp. 1–14). DEF Publishers (2018)
11. Zizek, S.S., Treven, S., Cancer, V.: Individual and organizational approaches–overcoming stress. Econ. Sci. 1–18 (n.d.)
12. Esia-Donkoh, K., Yelkpieri, D., Esia-Donkoh, K.: Coping with stress: strategies adopted by students at the Wenneba campus of university of education Wenneba, Ghana. US-China Educ. Rev. **B2**, 290–299 (2011)
13. Pant, I.: Stress management: strategies–cope up with job stress. Int. J. Res. Manag. Econ. Commer. 99–110 (2015)
14. Ogomegbunam, O.A.: Stress management strategies and employee performance: an application of correlational research design on manufacturing firms in Edo State, Nigeria. J. Econ. Financ. Manag. Stud. 678–690 (2023)

15. Harry, J.: Stress management and employee performance. Eur. J. Hum. Resour. Manag. Stud. 57–71 (2020)
16. Lian, S-Y., Tam, C.L.: Work stress, coping strategies and resilience: a study among working females. Asian Soc. Sci. 41–52 (2014)
17. Garg, S.: Role of work stress and coping strategies of employee's performance: an empirical study with reference–private university employees. Int. Reserach J. (2017)
18. Lembas, D., Starkowska, A., Mak, M., Konecka, M., Bikowska, M., Chlewicka, J.C., Korzonek, M.: Impact of demographic factors on usage of stress coping strategies chosen by elderly people. Fam. Med. Prim. Care Rev 34–38 (2017)
19. Goel, M., Verma, J.: Workplace stress and coping mechanism in a cohort of Indian service industry. Asia Pac. Manag. Rev. 113–119 (2021)
20. Jahangiri, F.R., Zeb, R., Arzeen, S.: Organizational stress and job performance: coping strategies as a mediator. Indian J. Econ. Bus. 701–712 (2022)
21. Akintunde-Adeyi, J.F., Akinbode, J.O., Emmanuel.: Stress management and employee performance. Int. J. Prof. Bus. Rev. 1–15 (2023)
22. Obiora, O.A., Ifeoma, I-OM., Uzochukwu, A.C.: Effect of stress management strategies on employee job performance in deposit money banks in Nigeria. Int. J. Adv. Multidiscip. Res. Stud. 200–208 (2023)
23. Sharma, P., Goswami, S.: Developing a sustainable workplace: a study on workplace stress and individual interventions–manage stress in Guwahati Traffic Police. Am. J. Soc. Dev. Am. J. Soc. Dev. 32–37 (2023)
24. Madhu, P., Singh, S.: Effect of stress management techniques on employee performance. Int. Res. J. Mod. Eng. Technol. Sci. 1962–1967 (2024)
25. Kofoworola, O.H., Alayode, A.M.: Strategies for managing stress for optimal job performance. Int. J. Psychol. Stud. 162–168 (2012)
26. Latif, M.A., Sultana, S.: Adaptation of occupational stress index. J. Life Earth Sci. 55–57 (2009)
27. Nekzada, N., Tekeste, S.F.: Stress causes and its management at the work place a qualitative study on the causes of stress and management mechanisms at Volvo Trucks AB, Umea. Umea Sch. Bus. Econ. 1–66 (2013)
28. Yusoff, R.B., Ali, A.M., Khan, A.: Assessing reliability and validity of job performance scale among university teachers. J. Basic Appl. Sci. Res. 35–41 (2014)

# Artificial Intelligence and Firm Performance: The Moderating Role of Circular Economy Intelligence in e-Commerce Firms

**Zahid Hussain and Arman Khan**

**Abstract** This chapter explores the interplay between artificial intelligence (AI) and firm performance (FP), adding a new dimension by examining the moderating effect of Circular economy intelligence (CEI). Using the analytical tools of SMART-PLS, our chapter systematically evaluates the collected data through measurement and SEM model. The chapter is based on a thorough inquiry to uncover the diverse impact of AI on FP. Through empirical testing and sophisticated statistical methods enabled by SMART-PLS, we reveal insights into the interrelated associations, highlighting the crucial role of CEI. The results of this chapter enrich the literature on AI and business outcomes and provide practical guidance for organizations operating at a crossroads of technology, sustainability, and economic performance. By integrating state-of-the-art analysis techniques focusing on CE, this chapter enhances our comprehension of how businesses can effectively leverage AI in changing economic and environmental scenarios. Lastly, this chapter offers a comprehensive view of the link between AI and FP, considering the moderating influence of CEI, and supports its insights through a rigorous analysis using SMART-PLS methodology.

**Keywords** Artificial intelligence · Circular economy intelligence · Firm performance · E-commerce · Pakistan

Z. Hussain (✉)
Khadim Hussain Shah Institute of Technology (KASBIT), Karachi, Pakistan
e-mail: Zahidhussain9341@gmail.com

A. Khan
Department of Business Administration, Shaheed Benazir Bhutto University, Shaheed Benazirabad, Pakistan
e-mail: armankhan@sbbusba.edu.pk



N. D. K. Nguyen et al. (eds.), *Digital Transformation and Sustainability: Methods and Applications*, Information Systems Engineering and Management 47,
https://doi.org/10.1007/978-3-031-91278-8_16

## 1 Introduction

AI can improve operational procedures, increasing efficiency and productivity across several organizational contexts [1]. AI enables data-driven decision-making, equipping organizations with efficient insights for strategic planning and resource allocations [2]. AI applications, including Chatbots and personalized recommendations, improve customer experiences, cultivating customer satisfaction and loyalty [3]. AI technologies are also associated with substantial business cost savings, especially in routine and repetitive tasks [4]. AI capacities in predictive analytics support effective risk mitigation strategies and robust fraud detection systems to improve organizational security [5]. AI integration supplements human capacities, leading to workforce enhancement rather than replacement, creating a more harmonious link between technology and human expertise [6]. AI-driven predictive maintenance models have shown value in optimizing machinery and equipment performance, reducing downtime, and lowering maintenance costs [7]. AI tools are known for their capacity to analyze extensive data sets rapidly and provide organizations with valuable insights into market trends and competitors' behavior for strategic positioning [8]. AI-driven analytics support creating personalized recommendation strategies and customizing content and promotions to individual customers' choices for higher engagement [5].

The adaptability of AI systems through machine learning algorithms allows organizations to learn from data continuously, adapt to shifting circumstances, and stay resilient in dynamic environments [9]. AI applications in SCM should enhance logistics, shorten lead times, and boost overall supply chain efficiency [4]. When implemented wisely, AI tools enhance employee satisfaction by automating mundane tasks and allowing human workers to focus on more meaningful creative aspects of their roles [10]. AI-driven dynamic pricing models enable organizations to adjust prices in actual time based on market demand, maximizing revenue and staying competitive [11]. AI tools in workplaces have been linked to improvements in health and safety, with predictive analytics helping identify potential hazards and prevent accidents [12].

Natural language processing (NLP) in analyzing customer feedback enables organizations to extract valuable insights and enhance products or services [8]. AI systems that facilitate cross-functional collaboration by integrating data across departments have been shown to break down silos and enhance communication with organizations [9]. AI should contribute to energy savings in operational procedures by optimizing resource usage and identifying possibilities for energy conservation [8]. AI-driven diagnostic tools in healthcare have demonstrated the capacity to improve accuracy in disease detection, leading to more precise and timely medical interventions [2]. Ethical considerations in AI deployment prompt organizations to prioritize responsible AI initiatives and address potential algorithm biases [5]. The rapidly shifting landscape of modern business has brought the link between AI and FP to the forefront of academic inquiry.

This study aims to provide a comprehensive understanding of a new factor, the moderating effect of CEI. The motivation for this chapter stems from the need to understand how organizations strategically leverage AI within dynamic economic and environmental contexts. The suitable interplay between AI and FP becomes more critical as technology shifts industries. This chapter adopts an innovative approach by incorporating CEI as a moderating factor, acknowledging its potential to trigger impactful changes in business operations. This chapter involves a rigorous analysis of collected data using the SMART-PLS methodology. This technique improves the accuracy and demonstrates the commitment to methodological rigor, ensuring a solid examination of measurement and SEM models. The literature synthesis situated this chapter at the AI, CEI, and FP links. A comprehensive review of scholarly works has revealed insights from different sources that significantly contribute to our understanding of the impactful implications of AI in organizational settings. The remarkable findings of this chapter provide valuable insights, bringing a refined and informed perspective to the forefront.

## 2 Literature Review

### *2.1 Artificial Intelligence and Firm Performance*

AI is the art and science of creating intelligent machines [3]. AI is the science of making machines capable of things that would require intelligence if done by humans [4]. AI is the study of agents that perceive the environment and act upon it [6]. AI is a collection of algorithms and intelligence that empower computers to learn and make decisions [13]. AI is a system that adapts its behavior based on data [8]. FP is how a firm accomplishes its objectives [9]. FP is how a firm's actions generate superior financial returns [11]. FP is the total stakeholders' return and market share growth [1]. It integrates both financial metrics (stakeholder return) and market-related metrics (market share growth) to evaluate performance [8]. AI integration increases operational efficiency, lowering costs regarding manual procedures and resource allocations [2]. AI creates strategic decision support systems, equipping executives with data-driven insights for more informed and strategic decision-making [14].

The effect of AI on CRM systems demonstrates that personalized customer engagements and targeted marketing strategies enhance customer satisfaction and loyalty [1]. Organizations that adopt AI-driven predictive analytics achieve significant revenue growth by anticipating market trends and customer behavior [15]. AI can enhance organizational adaptability by swiftly analyzing market shifts and adjusting real-time strategies [16]. AI applications in supply chain management enhance logistics, shorten lead times, and boost overall supply chain efficiency [17]. AI systems foster organizations' continuous learning culture by adapting to new information, trends, and technologies [7]. AI supports sustainable practices by optimizing resource allocations, reducing waste, and aligning business operations with environmental goals [6].

AI-driven forecasting models greatly enhance accuracy, enabling organizations to make more dependable predictions about demand, sales, and market trends [18]. AI-driven analytics efficiently predict customer churn, allowing organizations to implement retention strategies and maintain customer loyalty [19]. Effects of AI in project management, optimizing workflows, resource allocations, and task prioritization for successful project outcomes [20]. AI facilitates the alignment of innovation strategies with organizational goals, ensuring that technological advancements contribute to overall business objectives [21]. AI on learning and development initiatives provide personalized Training programs tailored to individual employees' needs [22]. AI algorithms that excel in customer segmentation allow organizations to target specific demographics with tailored marketing strategies [20]. AI in HR analytics enables organizations to make data-driven decisions about talent acquisition and performance management. AI can perform real-time data analysis and provide organizations with timely insights to respond swiftly to shifting market conditions. However, AI influences innovation in product design, aiding in product creation that aligns with evolving customer preferences and technological trends [23]. In risk management, AI provides organizations with predictive analytics to identify and mitigate risks before they escalate [24]. When ethically deployed, AI aligns with CSR goals and contributes to socially conscious business practices [25].

H1: Artificial Intelligence positively impacts Firm Performance in E-commerce firms.

## 2.2 *Circular Economy Intelligence and Firm Performance*

CEI is a factor that refers to the use of data, AI, and analytics to support the transition to CE, which is a model of production and consumption that overcomes waste and increases resource efficiency [26]. CEI is the capacity to collect and analyze the data procedures related to the circularity of products, materials, and systems and utilize this information to design, operate, and evaluate CE solutions [27]. CEI is the application of AI to improve and accelerate the development, adoption, and evaluation of CE initiatives across industries [7]. CEI is a data integration, analytics, and AI into the CE framework, optimizing resource flows, environmental reduction, and stakeholder value creation [28]. CEI is data Usage, analytics, and AI to support the CE principles of designing out waste and pollution, keeping usage of products and materials, and regenerating natural systems [24].

CEI is the capacity of data leverage, analytics, and AI to facilitate the transition to CE, which is a model of production and consumption that decouples growth from the consumption of finite resources [29]. CEI is data utilization, analytics, and AI to enable CE innovations, which is the procedure of creating new products or services and business models that contribute to the circularity of resources and environmental sustainability [30]. CEI refers to using advanced analytics, data-driven insights, and intelligent systems to support and enhance organizational CE initiatives

[31]. It involves using technology and business intelligence tools to improve resource efficiency and wastage and support sustainable business initiatives [29].

Nowadays, organizations using CEI significantly reduce resource consumption through advanced analytics, resulting in cost savings and environmental benefits [30]. CEI helps businesses identify and adopt efficient waste reduction strategies by analyzing data throughout the product lifecycle [32]. CEI enables comprehensive lifecycle assessments to provide insights into the environmental effects of products and inform decisions for eco-design [33]. CEI enhances supply chain transparency, allowing organizations to track materials, monitor suppliers, and ensure compliance with circular principles [4]. CEI stimulates innovation in circular product design by analyzing market trends, customer choices, and materials lifecycles [34]. CEI leads to economic benefits by identifying sustainable practices that lower environmental effects and increase long-term profitability [35]. CEI supports organizations in effectively communicating their circular initiatives to customers and building trust and brand loyalty [36]. CEI helps businesses comply with changing environmental regulations by providing data on their sustainability efforts [37]. A set of performance metrics developed through CEI offers organizations measurable indicators to evaluate the success of their circular initiatives [38].

According to [39], CEI encourages collaboration within ecosystems, promoting partnerships that improve circular practices across industries. CEI enables organizations to compare their circular performance with industry standards and competitors [40]. CEI improves predictive maintenance in circular procedures, optimizes the lifespan of products, and reduces disruptions [6]. CEI supports talent development by detecting skill gaps and training needs for employees involved in circular initiatives [41]. CEI enables the transition to Circular business models focusing on revenue generation through sustainable and circular practices [34]. CEI improves organizational resilience by providing insights into risks and alternative circular supply chain options [38]. CEI analyzes customer behavior, identifying trends and choices that affect circular product adoption and purchasing decisions [27]. CEI should be customized for emerging markets, considering unique issues and possibilities for circular practices in these areas [4]. CEI enables lifecycle cost analysis, helping organizations assess the economic feasibility of circular initiatives over the long term [24]. CEI significantly reduces Carbon footprints by optimizing procedures and materials [4]. A circular product certification system supported by CEI provides customers with trustworthy information about a product's circular attributes [24]. CEI supports electronic waste management and provides strategies for effective recycling, refurbishment, and responsible disposal [29]. CEI usage in creating circular procurement strategies ensures that purchased goods comply with circular principles throughout their lifecycle [26].

The development of standardized reporting frameworks for circular performance enabled by CEI improves transparency and comparability [30]. CEI helps to reduce water footprints in production procedures, addressing a vital aspect of sustainable resource management [36]. CEI enabled organizational adoption patterns of circular key performance indicators (KPIs), examining their impact on decision-making [38]. CEI improves economic resilience in organizations following circular principles, especially amid economic uncertainties [11]. The application of CEI to urban planning optimizing city structures and resource flows for increased circularity and sustainability [29]. CEI impacts customer awareness and provides information to support the success of circular initiatives [32]. CEI affects policy development, enabling government regulations and incentives. CEI in global supply chains fosters economic circularity on a broader scale [28]. Cultural shots within organizations enabled CEI to influence employees' mindsets, values, and organizational norms toward circular practices [4].

H2: Circular Economy Intelligence significantly impacts Firm Performance in E-commerce firms.

## *2.3 Circular Economy Intelligence Mediates Between Artificial Intelligence and Firm Performance*

CEI Plays a vital moderating role in the link between AI adoption and overall FP. CEI acts as a moderator by enhancing resource efficiency through AI-driven insights, resulting in enhanced efficiency and cost savings, ultimately influencing FP positively. Through AI-enhanced analytics, CEI detects waste reduction possibilities across the value chain, moderating the effects of AI on reducing waste and contributing to improving FP [16]. CEI guided by AI algorithms stimulates innovation in circular product design, moderating the influence of AI on creating sustainable products that align with customers' choices and industry trends [14]. CEI integrated with AI-driven account-ability tools is a mediator by ensuring visibility and traceability in supply chain effects on AI adoption impact whole supply chain performance [19]. CEI, supported by AI analytics, mediates the development and implementation of circular performance metrics and provides organizations with an advanced framework to measure the achievements of circular initiatives [3]. CEI usage sentiment analysis via AI mediates the effect on customer trust and brand loyalty by ensuring that circular initiatives align positively with customers' values and expectations [21].

CEI with AI-driven compliance monitoring acts as a mediator, ensuring that AI adoption complies with regulatory requirements and mitigates risks related to noncompliance [42]. CEI's usage of AI-driven analytics mediates employee engagement by fostering a culture of innovation and environmental responsibility within organizations adopting AI for circular initiatives [4]. CEI leveraging AI algorithms mediates the effect of AI on procurement techniques by ensuring that circular initiatives are integrated into supplier selection and material sourcing decisions [25]. CEI guided by AI predictive analytics mediates the effect of AI on economic resilience by offering insights that enhance organizational adaptability to market shifts and uncertainties [19]. CEI supported by AI-driven lifecycle assessments mediates the effect of AI on conducting cost analyses throughout product lifecycles, affecting decisions regarding Circular initiatives [17]. CEI adapted for particular market dynamics utilizing AI insights mediates the effect of AI on promoting circular initiatives in emerging markets, addressing unique issues and possibilities [2]. CEI integrates social intelligence via AI and mediates the effect on social effect assessments, ensuring that AI-driven circular initiatives consider and improve social standards [10].

CEI acts as a mediator by facilitating collaboration within ecosystems via AI collaboration platforms, promoting partnerships that improve the circular advantages across sectors [9]. CEI usage of AI for predictive maintenance mediates the effect of AI on enhancing the lifespan of circular procedures and overcoming disruptions, enhancing overall operational efficiency [20]. CEI, supported by AI analysis of data during M&A activities, mediates the effect of AI on successful integration, ensuring that circular considerations align with strategic decision-making [28]. CEI with AI-driven real-time data analysis mediates effective crisis management by offering timely insights during critical situations and supporting informed decision-making [11]. CEI utilizing AI for real-time analysis mediates the impact of AI on providing organizations with timely insights, ensuring swift response to changing market conditions for improved FP [21]. CEI driven by AI algorithms mediates the impact of innovation in product design, ensuring that AI contributes to creating products aligned with circular principles and consumer expectations [6].

CEI utilizing AI for brand sentiment analysis acts as a mediator, ensuring that AI-driven circular initiatives positively contribute to brand reputation and management [37]. CEI with AI-driven analytics mediates the impact on eco-labeling effectiveness, ensuring that AI supports accurate and transparent communication to enhance consumer awareness [39]. CEI incorporating AI in-sights mediates the impact of AI on promoting circular tourism practices, ensuring sustainable travel, and minimizing the environmental impact of the tourism industry [40].

H3: Circular Economy Intelligence mediates the relationship between Artificial Intelligence and Firm Performance in e-commerce firms.

## 3 Conceptual Framework

See (Fig. 1).

## 4 Methodology

This study used a 5-point scale to examine the relationship between the proposed variables of the framework. The sample consisted of e-commerce firms in Pakistan. Randomly selected from the Pakistan directory that listed the operating firms under SMEs in the country. Given that 95% of private businesses are e-commerce firms, the researcher could obtain a diverse sample for suitable responses. The primary respondents were managers. Online surveys are a vital tool in research as they are fast, cheap, and simple to conduct. The data collection involved sending out 550 questionnaires, of which 305 were completed and usable for data analysis. The measurement scales

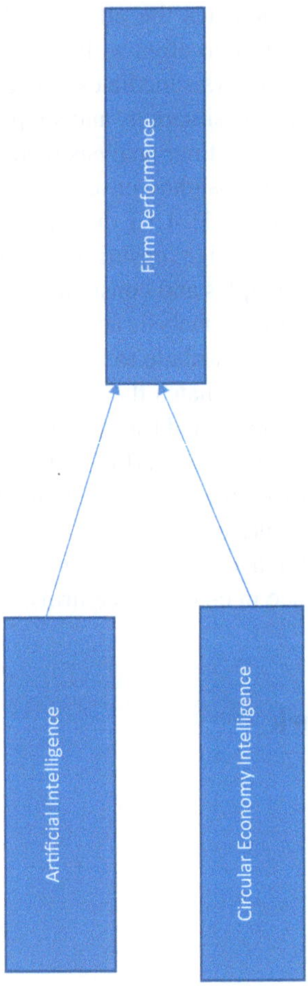


**Fig. 1** Conceptual framework. *Source* Author's work)

**Table 1** Measurement model (Reliability Test)

| | Cronbach's alpha | Composite | Reliability |
|---|---|---|---|
| *AVE* | | | |
| FP | 0.805 | 0.864 | 0.532 |
| AI | 0.835 | 0.878 | 0.548 |
| CEI | 0.942 | 0.958 | 0.846 |

*Source* Author's work

of the variables were adopted from previous studies. The 6-item scale for FP was adopted. The CEI based on competitive industry and competitive pressure was also adopted. The measurement scale to measure the AI applications was adapted from the industry, and the study used SMART-PLS for data analysis through measurement models and structural equation modeling techniques based on the collected data. The data was evaluated by checking its reliability, composite reliability, and average variance extract under the measurement model. The relationships between constructs were examined based on SEM through the collected data. This section shows the reliability test as the first step of the PLS algorithm (Table 1).

The reliability test results show that the three factors, FP, AI, and CEI, have high internal consistency and reliability. The Cronbach's alpha values are all above 0.8, indicating that each factor's items are closely related and measure the same construct. The composite reliability values are also above 0.8, which indicates that the factors are reliable and consistent indicators of the latent variables. The AVE values are above 0.5, which suggests that the factors explain more than half of the variance in their respective items. These results indicate that the measurement model is reliable and valid (Table 2).

This table shows the discriminant validity of the measurement model's three factors (FP, AI, and CEI). Discriminant validity refers to the extent to which the factors are distinct and not highly correlated. The diagonal elements of the table are the square roots of the AVE for each factor, which measure the amount of variance that the factor captures from its indicators relative to the measurement error. The off-diagonal elements are the correlations between the factors. A common rule of thumb is that the diagonal elements should be greater than the off-diagonal elements in the corresponding rows and columns. This indicates that the factors share more variance with their indicators than other factors. Based on this criterion, the table shows that the three factors have good discriminant validity, as the diagonal elements

**Table 2** Measurement model (Discriminant Validity)

| | FP | AI | CEI |
|---|---|---|---|
| FP | 0.729 | | |
| AI | 0.628 | 0.741 | |
| CEI | 0.531 | 0.648 | 0.921 |

*Source* Author's work

**Table 3** Measurement model

| Indicators | Loading | Uniqueness |
|---|---|---|
| AI1 | 0.854 | 0.524 |
| AI2 | 0.797 | |
| AI3 | 0.686 | |
| AI4 | 0.649 | |
| AI5 | 0.774 | |
| AI6 | 0.664 | |
| FP1 | 0.804 | 0.704 |
| FP2 | 0.868 | |
| FP3 | 0.806 | |
| FP4 | 0.756 | |
| FP5 | 0.296 | |
| FP6 | 0.672 | |
| CEI1 | 0.899 | 0.142 |
| CEI2 | 0.947 | |
| CEI3 | 0.937 | |
| CEI4 | 0.839 | |

*Source* Author's work

are more significant than the off-diagonal elements. This suggests that the factors are well differentiated and measure distinct constructs (Table 3).

The data in this table shows the reliability and validity of the data measurement model for the three factors: FP, AI, and CEI. Cronbach's alpha is a measure of internal consistency, which indicates how the items in each factor are related and measure the same construct. A common rule of thumb is that Cronbach's alpha should be above 0.7 for a reliable measurement. This table shows that all three factors have Cronbach's alpha values above 0.8, which indicates high internal consistency and reliability. Composite reliability is another measure of internal consistency, which indicates how reliable and consistent the factors are with the latent variables. A common rule of thumb is that composite reliability should be above 0.7 for a reliable measurement.

This table shows that all three factors have composite reliability values above 0.8, which indicates high internal consistency and reliability. AVE is a measure of convergent validity, which indicates how much variance the factors capture from its indicators relative to the measurement error. A common rule is that AVE should be above 0.5 for a valid measurement. This table shows that all three factors have AVE values above 0.5, which indicates high convergent validity. The data in this table shows that the measurement model has good reliability and validity for the three factors. This means that the measurement model is consistent, accurate, and reflective of the actual constructs of measuring (Table 4)

**Table 4** Hypothesis testing (Direct Relationship)

| | Original sample (0) | T. statistics (JO/STEDVI) | P-Value |
|---|---|---|---|
| AI > FP | 0.628 | 0.774 | 0.000 |
| CEI > FP | 0.531 | 0.789 | 0.000 |

*Source* Author's work

**Table 5** SEM (moderating effect)

| | Original sample (0) | T. statistics (JO/STEDVI) | P-Value |
|---|---|---|---|
| AI > FP | 0.503 | 5.936 | 0.000 |
| CEI > FP | 0.136 | 1.477 | 0.145 |
| Moderating effects | −0.074 | 0.947 | 0.347 |

*Source* Author's work

The table shows that AI and CEI positively and significantly affect FP, as the path coefficients are positive and the p-values are less than 0.05. This means that the higher the level of AI and CEI, the higher the level of FP. The table also shows that AI has a more substantial effect on FP than CEI, as the path coefficient of AI > FP is higher than that of CEI > FP. This means that AI has a more significant impact on FP than CEI (Table 5).

This table shows the moderating effect of variables. The original sample correlation coefficient between AI and FP is 0.503. The t-statistic (JO/STEDVI) is 5.986. The p-value is 0.000, less than the significance level (usually set at 0.05). The relationship between AI and AFP is statistically significant. AI has a positive impact on FP. The original sample correlation coefficient between CEI and FP is 0.136. The t-statistic (JO/STEDVI) is 1.477. The p-value is 0.145, which is greater than the significance level. The relationship between CEI and FP is not statistically significant at the conventional level. However, it's essential to consider the moderating effects. The moderating effect coefficient is -0.074. The t-statistic for moderating effects is 0.947. The p-value is 0.347, which is greater than the significance level. The moderating effect of CEI on the relationship between AI and FP is not statistically significant. While AI significantly influences FP, the role of CEI as a moderator requires further investigation. These findings contribute to our understanding of the complex dynamics in e-commerce firms.

## 5 Discussion and Findings

This study shows that AI significantly impacts FP. And CEI's moderating effect on the relationship between AI and CEI. This study explores the interplay between various factors, such as AI, CEI, and FP, in e-commerce firms with a specific focus on efficiency. By automating routine tasks, businesses can allocate busman resources strategically. This leads to cost savings and improved profitability. AI-driven personalization enhances customer satisfaction, encouraging repeat business and higher revenue. Customized recommendations and chatbots contribute to net profitability. AI algorithms optimize inventory management, supply chain logistics, and production schedules. Efficient resource allocation directly impacts the bottom line. AI automates tasks, reduces errors, and streamlines workflows. AI analyzes data to predict equipment failures, minimizing downtime. AI provides real-time insights for agile decision-making. AI leverages data from past customer behavior, browsing history, and purchase patterns to provide personalized product recommendations by understanding user preferences. AI suggests relevant items, increasing the likelihood of conversion. For instance, NLP-based AI can analyze language and images to match shoppers with desired products.

AI-powered chatbots handle customer inquiries, provide real-time assistance, and enhance user engagement. These virtual assistants can answer common queries, guide users through buying, and offer personalized recommendations. AI algorithms detect unusual patterns flagging potentially fraudulent transactions. By analyzing vast amounts of data, AI helps prevent unauthorized activities and protects customers and businesses. AI optimizes inventory levels by predicting demand patterns. It ensures that products are available when needed, minimizing stockouts and excess inventory. AI adjusts prices dynamically based on demand, competitor pricing, and other factors. Optimized pricing strategies lead to improved marketing ROI and competitiveness. AI analyzes customer behavior to predict churn. Businesses can proactively engage with at-risk customers, reducing churn rates. AI-generated ad creatives minimize manual effort and enhance relevance. Whether creating compelling visuals or writing persuasive copy, AI assists marketers in crafting effective ads. AI personalizes sales, engaging prospects with relevant messages at the right time, leading to higher conversions. By automating tasks, AI streamlines workflow reduces human effort and improves efficiency. AI enables precise targeting, ensuring that ad spending reaches high-potential segments. Automation and optimization lead to cost savings. AI doesn't replace jobs; it augments human capabilities. Collaborative efforts between humans and AI drive innovation and job growth. Exciting employees can upskill to work alongside AI systems, ensuring job security.

Thoughtful adoption of AI transforms e-commerce firms into strategic assets, driving financial performance and operational excellence. CEI encourages e-commerce companies to design products with longevity, reparability, and recyclability in mind. Creating durable and easily repairable items reduces the overall waste generated. E-commerce firms should prioritize product design that extends the lifespan of goods and facilitates repair. E-commerce companies can explore avenues like product refurbishment, remanufacturing, and second-hand markets. These practices promote resource reuse and reduce the need for new raw materials. Encourage customers to consider pre-owned or refurbished products. CEI ensures that materials are recycled efficiently. AI is crucial in optimizing recycling processes, identifying recyclable materials, and tracking their lifecycle. Implement AI-driven recycling systems to enhance material recovery rates. Life cycle assessments, carbon footprint calculations, and eco-labeling are essential tools integrated into CEI. AI helps evaluate the environmental impact of e-commerce operations. Regularly assess and communicate the environmental footprint of products and services.

AI algorithms optimize supply chains using predictive analytics, demand forecasting, and inventory management. This reduces excess stock, transportation costs, and energy consumption. Employ AI-driven supply chain management to minimize waste and enhance efficiency. CEI encourages e-commerce firms to adopt energy-efficient practices. AI can optimize energy usage in warehouses, distribution, and data centers. Implement bright lighting, HVAC systems, and real-time energy monitoring to conserve resources. AI simulations evaluate the environmental impact of different design choices. CEI informs sustainable product design by considering ecological factors. Use AI to assess design alternatives and select the most environmentally friendly options. AI assists in designing packaging that minimizes waste and is easy to recycle. Analyzing waste streams helps identify opportunities for recycling, composting, or repurposing. Opt for eco-friendly packaging materials and designs. Combined with AI-driven practices, CEI can significantly enhance resource efficiency, reduce waste, and contribute to a more sustainable e-commerce system. AI algorithms are pivotal in promoting sustainable practices within the e-commerce sector. By analyzing vast volumes of data, including market trends and customer choices, AI empowers businesses to optimize their operations and reduce carbon emissions.

AI provides real-time analysis of inventory levels, studying historical sales trends. By predicting demand accurately, AI optimizes supply chain management, ensuring the right products are available at the right time. This prevents overstocking and understocking, leading to more efficient operations and reduced waste. AI tools process data faster than humans, providing insights promptly. These tools continuously learn from data, becoming proficient at predicting outcomes and trends. CEI ensures that waste management aligns with circular economy principles. This involves minimizing waste, reusing materials, and promoting a sustainable product lifestyle. CEI considers the environmental and social impact of sourcing raw materials. AI assists in identifying ethical suppliers who align with sustainability goals. CEO guides sustainable production practices such as using renewable energy and reducing water usage. AI-powered supply chain transparency ensures responsible sourcing by tracking materials' origins and practices. CEI fosters collaboration with suppliers to improve sustainability practices. CEI enables transparent communication with customers, providing information about product regulations, while AI monitors changing regulations and helps e-commerce firms adapt. Dialogue with regulatory bodies shapes sustainable policies.

By integrating AI with CEI, businesses can thrive while minimizing their ecological footprint. Personalized recommendations, chatbots, fraud detection, and supply chain optimization enhance efficiency. Integrate CEI and consider environmental impact throughout the product lifecycle. Design for recyclability, reduce packaging waste and explore circular business models. Prioritize eco-friendly decisions alongside AI-driven profitability. Engage with industry peers, policymakers, and NGOs and collaborate to drive sustainable practices. The transformative potential of AI in e-commerce is immense, but responsible adoption guided by CEI is crucial.

Balancing technology, sustainability, and economic performance allows e-commerce firms to thrive in a dynamic landscape. AI and CEI pave the way for a greener, more sustainable e-commerce future.

## 6 Conclusion

The study aimed to explore the connections between AI adoption, CEI, and FP. A robust methodology was employed, utilizing a 5-point scale to measure variables. Data was collected from a diverse sample of e-commerce firms listed in the Pakistan Directory of SMEs. Managers were the primary respondents, and online surveys were an efficient data collection method. The reliability test results indicated high internal consistency and reliability across the three factors: FP, AI, and CEI. Cronbach's alpha values, composite reliability, and average variance extracted (AVE) exceeded recommended thresholds, confirming the model's reliability and validity. The discriminant validity analysis reinforced the distinctiveness of the factors. FP, AI, and CEI were well-differentiated and measured distinct constructs.

The detailed analysis of loadings and uniqueness further affirmed the measurement model's reliability and validity. All factors exhibited high internal consistency and reliability, meeting or exceeding commonly accepted thresholds. The study revealed significant positive relationships between AI, CEI, and FP. AI adoption had a more substantial impact on FP than CEI. This highlights the pivotal role of AI in enhancing e-commerce firms' overall performance. SEM with moderating effects provided nuanced insights. AI demonstrated a significant positive impact on FP. However, CEI's moderating effect on the relationship between AI and FP was not statistically significant. Further investigation is needed to understand the specific role of CEI as a moderator. E-commerce firms in Pakistan should strategically embrace AI to enhance their overall performance.

These findings emphasize the complex interplay between technological adoption, sustainable practices, and business outcomes. The study's robust methodology and comprehensive measurement model analysis enhance the methodological toolkit for research in this field. It underscores the significance of considering AI adoption and CEI when studying FP. Strategic emphasis on AI adoption can positively impact FP. While CEI didn't significantly moderate the relationship in this study, further exploration is warranted. E-commerce firms in Pakistan can benefit from these insights as they navigate the evolving landscape of technology and sustainability. The intricate interplay between AI, CEI, and FP demands ongoing exploration and strategic decision-making. This study lays a valuable groundwork for future research and provides actionable guidance for businesses in Pakistan's digital era.

## References


1. Fraga-Lamas, P., Lopes, S. I., Fernández-Caramés, T.M.: Green IoT and edge AI as key technological enablers for a sustainable digital transition towards a smart circular economy: an industry 5.0 use case. Sensors 21(17), 5745 (2021)
2. Arsawan, I.W.E., Atmaja, I.M.D.A.S., Hariyanti, N.K.D., Kariati, N.M., Yudistira, C.G.P., Darmayanti, P.A.: Circular economy business model: bibliography analysis and future Agenda. In International conference on applied science and technology on social science 2022 (iCAST-SS 2022) (pp. 143–147). Atlantis Press. (2022)
3. Di Vaio, A., Palladino, R., Hassan, R., Escobar, O.: Artificial intelligence and business models in the sustainable development goals perspective: a systematic literature review. J. Bus. Res. **121**, 283–314 (2020)
4. Pizzi, S., Corbo, L., Caputo, A.: Fintech and SMEs sustainable business models: reflections and considerations for a circular economy. J. Clean. Prod. **281**, 125217 (2021)
5. Abdallah, M., Talib, M.A., Feroz, S., Nasir, Q., Abdalla, H., Mahfood, B.: Artificial intelligence applications in solid waste management: a systematic research review. Waste Manag.Manag. **109**, 231–246 (2020)
6. Sun, T.Q., Medaglia, R.: Mapping the challenges of artificial intelligence in the public sector: evidence from public healthcare. Gov. Inf. Q. **36**(2), 368–383 (2019)
7. Bag, S., Pretorius, J.H.C., Gupta, S., Dwivedi, Y.K.: Role of institutional pressures and resources in adopting big data analytics powered artificial intelligence, sustainable manufacturing practices, and circular economy capabilities. Technol. Forecast. Soc. Chang. **163**, 120420 (2021)
8. Abioye, S.O., Oyedele, L.O., Akanbi, L., Ajayi, A., Delgado, J.M.D., Bilal, M., Ahmed, A.: Artificial intelligence in the construction industry: a review of present status, opportunities, and future challenges. J. Build. Eng. **44**, 103299 (2021)
9. Chauhan, C., Parida, V., Dhir, A.: Linking circular economy and digitalization technologies: a systematic literature review of past achievements and future promises. Technol. Forecast. Soc. Chang. **177**, 121508 (2022)
10. Alhawari, O., Awan, U., Bhutta, M.K.S., Ülkü, M.A.: Insights from circular economy literature: a review of extant definitions and unraveling paths to future research. Sustainability **13**(2), 859 (2021)
11. Gazzola, P., Pavione, E., Pezzetti, R., Grechi, D.: Trends in the fashion industry. The perception of sustainability and circular economy: a gender/generation quantitative approach. Sustainability 12(7), 2809 (2020)
12. Elliott, K., Price, R., Shaw, P., Spiliotopoulos, T., Ng, M., Coopamootoo, K., van Moorsel, A.: Towards an equitable digital society: artificial intelligence (AI) and corporate digital responsibility (CDR). Society 58(3), 179–188 (2021)
13. Bag, S., Wood, L.C., Mangla, S.K., Luthra, S.: Procurement 4.0 and its implications on business process performance in a circular economy. Resour. Conserv. Recycl. 152, 104502 (2020)
14. Bag, S., Pretorius, J.H.C.: Relationships between industry 4.0, sustainable manufacturing and circular economy: proposal of a research framework. Int. J. Organ. Anal. 30(4), 864–898 (2022)
15. Andronie, M., Lăzăroiu, G., Ştefănescu, R., Uţă, C., Dijmărescu, I.: Sustainable, thoughtful, and sensing technologies for cyber-physical manufacturing systems: a systematic literature review. Sustainability **13**(10), 5495 (2021)
16. Ciccullo, F., Cagliano, R., Bartezzaghi, G., Perego, A.: Implementing the circular economy paradigm in the agri-food supply chain: the role of food waste prevention technologies. Resour. Conserv. Recycl.. Conserv. Recycl. **164**, 105114 (2021)
17. Antikainen, M., Uusitalo, T., Kivikytö-Reponen, P.: Digitalization as an enabler of the circular economy. Procedia Cirp **73**, 45–49 (2018)
18. Denicolai, S., Zucchella, A., Magnani, G.: Internationalization, digitalization, and sustainability: are SMEs ready? A survey on synergies and substituting effects among growth paths. Technol. Forecast. Soc. Chang. **166**, 120650 (2021)

19. Chidepatil, A., Bindra, P., Kulkarni, D., Qazi, M., Kshirsagar, M., Sankaran, K.: From trash to cash: how can blockchain and multi-sensor-driven artificial intelligence transform the circular economy of plastic waste? Adm. Sci. **10**(2), 23 (2020)
20. Cannas, V.G., Ciano, M.P., Saltalamacchia, M., Secchi, R.: Artificial intelligence in supply chain and operations management: a multiple case study research. Int. J. Prod. Res. 1–28 (2023)
21. Esposito, M., Tse, T., Soufani, K.: Introducing a circular economy: new thinking with new managerial and policy implications. Calif. Manage. Rev. **60**(3), 5–19 (2018)
22. Khan, I.S., Ahmad, M.O., Majava, J.: Industry 4.0 and sustainable development: a systematic mapping of triple bottom line, circular economy and sustainable business models perspectives. J. Clean. Prod. 297, 126655 (2021)
23. Rajput, S., Singh, S.P.: Connecting circular economy and industry 4.0. Int. J. Inf. Manage. **49**, 98–113 (2019)
24. Javaid, M., Haleem, A., Singh, R.P., Suman, R.: Artificial intelligence applications for industry 4.0: A literature-based study. J. Ind. Integr. Manag. 7(01), 83–111 (2022)
25. Lacy, P., Long, J., Spindler, W.: The circular economy handbook (Vol. 259). London, UK: Palgrave Macmillan UK (2020)
26. Kristofferson, E., Blomsma, F., Mikalef, P., Li, J.: The smart circular economy: a digital-enabled circular strategies framework for manufacturing companies. J. Bus. Res. **120**, 241–261 (2020)
27. Rehman Khan, S.A., Yu, Z., Sarwat, S., Godil, D.I., Amin, S., Shujaat, S.: The role of block chain technology in circular economy practices to improve organisational performance. Int J Log Res Appl J Log Res Appl **25**(4–5), 605–622 (2022)
28. Jabbour, C.J.C., Fiorini, P.D.C., Wong, C.W., Jugend, D., Jabbour, A.B.L.D.S., Seles, B.M.R.P., da Silva, H.M.R.: First-mover firms in the transition towards the sharing economy in metallic natural resource-intensive industries: implications for the circular economy and emerging industry 4.0 technologies. Resour. Policy 66, 101596 (2020)
29. Ranta, V., Aarikka-Stenroos, L., Väisänen, J.M.: Digital technologies catalyzing business model innovation for circular economy—multiple case study. Resour. Conserv. Recycl.. Conserv. Recycl. **164**, 105155 (2021)
30. Kumar, S., Raut, R.D., Narkhede, B.E.: A proposed collaborative framework for healthcare workers using artificial intelligence of things (AI-IoT) during the COVID-19 pandemic. Int. J. Healthc. Manag. **13**(4), 337–345 (2020)
31. Leal Filho, W., Yang, P., Eustachio, J.H.P.P., Azul, A.M., Gellers, J.C., Gielczyk, A., Kozlova, V.: Deploying digitalization and artificial intelligence in sustainable development research. Environ. Dev. Sustain. **25**(6), 4957–4988 (2023)
32. Nandi, S., Sarkis, J., Hervani, A.A., Helms, M.M.: Redesigning supply chains using blockchain-enabled circular economy and COVID-19 experiences. Sustain. Prod. Consum. **27**, 10–22 (2021)
33. Persis, D.J., Venkatesh, V.G., Sreedharan, V.R., Shi, Y., Sankaranarayanan, B.: Modeling and analysing the impact of circular economy; internet of things and ethical business practices in the VUCA world: evidence from the food processing industry. J. Clean. Prod. **301**, 126871 (2021)
34. Okorie, O., Salonitis, K., Charnley, F., Moreno, M., Turner, C., Tiwari, A.: Digitization and the circular economy: a review of current research and future trends. Energies **11**(11), 3009 (2018)
35. Sarc, R., Curtis, A., Kandlbauer, L., Khodier, K., Lorber, K.E., Pomberger, R.: Digitalization and intelligent robotics in value chain of circular economy oriented waste management–a review. Waste Manag.Manag. **95**, 476–492 (2019)
36. Stahl, B.C., Andreou, A., Brey, P., Hatzakis, T., Kirichenko, A., Macnish, K., Wright, D.: Artificial intelligence for human flourishing–beyond principles for machine learning. J. Bus. Res. **124**, 374–388 (2021)
37. Tseng, M.L., Tran, T.P.T., Ha, H.M., Bui, T.D., Lim, M.K.: Sustainable industrial and operation engineering trends and challenges toward industry 4.0: a data driven analysis. J. Ind. Prod. Eng. 38(8), 581–598 (2021)

38. Sousa, M.J., Dal Mas, F., Osório de Barros, G., Tavares, N.: Artificial intelligence: technologies, applications, and policy perspectives. Insights from Portugal. Decis. Intell. Anal. Implement. Strat. Bus. Manag. 69–84 (2022)
39. Vinuesa, R., Azizpour, H., Leite, I., Balaam, M., Dignum, V., Domisch, S., Fuso Nerini, F.: The role of artificial intelligence in achieving the sustainable development goals. Nat. Commun.Commun. **11**(1), 1–10 (2020)
40. Zhang, A., Venkatesh, V.G., Liu, Y., Wan, M., Qu, T., Huisingh, D.: Barriers to smart waste management for a circular economy in China. J. Clean. Prod. **240**, 118198 (2019)
41. Yigitcanlar, T., Butler, L., Windle, E., Desouza, K.C., Mehmood, R., Corchado, J.M.: Can building artificially intelligent cities safeguard humanity from natural dis-asters, pandemics, and other catastrophes? An urban scholar' s perspective. Sensors **20**(10), 2988 (2020)
42. Garcia-Muiña, F.E., González-Sánchez, R., Ferrari, A.M., Settembre-Blundo, D.: The paradigms of industry 4.0 and circular economy as enabling drivers for the competitiveness of businesses and territories: the case of an Italian ceramic tiles manufacturing company. Soc Sci 7(12), 255 (2018)

36. Sousa, M.J., Dal Mas, F., Osório de Barros, G., Tavares, N.: Artificial intelligence technologies, applications, and policy perspectives. Insights from Portugal. Decis. Intell. Anal. Eng. Strat. Bus. Manag. 69–84 (2022)
37. Vlačić, E., [illegible]
38. Zhang, [illegible]
39. [illegible]
40. [illegible]

# Technological Innovation in Public Service Delivery: Performance and Uptake of OPGRS in the Landscape of Digital Governance

**Ahmad Raza Akhtar, Ghulam Murtaza, and Muhammad Luqman**

**Abstract** Technological innovation is essential for enhancing service accessibility, affordability, and user experience. Private organizations across various sectors, including health, education, and supply chain, are leveraging technology to transform operations. Likewise, public organizations, including municipal administrations, utility services, law enforcement agencies, and public grievance redressal bodies, are adopting digital technologies to improve public service delivery, enhancing governance efficiency. However, despite these developments, the full potential of digital technology in governance remains unrealized. This chapter examines the performance and uptake of Online Public Grievance Redressal Systems (OPGRS): Pakistan Citizen Portal (PCP) and Wafaqi Mohtasib (WM). These OPGRS represent a landmark initiative by the Government of Pakistan in digital governance and are ranked highly developed, integrating multiple government departments and agencies, engaging citizens nationwide, enabling grievance tracking, and supporting efficient resolution of public grievances related to governance. For analysing the performance and uptake of OPGRS, the data from the Prime Minister Delivery Unit, Prime Minister Office of Pakistan, and Wafaqi Mohtasib (Ombudsmen) Secretariat, Islamabad, were taken on public complaints. It covers four provinces and Azad Jammu & Kashmir; 38 districts; 14 professions; 43 complaint types; critical sectors like energy, municipal services, education, and health; 10 ministries; 20 state departments; and 20 government offices. Evaluation metrics include grievance resolution (received, resolved, pending), complaint timelines, citizen feedback (satisfaction level), and

A. R. Akhtar
Department of Economics, S3H, National University of Sciences & Technology (NUST), Islamabad, Pakistan
e-mail: ahmadr.mse21s3h@student.nust.edu.pk

G. Murtaza (✉)
Faculty of Arts and Society, Charles Darwin University, Darwin, Australia
e-mail: ghulam.murtaza@cdu.edu.au

M. Luqman
Kashmir Institute of Economics, The University of Azad Jammu & Kashmir, Muzaffarabad, Pakistan
e-mail: luqman.khan@ajku.edu.pk



N. D. K. Nguyen et al. (eds.), *Digital Transformation and Sustainability: Methods and Applications*, Information Systems Engineering and Management 47,
https://doi.org/10.1007/978-3-031-91278-8_17

departmental responsiveness. Our findings identify strengths and areas for improvement, illustrating how Pakistan's digital governance initiatives, such as OPGRS, can guide resource-limited developing countries in achieving transparent and efficient public administration systems for improved public service delivery.

**Keywords** Technological innovation · Public service delivery · Digital governance · Online public grievances redressal systems (OPGRS) · Operational management · Pakistan

**JEL Classifications** O33 · H83 · D73 · L86

## 1 Introduction

Organizations are increasingly embracing digital technologies across various sectors, such as healthcare, education, and supply chains, to improve accessibility, affordability, and service quality [5]. This shift is because integrating digital tools within traditional service delivery models has proven transformative, significantly enhancing organizational performance, productivity, resilience, and societal impact [28]. Particularly within public sector organizations, digital transformation is vital for building adaptable, efficient, and citizen-focused public administrative systems that meet contemporary citizens' demands. For the emerging markets and developing economies (EMDEs) facing challenges such as socioeconomic disparities, governance inefficiencies, and resource constraints, technological innovation is critical for improved public service delivery addressing citizen needs, fostering engagement, optimising workflows, and ensuring equitable access [6, 14].

In the case of EMDEs, the traditional public service model remains prevalent, characterized by a hierarchical governance structure that prioritizes procedural consistency but often lacks the flexibility to respond swiftly to evolving public demands, limiting responsivenesss. In contrast, many developed countries are adopting more responsive public service models. For example, the integrated service delivery model consolidates multiple services into a single access point, streamlining interactions, improving coordination, and enhancing user experience by providing seamless access to various public services [14]. Similarly, the self-service and empowerment model allow citizens direct digital access to services, reducing reliance on in-person assistance and fostering autonomy in handling routine administrative tasks [28]. Likewise, the citizen-centric model prioritises citizen feedback through digital platforms, enabling services to evolve in response to real-time feedback and fostering a more responsive public sector [14].

In recent years, digital governance models have emerged as transformative approaches to public service delivery, leveraging automation and digitization to enhance transparency, accessibility, and responsiveness. These models address traditional bureaucratic inefficiencies by streamlining workflows, reducing delays, and eliminating bottlenecks, thereby expanding access to critical services such as healthcare, education, and social welfare [12]. Through digital integration, these models cater to citizens' growing expectations for efficient and transparent public administration, fostering societal equity and well-being.

Digital governance represents the convergence of innovative service delivery frameworks, integrating advanced technologies to improve quality and responsiveness [1]. This paradigm directly supports the 2030 global development agenda, particularly the Sustainable Development Goals (SDGs). For instance, SDG 16 emphasizes the need for peace, justice, and strong institutions, SDG 9 advocates for resilient infrastructure and innovation, while SDG 17 highlights partnerships as vital for achieving effective service delivery [24]. By aligning with these goals, digital governance can contribute to sustainable development by fostering inclusivity, efficiency, and transparency in public administration systems [3]. Adopting digital governance thus represents a critical pathway for nations to modernize public services, address governance challenges, and promote long-term sustainability. This approach demonstrates its potential to create more accessible and equitable systems, meeting the evolving needs of citizens in a digital age.

Public service delivery refers to the mechanisms through which government agencies and public sector organizations provide essential services to citizens, aiming to address their social and economic needs while enhancing living standards. These services encompass various critical domains, including social welfare, healthcare, education, public safety, utilities, infrastructure, and administrative functions. For example, healthcare services provided through hospitals, clinics, and emergency care systems ensure citizens have access to preventative and urgent medical attention, safeguarding public health. Similarly, educational institutions such as schools, universities, and vocational training centers contribute to personal development, economic growth, and social advancement. Public safety services, including police, fire departments, and disaster response units, are crucial for maintaining security and protecting communities [5]. Social welfare programs, offering pensions, unemployment benefits, and disability support, aim to assist vulnerable populations and uphold a stable standard of living [15]. Utilities such as water supply, sanitation, electricity, and waste management provide indispensable resources essential for public health and daily life [26]. Furthermore, infrastructure, including roads, bridges, and transportation systems, facilitates connectivity, economic activities, and mobility [1]. Administrative services, such as licensing, civil registration, permits, and grievance redressal mechanisms, simplify interactions between citizens and government bodies, enhancing accessibility and efficiency. The overarching goal of public service delivery is to improve quality of life, ensure equitable resource access, and promote societal well-being.

Although technological innovation in public service delivery has significantly improved accessibility, efficiency, transparency, and equity, enabling governments to address complex public needs more comprehensively [10], the full transformative potential of digital tools remains underexplored, particularly in resource-limited EMDEs. In many such countries, traditional service delivery models still dominate public organizations, limiting the impact of digital innovation [12]. Public service delivery in these regions faces numerous challenges. For instance, in Pakistan, systemic issues such as corruption, bureaucratic inefficiencies, and infrastructure neglect hinder service quality and fuel public dissatisfaction [18]. Reports indicate that corruption in critical institutions like the police and judiciary undermines public trust and disrupts effective service delivery, leading to economic losses and social unrest in Pakistan [21, 22]. Bureaucratic inefficiencies further exacerbate delays and resource mismanagement, while neglected infrastructure impedes equitable access to essential services such as healthcare, education, and utilities in Pakistan [6]. Disruptive innovation theory also underscores the importance of evaluating digital platforms for their ability to serve diverse societal groups, particularly underserved populations, to maximize their potential in transforming public service delivery [3, 4]. By focusing on inclusivity and addressing systemic barriers, digital innovations can play a pivotal role in improving governance outcomes, fostering public trust, and enhancing service equity and quality [1].

To address the public administration challenges, such as corruption, bureaucratic inefficiencies, and lack of infrastructure, Pakistan has recently implemented Online Public Grievance Redressal Systems (OPGRS), particularly the Pakistan Citizens Portal (PCP) and the Wafaqi Mohtasib (WM) or Federal Ombudsman, as part of its digital governance model to streamline grievance redressal in public service delivery. These OPGRS platforms are highly developed and support cross-departmental integration, real-time tracking, and structured resolution timelines, promoting transparency and efficiency in governance and thereby helping improve public service delivery [8]. For instance, since its launch in 2018, the PCP has handled over 3.5 million complaints with a resolution rate of approximately 89%, demonstrating digital tool capacity to improve public trust by enhancing administrative responsiveness and accountability [16].

Given the above perspective, this chapter examines Pakistan's initiatives in digital governance, such as OPRGs, providing insights into effective strategies for building a responsive and citizen-centric public administration system that improves public service delivery. In particular, within the digital governance landscape of Pakistan, the study evaluates the performance and uptake of OPGRS using data sourced from the Prime Minister Delivery Unit (PMDU), Prime Minister Office, and Wafaqi Mohtasib Secretariat, Islamabad, across four provinces (including Azad Jammu and Kashmir), 38 administrative districts, and multiple professional categories and complaint types. The evaluation spans key areas, including 10 ministries/divisions, 20 state departments, and 20 government offices, and examines grievance resolution rates, timelines, citizen feedback, and departmental responsiveness. Furthermore, this comprehensive assessment categorises data by administrative region, profession, and complaint type. The study's findings highlight the strengths and areas for improvement within Pakistan's digital governance. By identifying challenges and leveraging opportunities in public service delivery that align with the effectiveness of OPGRS, the study presents how Pakistan's discourse on digital governance can serve as a model for resource-constrained EMDEs in establishing transparent and efficient public administration systems. The importance of this study, as reinforced by the World Bank [27], lies in the emphasis on digital governance to enhance service efficiency, increase citizen satisfaction, and provide valuable insights for other countries that need similar digital advancements in governance.

The chapter is structured as follows: Section 2 examines governance inefficiencies and administrative challenges in EMDEs, discussing barriers to public service delivery, such as corruption, through indicators like the corruption perception index, as well as enablers, including digital infrastructure, supported by e-governance capabilities and digital literacy levels. Section 3 explores Pakistan's digital initiatives, focusing on OPGRS, particularly the Pakistan Citizen Portal and Wafaqi Mohtasib. Section 4 assesses the uptake and performance of OPGRS, using evaluation metrics such as complaint receipt, disposal and resolution rates, and citizen satisfaction. Finally, Section 5 presents conclusions, underscoring the importance of data-driven evaluations to refine policymaking, enhance OPGRS effectiveness in public service delivery, and provide insights for resource-constrained EMDEs.

## 2 Public Service Delivery: Challenges and Opportunities for EMDEs

In EMDEs, including Pakistan, effective public service delivery is influenced by a combination of governance challenges and a lack of digital infrastructure. The following sub-sections delve into the core barriers and potential enablers, respectively.

### *2.1 Barriers to Public Service Delivery: Governance Inefficiencies and Administrative Challenges*

Figure 1 highlights governance inefficiencies and administrative challenges in public systems across EMDEs, presenting an average overview from 2010 to 2023. It provides insights into administrative performance, public sector corruption, and governance efficiency. For instance, the Government Effectiveness Index (GEI) indicates how well governments manage public services, ensure civil service quality, and maintain independence from political influence. Countries like China, South Korea, and the UAE demonstrate higher GEI scores, reflecting their strong policymaking capabilities, effective implementation, and consistent public service delivery. These nations have increasingly leveraged technology and digital governance tools to boost administrative efficiency, combat corruption, and improve service quality, contributing to their higher GEI scores. In contrast, countries like Pakistan, Egypt, and Argentina exhibit lower GEI scores, indicating obstacles in policy implementation, public service quality, and political interference. These scores signal notable governance challenges, including limited administrative capacity, ineffective service delivery, and fragile policy enforcement mechanisms.

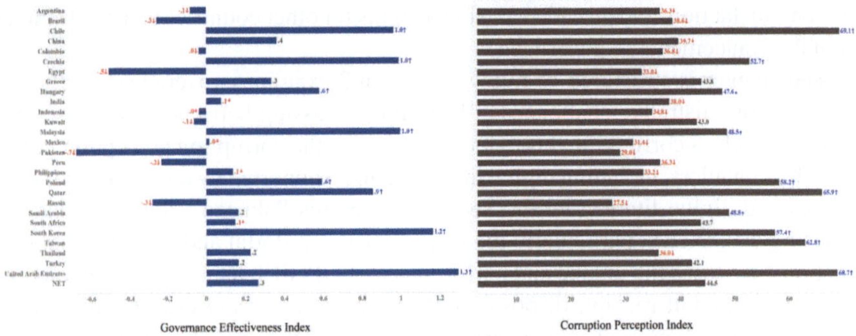


**Fig. 1** Institutional and governance effectiveness across EMDEs, average scores 2010–2023. *Note* **Authors' own**. The visualisation represents the comparison of emerging economies of the world. The data on the Government Effectiveness Index (−2.5 weak; 2.5 strong), which captures perceptions of the quality of public services, the quality of the civil service and the degree of its independence from political pressures, the quality of policy formulation and implementation, and the credibility of the government's commitment to such policies, is taken from (www.theglobaleconomy.com). Corruption Perceptions Index (100 = no corruption) indicates perceptions of public sector corruption, i.e., administrative and political corruption. The indicator values are determined by using information from surveys and assessments of corruption collected by various reputable institutions and taken from (www.theglobaleconomy.com). All data spans 2010–2023, and average values are visualised

Figure 1 also reveals that high levels of public sector corruption in countries such as Russia, Brazil, and Mexico create significant obstacles to effective governance by weakening institutional integrity. Corruption disrupts essential services and diminishes public trust in government institutions, further impairing service accessibility, reliability, and fairness, especially in settings where administrative capacity is already compromised. Trends indicate that EMDEs with higher corruption levels typically demonstrate lower government effectiveness, as corruption undermines institutional transparency and operational integrity. For instance, China's strong government effectiveness and relatively low corruption levels indicate a focused administrative approach to effective policy execution and efficient service delivery. Conversely, Pakistan's GEI score of −0.7, coupled with a low Corruption Perceptions Index (CPI), underscores the dual challenges of governance inefficiencies and pervasive corruption. This combination undermines service accessibility, efficiency, and quality, impeding Pakistan's development trajectory. The governance and administrative challenges illustrated in Fig. 1 emphasise the complexity EMDEs face in advancing public service delivery. EMDEs with lower governance effectiveness and higher corruption rates must build administrative capacity and implement robust anti-corruption strategies to improve service accessibility, equity, and quality. Digital governance reforms such as OPGRS are pivotal in addressing these challenges, fostering transparent, accountable, and efficient public service systems, and contributing to broader sustainable and inclusive development goals.

## 2.2 *Enablers of Public Service Delivery: Digital Infrastructure*

Digital transformation is increasingly recognised as a key enabler of public service delivery, enabling governments to streamline resources, expand access, and reduce bureaucratic inefficiencies. For EMDEs, digital governance holds great potential to enhance and modernise public services; however, these countries face unique constraints compared to developed economies. Nations such as the United States, the European Union, and Singapore lead globally in technology-driven governance, supported by robust digital infrastructure and highly skilled populations. Conversely, EMDEs across Asia, Africa, and Latin America encounter challenges like limited infrastructure, digital skills gaps, and budget constraints, which slow their progress toward digital transformation. Figure 2 highlights the digital infrastructure and the disparities within EMDEs for the year 2022, measured by indicators including the (a) e-Government Development Index, (b) Digital Skills Index, (c) Internet Users, and (d) ICT Development Index.

The e-Government Development Index (EGDI), measured at a scale of 100–0, with 100 highest and 0 lowest, provides a holistic picture of digital governance capacity across countries, revealing notable differences in e-governance maturity. For example, South Korea, Qatar, and the United Arab Emirates rank at the top with EGDI scores of 0.9283, 0.8706, and 0.9012, respectively, placing them in the 'Very High' category. These countries exemplify advanced digital infrastructure, robust online services, and strong digital literacy, creating an environment for effective

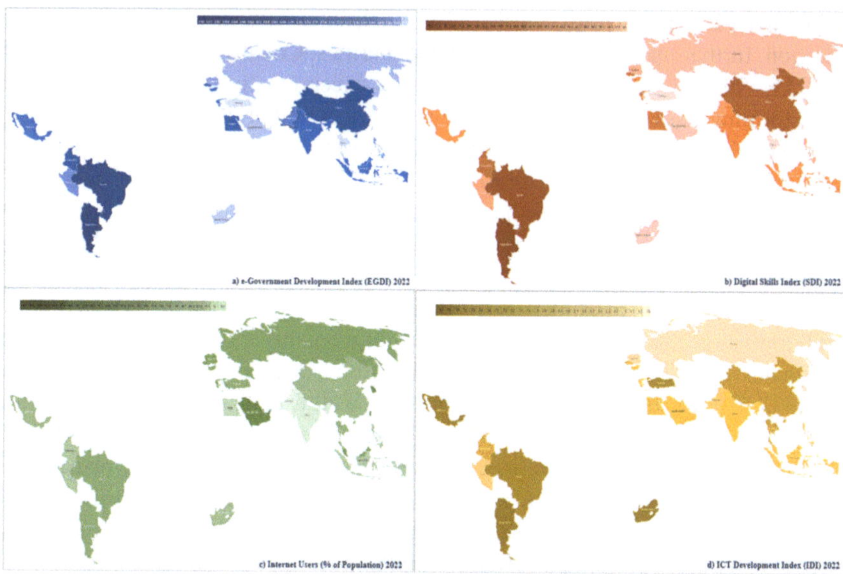


**Fig. 2** Digital infrastructure across EMDEs. *Source* Authors' own

and inclusive public service delivery. Their high rankings reflect extensive investments in digital tools and policy frameworks that prioritise accessibility and improve government-citizen interactions, setting global benchmarks for e-governance. In contrast, countries like Egypt (0.5723), Peru (0.6214), and Pakistan (0.4022) rank significantly lower, underscoring the challenges faced by EMDEs. Pakistan's low EGDI classification highlights critical gaps in digital infrastructure, online service provision, and digital literacy. This lower scoring indicates broader issues facing many EMDEs, often constrained by socioeconomic limitations and governance inefficiencies that hinder the development of effective e-governance frameworks. The digitisation of public services in these regions is often impeded by limited internet access, insufficient ICT investment, and significant disparities in digital skills. These contrasts between high- and low-ranking countries underscore the necessity of targeted investments in ICT and human capital development as foundational pillars for successful digital governance.

The Digital Skills Index (DSI), measured on a scale of 100–0, with 100 highest and 0 lowest, varies significantly across regions and is vital for economic participation and public engagement. High DSI scores—such as South Korea's 79.2, Qatar's 76.0, and Taiwan's 77.5—indicate a digitally capable population able to create and interact with sophisticated digital services. In contrast, lower DSI scores in Pakistan (46.0) and Egypt (50.1) highlight the challenges in building a digitally literate workforce. For EMDEs with low digital skills, strengthening digital education and literacy is essential for economic competitiveness and fostering the effective use of digital government platforms like OPGRS. Without sufficient digital skills, even well-designed e-government services may remain underutilised, limiting their reach and impact.

Internet penetration rates, measured through internet users as a percentage of pollution, further underscore these digital disparities. Countries like Qatar (99.7%), the UAE (99.5%), and Kuwait (99.6%) enjoy exceptionally high rates of internet users, forming a solid foundation for digital services and promoting citizen engagement in public governance. In contrast, internet users are comparatively low in countries such as Pakistan (36.5%) and India (43%), highlighting structural barriers to connectivity that restrict the reach of digital services, especially in rural and underserved areas. Addressing these connectivity gaps is crucial for more equitable digital governance.

The ICT Development Index (IDI), measured at a scale of 10–0, with 10 highest and 0 lowest, evaluating the overall maturity of a country's ICT ecosystem, further illuminates these differences. Countries like South Korea (8.1), Qatar (7.6), and Poland (7.5) score highly on the IDI, reflecting well-established ICT infrastructure that supports advanced public service platforms and promotes a digitally integrated society. Conversely, low IDI scores in Pakistan (4.8) and Egypt (5.5) reveal gaps in ICT access, usage, and digital skills, which are essential for an inclusive digital ecosystem. For these countries, prioritizing ICT infrastructure and enhancing digital usage is essential to creating an effective digital governance framework.

The above-explained statistics on digital infrastructure across EMDEs reveal that countries with limited digital infrastructure, such as Pakistan, can still achieve significant progress in digital governance by focusing on strategic priorities [19]. Case studies from countries like Singapore, Estonia, and India demonstrate that citizen-centric approaches, strategic partnerships, and interdepartmental integration are key to successful digital transformation, even under resource constraints [11, 15].

In Pakistan, the Online Public Grievance Redressal Systems (OPGRS) exemplify the potential of such digital transformation in governance, leveraging existing mobile and internet networks to deliver accessible grievance redressal mechanisms [9]. OPGRS has been instrumental in reducing corruption, enhancing responsiveness, and fostering public trust—key elements in modernising public administration and promoting inclusive and sustainable governance [2, 7].

Building on the barriers and enablers of public service delivery identified in EMDEs, the following sections examine Pakistan's digital governance initiatives, particularly OPGRS, and critically analyse their performance and uptake. The analysis will highlight how digital governance initiatives, i.e., OPGRS, can significantly enhance citizen satisfaction and engagement, ultimately improving governance and contributing to more effective and equitable public service delivery.

## 3 Pakistan's Initiatives in the Landscape of Digital Governance: Online Public Grievances Redressal Systems (OPGRS)

The global rise in caseloads and delays in judicial systems has prompted many countries to adopt online dispute resolution (ODR) platforms to expedite minor case resolutions. In the UK, for instance, ODR systems have successfully alleviated court backlogs by efficiently managing small claims and civil disputes, enhancing citizens' access to justice. Pakistan faces similar issues within its judicial system, where economic instability, rapid population growth, and complex legal processes have limited access to timely justice, particularly for marginalised communities. With over 2.26 million cases pending nationwide, delays in delivering justice are a pressing concern. Recognising the urgent need for modernisation, Pakistan has embarked on a digital transformation of its governance framework. By adopting ODR, Pakistan aims to introduce accessible and timely grievance resolution.

Traditionally, Pakistan has managed public grievances through judicial courts, specialised anti-corruption agencies like the National Accountability Bureau (NAB), administrative reviews, and public consultations to uphold transparency and accountabilityy. Complementing these formal avenues are alternative dispute resolution (ADR) methods, such as mediation and arbitration, providing additional options for citizens who face barriers in the formal judicial process. Based on these existing mechanisms, digital governance platforms, such as OPGRS, have the potential to make grievance redressal more accessible, efficient, and transparent.

OPGRS represents a significant advancement in the broader landscape of digital governance in Pakistan. Accessible to citizens via smartphone apps or web portals, OPGRS are akin to ODR systems and fall within the ADR framework while operating in the broader digital governance landscape. By leveraging technology, OPGRS enhances transparency and efficiency in grievance management, representing a crucial step toward modernising the public administration system. Countries like India and China have pioneered similar models, with India's Centralized Public Grievance Redress and Monitoring System (CPGRAMS) and China's 12345 Government Service Hotline achieving high-resolution rates and streamlined grievance handling.

This study evaluates the OPGRS initiative taken by the Government of Pakistan, focusing on the performance and uptake of these digital platforms. However, numerous digital platforms operate in Pakistan in the broader framework of digital governments. Nonetheless, OPGRS are particularly designed to address citizens' grievances related to public services and foster trust between citizens and the state, ultimately strengthening public administration systems.

Applying PRISMA guidelines, which likely can enhance transparency and reproducibility in systematic reviews [13], we thoroughly reviewed available digital governance platforms to differentiate OPGRS platforms from other digital governance platforms available in Pakistan. The initial step identified 161 digital governance platforms across various sectors. Following this, 23 platforms were screened as potential OPGRS. Using specific inclusion and exclusion criteria, such as the platform's developmental stage and accessibility via smartphone apps, the review process continued through the PRISMA stages, culminating in the selection of two key OPGRS: the Pakistan Citizen Portal (PCP) and the Wafaqi Mohtasib (WM) (or Federal Ombudsman). These digital platforms with the highest stage of development (stage 4) were chosen to analyse their performance and user engagement in-depth. A detailed application of each PRISMA stage is depicted in Fig. 3, while OPRGS with their stages of development are explained in Table 1.

The PCP and WM are recognised as Pakistan's most advanced OPGRS (see Table 1). The PCP integrates various governmental departments, allowing citizens to raise grievances across various public service areas, from healthcare and education to utilities. In contrast, the WM portal provides specialised redressal for administrative grievances, safeguarding citizens' rights against government mismanagement. Through these digital grievance redressal platforms, Pakistan is making strides in modernising public service delivery. The PCP and WM enhance administrative efficiency and accountability and foster trust in state institutions by enabling faster, more transparent grievance resolution processes. These platforms also highlight Pakistan's commitment to leveraging digital governance as a catalyst for improved public service delivery, reflecting a broader trend of technological innovation in governance that supports inclusive, responsive, and sustainable development.

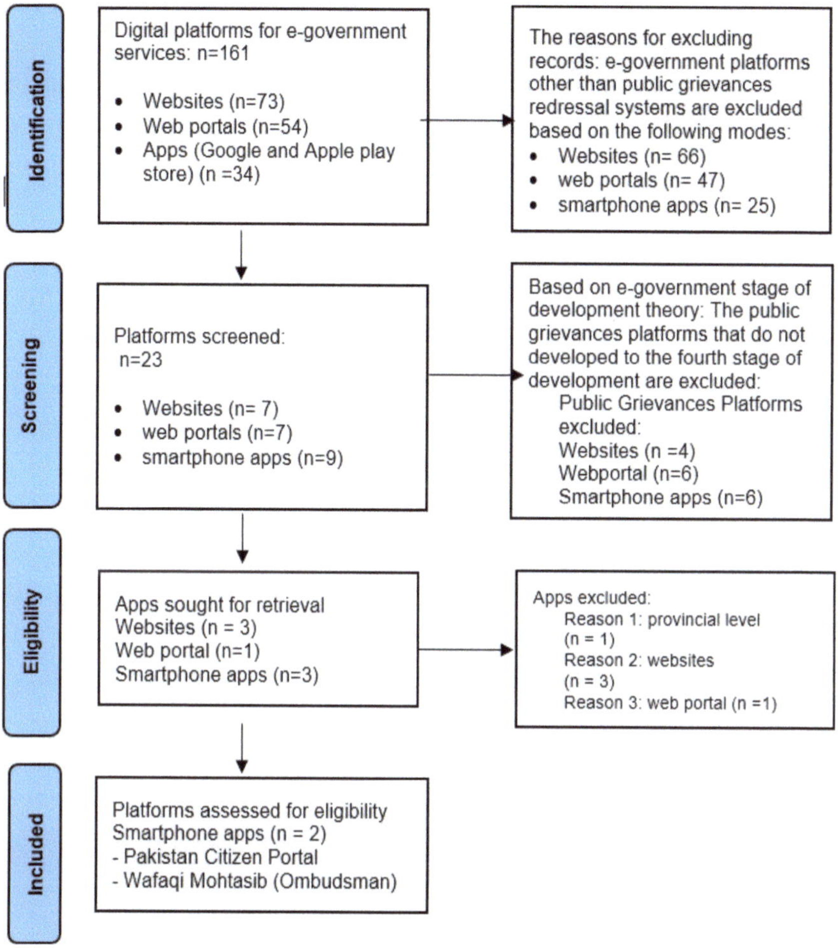


**Fig. 3** Selection of OPGRS. *Source* Identified by the authors from UN e-Government Survey (Pakistan) 2022, [20], following PRISMA guidelines, available at http://www.prisma-statement.org

## 4 Performance and Uptake of OPGRS

This section provides an in-depth performance analysis and uptake of OPGRS, focusing on PCP and WM. Data obtained from the Prime Minister Delivery Unit (PMDU), Prime Minister Office, Islamabad and the Wafaqi Mohtasib Secretariat, Islamabad, from 2018 to 2023 for PCP and 2019–2023 for WM sheds light on these platforms in facilitating efficient communication between citizens and government institutions for grievance resolution to improve public service delivery.

With Pakistan's population at approximately 241 million, significant demographic variations impact OPGRS use and performance. Punjab, with 127.6 million people, is the most populous province, representing 52.96% of the national population. This

**Table 1** OPGRS with their stages of development

| Stages of development (Layne and Lee, 2001) | Description | Key criterion for development stage inclusion | Level | OPGRS | | | Summary |
|---|---|---|---|---|---|---|---|
| | | | | Websites | Online portal | Smartphone apps | |
| Stage 1 (Least developed): *Cataloguing* | At this stage, governments do not have much Internet expertise and prefer to minimise the risk by doing a small project. Parts of the government's non-transactional information are put on the site | • Online presence<br>• Catalogue presentation<br>• Downloadable forms | – | – | – | – | Online<br>Websites: 0<br>Online web portal: 0<br>Smartphone apps: 0 |
| Stage 2 (Less developed): *Transaction* | This stage empowers citizens to deal with their governments online anytime, saving hours of paperwork, the inconvenience of travelling to a government office, and time spent waiting in line | • Service and forms online<br>• Working database supporting online transaction | – | – | – | – | Online<br>Websites: 0<br>Online web portal: 0<br>Smartphone apps: 0 |

(continued)

**Table 1** (continued)

| Stages of development (Layne and Lee, 2001) | Description | Key criterion for development stage inclusion | Level | OPGRS | | | Summary |
|---|---|---|---|---|---|---|---|
| | | | | Websites | Online portal | Smartphone apps | |
| Stage 3 (Medium developed): *Vertical Integration* | At this stage, the focus is moving toward transforming government services rather than automating and digitising existing processes | • Local systems linked to higher-level systems<br>• Within similar functionalities | National | Federal Tax Ombudsman (Government of Pakistan): https://fto.gov.pk/Default#close | – | FTO Portal (Federal Tax Ombudsman Pakistan) Complaint Management App: https://play.google.com/store/apps/details?id=com.pral.ftocmis&hl=en_US | Online Websites: 4 Online web portal: 6 Smartphone apps: 6 |
| | | | | Pakistan Telecommunication Authority (PTA): https://www.pta.gov.pk/en/consumer-support/complaints/complaints | – | PTA CMS Pakistan Telecommunication Authority—PTA https://play.google.com/store/apps/details?id=pk.gov.pta.cms&hl=en&gl=US | |

(continued)

**Table 1** (continued)

| Stages of development (Layne and Lee, 2001) | Description | Key criterion for development stage inclusion | Level | OPGRS | | | Summary |
|---|---|---|---|---|---|---|---|
| | | | | Websites | Online portal | Smartphone apps | |
| | | | | – | – | Foreign Minister's Portal: https://play.google.com/store/apps/details?id=com.nitb.foreignministry.foreign_ministry_portal | |
| | | | | – | SUNWAI (Customer Complaint Service by State Bank of Pakistan): https://complaint.sbp.org.pk/index.html | Sunwai Customer Complaint (State Bank of Pakistan): https://play.google.com/store/apps/details?id=com.sbp.sbp_complaints_management&hl=en&gl=US | |

(continued)

**Table 1** (continued)

| Stages of development (Layne and Lee, 2001) | Description | Key criterion for development stage inclusion | Level | OPGRS | | | Summary |
|---|---|---|---|---|---|---|---|
| | | | | Websites | Online portal | Smartphone apps | |
| | | | | Securities and Exchange Commission Pakistan (SECP)—Service Desk Management System: https://sdms.secp.gov.pk/~sdmsadmn/index.php?action=complaint | – | SECP's Service Desk Management Securities Exchange Commission Pakistan (SECP) https://play.google.com/store/apps/details?id=com.secp.sdms&hl=en&gl=US | |

(continued)

**Table 1** (continued)

| Stages of development (Layne and Lee, 2001) | Description | Key criterion for development stage inclusion | Level | OPGRS | | | Summary |
|---|---|---|---|---|---|---|---|
| | | | | Websites | Online portal | Smartphone apps | |
| | | | | – | Power Information Technology Company (PITC)—Customer Complaint Management System (CCMS), Ministry of Energy (Power Division) http://ccms.pitc.com.pk/login | Roshan Pakistan—Online Elect, Ministry of Energy (Power Division) https://play.google.com/store/apps/details?id=com.ssy.roshanpakistan&hl=en&gl=US | |
| | | | Provincial | Punjab Electricity Complaints Registration and Dispute Resolution: www.energy.punjab.gov.pk | Portal of Punjab Electricity Complaints Registration and Dispute Resolution: http://103.8.114.218/ecpp/intro.aspx | – | |

(continued)

**Table 1** (continued)

| Stages of development (Layne and Lee, 2001) | Description | Key criterion for development stage inclusion | Level | OPGRS | | | Summary |
|---|---|---|---|---|---|---|---|
| | | | | Websites | Online portal | Smartphone apps | |
| | | | | – | Overseas Pakistani Complaint Portal: https://ospc.punjab.gov.pk/ | – | |
| | | | | – | Inspector General of Police Sindh (Complaint Management System) https://igpcms.sindhpolice.gov.pk/ | – | |
| | | | | – | Women & Children Protection Cell (https://wpc.sindhpolice.gov.pk) | – | |

(continued)

**Table 1** (continued)

| Stages of development (Layne and Lee, 2001) | Description | Key criterion for development stage inclusion | Level | OPGRS | | | Summary |
|---|---|---|---|---|---|---|---|
| | | | | Websites | Online portal | Smartphone apps | |
| Stage 4 (Highly developed): *Horizontal Integration* | Databases across different functional areas will communicate with each other and ideally share information so that information obtained by one agency will propagate throughout all government functions | • Systems integrated across different functions | National | Prime Minister Performance Delivery Unit (PMDU): PM's Performance Delivery Unit (citizenportal.gov.pk) | Pakistan Citizen's Portal: Pakistan Citizen's Portal (citizenportal.gov.pk) | Pakistan Citizen Portal (National IT Board, Government Of Pakistan): https://play.google.com/store/apps/details?id=com.govpk.citizensportal | Online Websites: 3 Online web portal: 1 Smartphone apps: 3 |
| | | | | Wafaqi Mohtasib (Ombudsman)'s Secretariat (Government of Pakistan): https://mohtasib.gov.pk/ | – | Wafaqi Mohtasib CMIS (Wafaqi Mohtasib Secretariat): https://play.google.com/store/apps/details?id=com.pk.gov.wafaqimohtasib&hl=en_US | |

(continued)

**Table 1** (continued)

| Stages of development (Layne and Lee, 2001) | Description | Key criterion for development stage inclusion | Level | OPGRS | | | Summary |
|---|---|---|---|---|---|---|---|
| | | | | Websites | Online portal | Smartphone apps | |
| | | | Provincial | CM Secretariat (Complaint Cell): https://cm.sindh.gov.pk/complaint | – | – | |
| | | | | – | – | Ombudsman Punjab (OPMIS), Office of Ombudsman (Mohtasib) Punjab: https://play.google.com/store/apps/details?id=com.opmispunjab.OPMISERP&hl=en&gl=US | |
| Total | | | | 7 | 7 | 9 | 23 |

*Source* Authors' Compilation, United Nations Department of Economic and Social Affairs [20]

is followed by Sindh (23.04%), Khyber Pakhtunkhwa (15.42%), and Balochistan (5.94%), while Azad Jammu and Kashmir (AJK) and Gilgit-Baltistan together account for around 2.5% of the population. These population dynamics reflect varying demand levels for grievance redressal services across regions, as seen in the higher complaint volumes from populous areas like Punjab and Sindh, which present unique administrative challenges. In contrast, smaller regions face specific constraints related to digital infrastructure.

The study breaks down OPGRS performance into key domains: status of grievances, analysing the number of complaints received, resolved, and pending; citizen feedback, with satisfaction levels categorised as satisfied, unsatisfied, or neutral; departmental responsiveness, examining complaint volumes and resolutions for various different categories; resolution time profile, detailing complaint handling from submission to resolution; and department-specific dependency and overdue status. Grievance status reveals complaint volumes and departments' capacity to address them, pinpointing where targeted interventions could alleviate bottlenecks and enhance response efficiency.

The analysis of citizen feedback offers insights into public perception of service quality, emphasising areas where grievance handling and communication may be improved to boost satisfaction. Departmental responsiveness is essential in a multi-departmental system like OPGRS, as metrics such as response times and interdepartmental coordination reflect where structural improvements could strengthen overall performance. The resolution time profile provides a closer look at efficiency, revealing potential delays within certain departments and suggesting process automation to enhance responsiveness. The following sections provided a detailed performance and uptake analysis related to PCP and WM.

## *4.1 Pakistan Citizens Portal (PCP)*

The PCP is a government-owned smartphone application (available on Android and iOS) designed to connect citizens with government agencies and departments across Pakistan. It provides a nationwide platform for citizens to report public service delivery grievances and offer suggestions for improving public services. PCP covers data across Pakistan's four provinces, Punjab, Khyber Pakhtunkhwa, Sindh, and Balochistan, including Azad Jammu & Kashmir and the federal government. The PCP data comprehensively includes 38 administrative districts, 14 professional categories, gender, and 43 complaint categories, such as Energy & Power, Municipal Services, Education, Health, and Corruption/Malpractice. Additionally, it covers 10 ministries/divisions, 20 state departments, and 20 government offices.

Table 2 presents registered user membership and complaint trends on the PCP from 2018 to 2023, focusing on user distribution by location and overall engagement, showing that most PCP users are domestic, with inland members comprising 92.5% of the total, underscoring the platform's strong focus within Pakistan. Overseas members account for 7.1%, and foreign users make up a minimal 0.4%, reflecting limited international engagement. The data also highlights complaint registration and resolution rates across provinces, with inland complaints dominating at 94.8%, further reinforcing PCP's extensive use within Pakistan. Overseas complaints comprise 4.9%, and those from foreign users remain minimal at 0.3%. The province-wise breakdown shows that the federal government and Punjab receive the highest complaints. Furthermore, Table 2 shows the complaints receipts and disposal and measures the complaint resolution rates (in percentages) from 2019 to 2023 on the PCP, showing considerable variation across provinces, with each region exhibiting unique patterns in high, moderate, and low-performing districts. This province-wise analysis highlights the shades of complaint-handling efficiency across Pakistan. Further details show that in Punjab, most districts report high-resolution rates, with notable top performers. Khyber Pakhtunkhwa also demonstrates a high-resolution rate, suggesting efficient handling of grievances. However, Sindh shows a gap between received and resolved complaints, indicating potential for improvement in its resolution process. Lower complaint volumes are noted in Balochistan, Azad Jammu and Kashmir (AJK), and Gilgit-Baltistan, likely due to smaller populations and varying levels of digital access. This distribution illustrates PCP's broad reach across Pakistan, highlighting regional engagement and complaint-handling effectiveness disparities.

Figure 4 presents the resolution rate by distinct and province. Khyber Pakhtunkhwa (KP) consistently exhibits high-resolution rates across its districts, positioning it as one of the stronger provinces in complaint resolution. Districts like Charsadda and Dir Lower showcase close to 100% exemplary resolution rates, highlighting KP's effective administrative processes and possibly more localised, manageable complaint volumes. Swat (62%) and Lakki Marwat (60%) also reflect strong resolution rates, reinforcing KP's efficient handling across rural and more populated districts. However, a few districts, such as South Waziristan Lower (40%), present moderate resolution rates, suggesting that while KP is overall effective, certain areas may benefit from additional support to align with the provincial average.

In Sindh, the pattern reveals significant variability, with urban centers like Karachi Central and Karachi West displaying some of the lowest resolution rates in the country, at 32.1% and 30.2%, respectively. This low performance contrasts sharply with the generally high rates in rural districts. The lower rates in Karachi districts highlight the challenges associated with high complaint volumes and complex urban issues, potentially requiring more targeted resources and administrative interventions. Outside Karachi, Hyderabad (34%) also reports moderate resolution levels, reflecting administrative or infrastructural challenges that impact complaint resolution in these urban centers. Sindh's rural areas generally show better performance, indicating that the complexities of urban districts primarily drive the variability in complaint resolution within the province. Balochistan shows a range of resolution rates but generally leans towards high performance in many districts. Awaran stands out with an exceptionally high-resolution rate of 86%, likely due to lower complaint volumes and the resulting ability of local administrations to handle complaints effectively. Other districts, such as Quetta (37%), maintain moderate rates consistent with the provincial average but reflect challenges in addressing complaints in

**Table 2** Member registration and overall resolution rate, 2018–2023

| **Total registered members/citizens** | 4,167,932 | **In percentage** |
|---|---|---|
| Inland | 3,854,173 | 92.47 |
| Overseas | 297,246 | 7.13 |
| Foreigners | 16,513 | 0.40 |
| **Province wise registered members/citizens** | | |
| Punjab | 2,372,201 | 56.92 |
| Khyber Pakhtunkhwa | 692,043 | 16.60 |
| Sindh | 561,058 | 13.46 |
| Federal Govt. | 93,755 | 2.25 |
| Balochistan | 59,743 | 1.43 |
| Azad Jammu & Kashmir | 46,489 | 1.12 |
| Gilgit-Baltistan | 17,677 | 0.42 |
| **Overall resolution rate** | 5,332,973 | 98.67 |
| Federal Govt. | 2,731,516 | 99.39 |
| Punjab | 1,703,155 | 99.07 |
| Khyber Pakhtunkhwa | 541,617 | 99.01 |
| Sindh | 282,157 | 88.93 |
| Balochistan | 45,341 | 98.50 |
| Azad Jammu and Kashmir | 22,404 | 97.66 |
| Gilgit Baltistan | 6,783 | 94.93 |

*Source* Pakistan Citizen Portal [16]

the capital city. Districts like Panjgur (20%) report some of the lowest resolution rates in Balochistan, likely due to a combination of limited administrative resources and geographical remoteness that complicates efficient complaint handling. While Balochistan faces logistical challenges, it generally maintains strong resolution rates in several districts, indicating a concerted effort to handle complaints effectively. In Azad Jammu and Kashmir (AJK), most districts display moderate to high-resolution rates. Kotli and Mirpur exhibit solid rates around the provincial average, while Muzaffarabad (32%) has a slightly lower rate, reflecting some inefficiencies in complaint handling in the AJK's capital. A notable outlier is Neelum, with a relatively higher rate of 58%, suggesting that local administrations can better resolve complaints effectively in smaller or less populated districts. However, most districts in AJK remain within a moderate satisfaction range, indicating ongoing efforts to address complaints efficiently despite occasional limitations in resources or infrastructure. Gilgit-Baltistan shows moderate to high-resolution rates across its districts. Gilgit and Skardu maintain rates around 44–47%, reflecting a moderately effective approach to complaint resolution. While these rates are lower than those in some other provinces, they are consistent across Gilgit-Baltistan, suggesting a stable but moderate level of effectiveness in addressing citizen complaints. The province's unique geographical

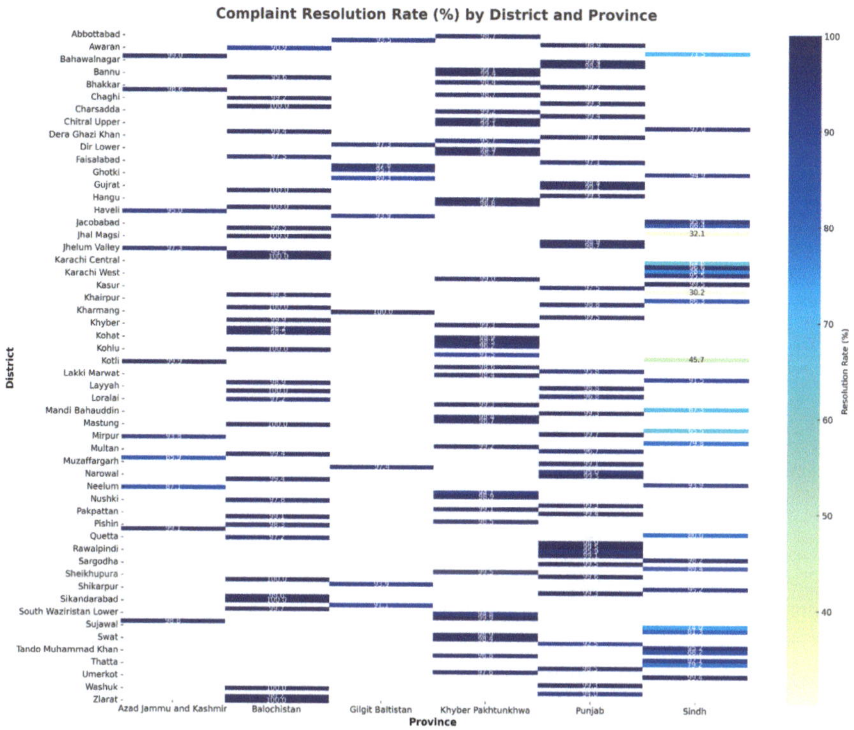


**Fig. 4** Complainants resolution rate by district and province. Authors' own

challenges and dispersed population may contribute to the moderate rates observed, indicating that further administrative support could enhance resolution outcomes.

These trends indicate that the PCP's complaint resolution rates reveal high effectiveness in provinces such as KP and Punjab, where most districts achieve rates close to or at 100%. Balochistan also demonstrates strong resolution rates in several districts despite challenges related to remoteness. In contrast, Sindh, particularly in Karachi, exhibits significant variability, with urban centers reporting some of the lowest rates in the country, highlighting the need for targeted improvements. AJK and Gilgit-Baltistan display moderate resolution rates, suggesting steady performance and opportunities for enhanced support. This province-wise analysis underscores the importance of a region-specific approach to complaint resolution, focusing on areas with low-resolution rates to achieve consistent improvements across Pakistan.

Figure 5 explains the citizens' satisfaction levels on the PCP that vary significantly across different complaint categories, highlighting areas of effective resolution and those needing improvement. Analysing satisfaction by profession, ministry, and type of complaint reveals distinct trends, showcasing the PCP's strengths in handling specific issues and identifying challenges in more complex or specialised areas.

Concerning profession, students and teachers show relatively high satisfaction levels, with 50.38% and 48%, respectively. This high engagement indicates that

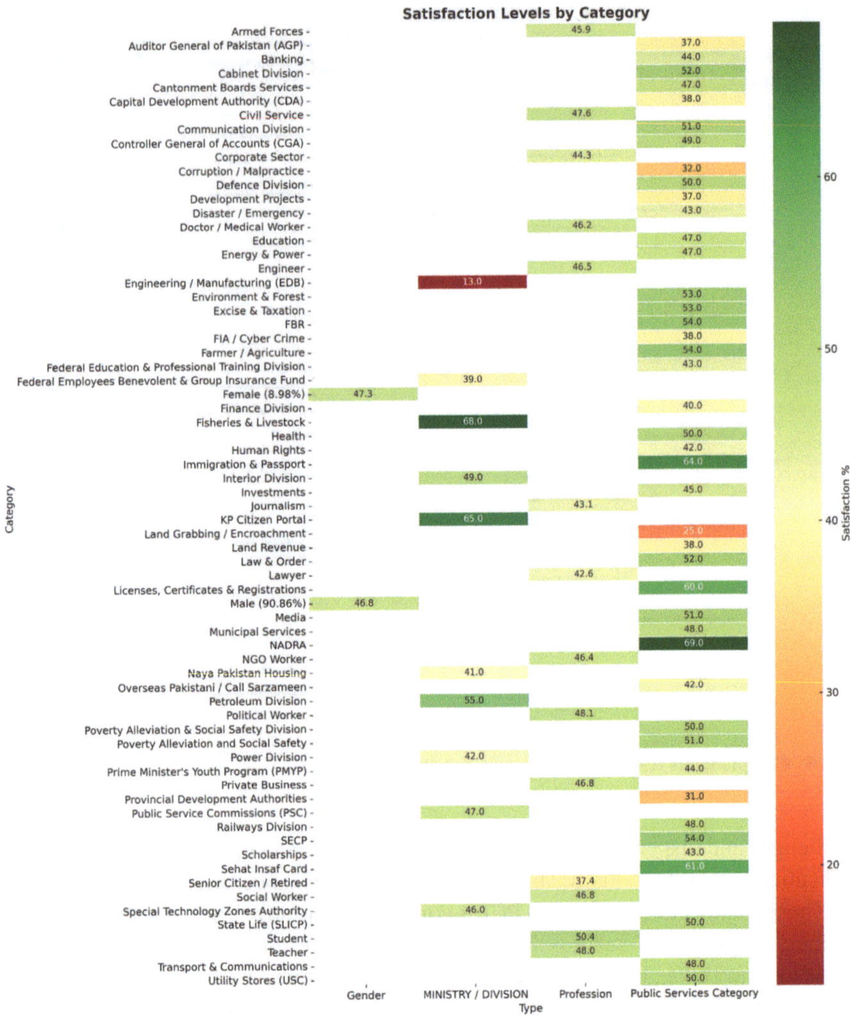


**Fig. 5** Citizens satisfaction levels by complaint category, 2019–2023. Authors' own

educational concerns are consistently addressed, reflecting the platform's efficiency in managing straightforward administrative issues within the educational sector. Political workers also display a comparable satisfaction level at 48.1%, suggesting that the PCP effectively resolves politically oriented complaints. However, certain professions, such as senior citizens and retirees, report lower satisfaction, with only 37.36% satisfaction. This demographic may require specialised support for welfare, healthcare, and retirement-related issues, which can be challenging to address through standard complaint-handling processes. Similarly, corporate sector professionals and lawyers report lower satisfaction levels of 44.31% and 42.61%, respectively, likely due to the complexity of legal and regulatory matters. When analysing satisfaction by ministry or division, certain areas emerge as highly effective. For example, the Interior Division and Cabinet Division display strong satisfaction rates of 49% and 52%, respectively, indicating high performance in handling complaints related to government administration. Conversely, divisions such as Engineering/ Manufacturing (EDB) and Provincial Development Authorities report some of the lowest satisfaction levels, at 13% and 31%, respectively. The particularly low satisfaction in the Engineering/Manufacturing sector highlights the need for specialised expertise and faster resolution processes to improve user outcomes in technical areas. Similarly, Provincial Development Authorities, which manage infrastructure and development projects at the regional level, may need more support and resource constraints, contributing to low satisfaction. Regarding public services categories, areas like NADRA (National Database and Registration Authority) and Immigration & Passport services show high satisfaction levels of 69% and 64%, respectively. These high levels reflect efficient management of identification, registration, and passport-related issues, likely facilitated by streamlined processes in these departments. In contrast, land-related issues, such as Land Grabbing/Encroachment, exhibit one of the lowest satisfaction levels at 25%. This area presents unique challenges, as land disputes often involve complex legal proceedings and enforcement difficulties, leading to significant dissatisfaction among citizens. Corruption/Malpractice complaints also show low satisfaction, at 32%, likely due to the sensitivity and complexity of handling corruption cases, which often require extended investigations.

The PCP's yearly trends in satisfaction levels indicate that areas with high complaint volumes generally maintain higher resolution efficiency. However, certain areas consistently report moderate satisfaction levels, such as students (50.38%) and health (50%), indicating that while complaints are resolved effectively, there remains room for improvement to meet user expectations fully. High-volume areas like Energy & Power, with a 47% satisfaction rate, demonstrate the PCP's ability to handle complex complaints reasonably. However, further refinement in these processes could improve user experiences. Overall, areas with high satisfaction, such as NADRA, Immigration & Passport, and the Sehat Insaf Card program, indicate the PCP's strength in handling specialised services efficiently. Conversely, areas like Engineering/Manufacturing, Provincial Development Authorities, and land-related issues require targeted improvements to address inherent complexities and enhance citizen satisfaction. The PCP's overall performance reflects its adaptability and effectiveness in managing various complaints. Still, specific areas, particularly those with lower satisfaction, could benefit from enhanced resource allocation, streamlined processes, and specialised support to better serve citizens' needs.

Figure 6 visualised the satisfaction levels across Pakistan's districts and provinces, revealing varied effectiveness in complaint resolution on the Pakistan Citizen's Portal (PCP). The variations show the differences in complaint-handling efficiency across regions, with distinct trends emerging in urban centres, rural areas, and remote districts.

In Punjab, there is a noticeable moderate to low satisfaction trend across many districts. Lahore, with the highest complaint volume of 36,888, shows a moderate satisfaction rate of 46%, indicating that while complaints are being addressed,

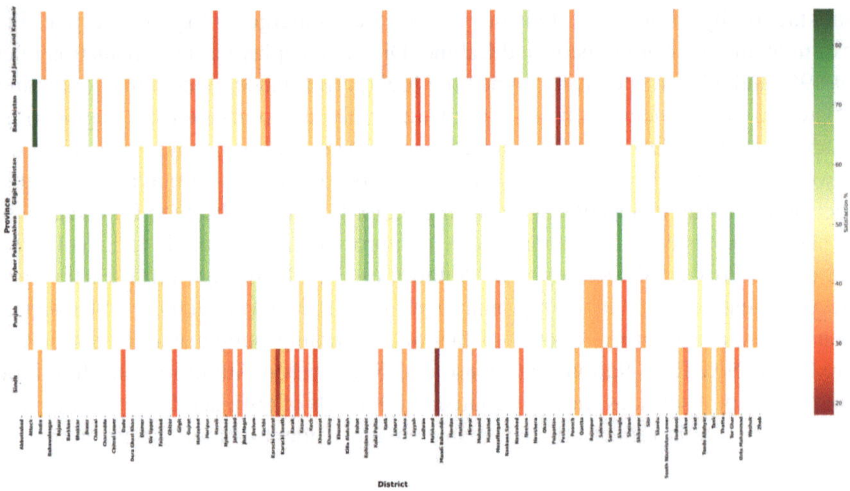


**Fig. 6** Complainant satisfaction level (%) by province and district. Authors' own

the level of effectiveness falls short of citizen expectations. Similarly, other high-complaint districts like Rawalpindi (37%), Kasur (45%), and Faisalabad (45%) reflect consistent yet moderate resolution outcomes. Some areas, however, show notably lower satisfaction, such as Sheikhupura (29%) and Layyah (31%), where challenges in addressing complaints seem more pronounced. Khyber Pakhtunkhwa (KP) demonstrates relatively higher satisfaction levels than other regions, reflecting a stronger complaint resolution performance. Districts like Malakand (68%) and Shangla (76%) stand out with high satisfaction rates, highlighting effective complaint handling. Major cities in KP, such as Peshawar (60%) and Mardan (57%), also display above-average satisfaction, indicating that KP's administrative processes might be well-equipped to handle citizen complaints effectively. However, certain areas, like South Waziristan Lower (40%), still report moderate satisfaction, suggesting that localised improvements could further boost satisfaction levels. In Sindh, satisfaction levels are generally lower, especially in urban centres like Karachi. Malir and Karachi East have some of the lowest satisfaction rates in the country, at 18% and 22%, respectively, indicating difficulties in managing the complex needs of densely populated urban areas. Other districts, including Hyderabad (34%) and Sukkur (32%), also show moderate satisfaction, reflecting potential administrative or infrastructure limitations that impact the effectiveness of complaint resolution in Sindh's urban areas. Balochistan shows a mix of satisfaction levels, with an overall trend toward moderate to low satisfaction. The provincial capital, Quetta (37%), reflects this moderate trend, which is in line with other areas of Balochistan. Some remote districts, such as Panjgur (20%), report low satisfaction, possibly due to limited resources or logistical challenges. In contrast, Awaran stands out with a high satisfaction rate of 86%, likely due to lower complaint volumes, which may allow for more focused and efficient complaint handling by local authorities. Azad Jammu and Kashmir (AJK) displays mostly moderate satisfaction levels, with major districts like Mirpur (32%) and Muzaffarabad (32%) showing average satisfaction. The region's overall trend indicates challenges in meeting citizen expectations comprehensively. However, Neelum has a higher satisfaction rate of 58%, suggesting that certain districts in AJK may have more effective local administration. However, most districts remain in the moderate range, indicating a need for strengthened complaint resolution system. A similar pattern emerges in Gilgit-Baltistan, with moderate satisfaction levels across most districts. Gilgit shows a satisfaction rate of 44%, while Skardu (46%) and Diamer (47%) report slightly higher satisfaction. This consistency suggests that while complaint resolution is generally effective, there is room for further improvement to better meet citizen expectations. The pattern indicates that localised administrative structures may contribute positively to complaint resolution, although resource enhancements could improve satisfaction.

Overall, Fig. 6 reflects significant regional variation in satisfaction with complaint resolution on the PCP. Urban centres, particularly in Karachi and Lahore, face challenges in achieving high satisfaction levels, likely due to high complaint volumes and the complex nature of issues in these populous regions. In contrast, some rural and remote districts, particularly in KP and certain areas of Balochistan, report relatively higher satisfaction. This suggests that localised governance structures and fewer complaints may contribute to more effective complaint handling in less densely populated areas. Analyzing citizens' satisfaction at district levels underscores the importance of tailored approaches to complaint resolution. Regions like KP, which demonstrate consistently high satisfaction, may serve as models for effective complaint handling. However, systemic challenges in Sindh's urban centres and specific areas of Balochistan indicate a need for targeted improvements. Remote or low-complaint districts sometimes exhibit high satisfaction due to localised, focused administration, though these approaches may not be scalable to higher-volume areas. AJK and Gilgit-Baltistan show mixed results, reflecting unique regional governance needs and indicating that improvements in administrative support could further enhance satisfaction. Together, these trends highlight the need for region-specific strategies to maximise the effectiveness of complaint resolution across Pakistan's diverse landscape.

## 4.2 *Wafaqi Mohtasib (WM)*

The WM is a government-owned smartphone application representing significant technological advancements in Pakistan's public grievance redressal infrastructure. The WM integrates many governmental departments, allowing citizens to submit complaints on various public service delivery grievances, such as healthcare and utility services. It also serves a specialised role in handling administrative grievances and safeguarding citizens' rights against governmental misconduct.

Figure 7 shows a strong upward trend in complaint volumes from 2019 to 2023, reflecting a steady increase in citizen engagement. Major urban centres, such as

Islamabad, Lahore, and Karachi, have consistently recorded the highest complaint volumes. For example, complaints in Islamabad rose from 15,764 in 2019 to 23,775 in 2023, resulting in a five-year total of 106,162 complaints. Similarly, Lahore and Karachi saw cumulative complaints of 116,550 and 96,483, respectively. This trend suggests that the WM has become an essential tool for addressing public grievances, especially in densely populated areas with higher public awareness and accessibility to the platform. Figure 7 also reveals that major cities with high complaint volumes have maintained excellent resolution efficiency (Fig. 9). In Islamabad, the total disposal of complaints slightly exceeded the receipt total, with 106,371 complaints resolved out of 106,162 received, demonstrating a resolution efficiency of nearly 100%. Similarly, Lahore disposed of 113,913 out of 116,550 complaints (about 98% efficiency), and Karachi managed 95,927 out of 96,483 complaints received, maintaining close alignment between receipt and disposal rates. These figures indicate that these cities are well-equipped to handle high complaint volumes, supported by robust complaint resolution systems and administrative capacity. Other high-volume cities like Multan and Peshawar also demonstrated strong resolution efficiency, with 98% and 95% disposal rates, respectively, showing that their complaint management frameworks are similarly effective. Moderate-volume cities like Hyderabad and Quetta also show high-resolution efficiency despite handling fewer complaints. Hyderabad received 46,608 complaints and disposed of 45,823 (98% efficiency), while Quetta handled 36,210 complaints, with 34,377 resolved (approximately 95% efficiency). This indicates that mid-sized cities maintain high standards in complaint handling, closely aligning disposals with receipts, which further emphasises the effectiveness of the PCP across different administrative levels. On the other hand, smaller and remote areas show lower resolution efficiency, likely due to limited resources or logistical challenges. For instance, Khuzdar received 3,837 complaints but disposed of 3,345, reflecting an efficiency of about 87%. Kharan saw 6,386 complaints, with 5,988 resolved (94% efficiency), while Wana (Collection) handled 2,256 complaints and resolved 2,068 (92% efficiency). These gaps between receipt and disposal rates suggest that smaller localities may face challenges in promptly addressing complaints. Such regions may benefit from targeted support, such as resource allocation or administrative assistance, to improve resolution rates.

Examining yearly trends provides further insight into the WM's uptake. In 2019, complaint volumes were relatively low, allowing for nearly perfect resolution efficiency across most regions. However, 2020 saw a significant surge in complaints, with major cities like Lahore, Karachi, and Islamabad experiencing the largest increases. Despite this, resolution efficiency remained high, indicating that the WM successfully scaled its resources to meet the increased demand. This efficiency continued into 2021 and 2022, where complaint volumes remained high and disposal rates closely aligned with receipts. Minor gaps in some localities during peak periods in 2022 suggest temporary backlogs but did not significantly impact overall resolution efficiency. By 2023, complaint volumes had stabilised at high levels, and the WM maintained strong resolution efficiency, demonstrating well-established processes to handle sustained public demand.

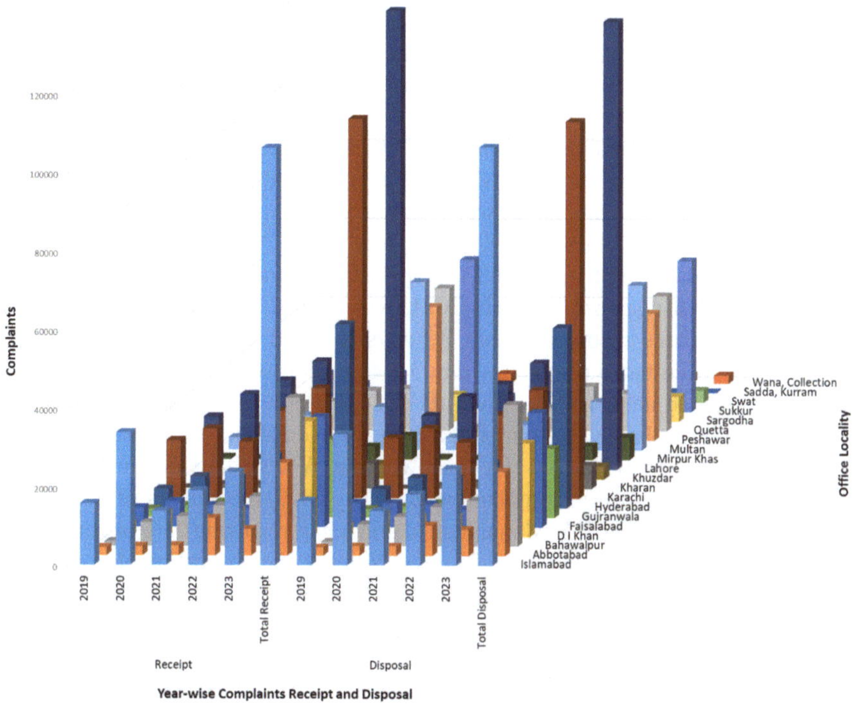


**Fig. 7** Annual complaints disposal by location. Authors' own

Conclusively, the WM has maintained high-resolution efficiency over five years, with major cities excelling in both complaint volume and resolution rates. Moderate cities also performed well, aligning disposal rates closely with receipt rates and demonstrating effective complaint management at various levels. Smaller or remote regions with lower resolution efficiency, such as Khuzdar and Kharan, may benefit from additional resources to enhance complaint handling. Increasing awareness and engagement in low-complaint regions like Wana and Sargodha could also ensure broader access to the platform's services. Overall, the WM's resolution trends indicate a robust system capable of addressing rising complaint volumes effectively, especially in urban centres. By sustaining high public satisfaction through efficient complaint handling, the PCP has established itself as an essential platform for promoting accountability and responsiveness in Pakistan's public service system.

Figure 8 illustrates the time profile of the disposal of complaints and provides an in-depth view of the efficiency with which the WM handled complaints from 2019 to 2023. During this period, the volume of complaints rose significantly, yet WM's resolution timelines remained largely consistent and, in some years, even improved.

The data categorises complaint resolutions by time intervals, from within 30 days to over a year, offering insights into how quickly citizens' grievances were addressed. In 2019, WM processed 74,892 complaints, with a strong initial response reflected by 36% of cases resolved within 30 days (26,933 complaints). As the resolution period extended, the percentage of cases resolved decreased: 29% were resolved within 31–45 days, 22% within 46–60 days, and only 12% took 61–90 days. Less than 1% extended beyond 120 days.

The volume of complaints rose sharply in 2020, reaching 130,164. WM improved its efficiency despite this increase, with 45% of complaints resolved within 30 days. Additionally, 20% of complaints were resolved within 31–45 days, 18% within 46–60 days, and 15% within 61–90 days. Only about 2% of complaints required more than 120 days, highlighting WM's capability to handle larger caseloads while maintaining swift response times. In 2021, with 106,824 complaints handled, the percentage of cases resolved within 30 days dropped slightly to 32% (33,961 complaints), while those resolved within 31–45 days rose to 25%. The 46–60-day category remained stable at 22%, while 19% of complaints were resolved within 61–90 days. Although there was a minor decrease in rapid resolutions, most complaints were still addressed within 60 days, showcasing WM's focus on timely resolutions despite fluctuations in complaint volume. By 2022, the number of complaints increased again to 157,803. WM maintained efficiency, resolving 35% of complaints within 30 days and 27% within 31–45 days. Notably, the proportion of complaints resolved within 46–60 days rose to 24%, while those resolved in 61–90 days dropped to 13%, reflecting improved handling of mid-range resolution times. This year underscored WM's commitment to prompt responses even as complaint volumes grew. In 2023, WM faced the highest volume of complaints yet, with 193,033 cases. Nonetheless, the WM sustained its efficiency, resolving 36% of complaints within 30 days and 27% within 31–45 days. A further 25% were resolved within 46–60 days, while

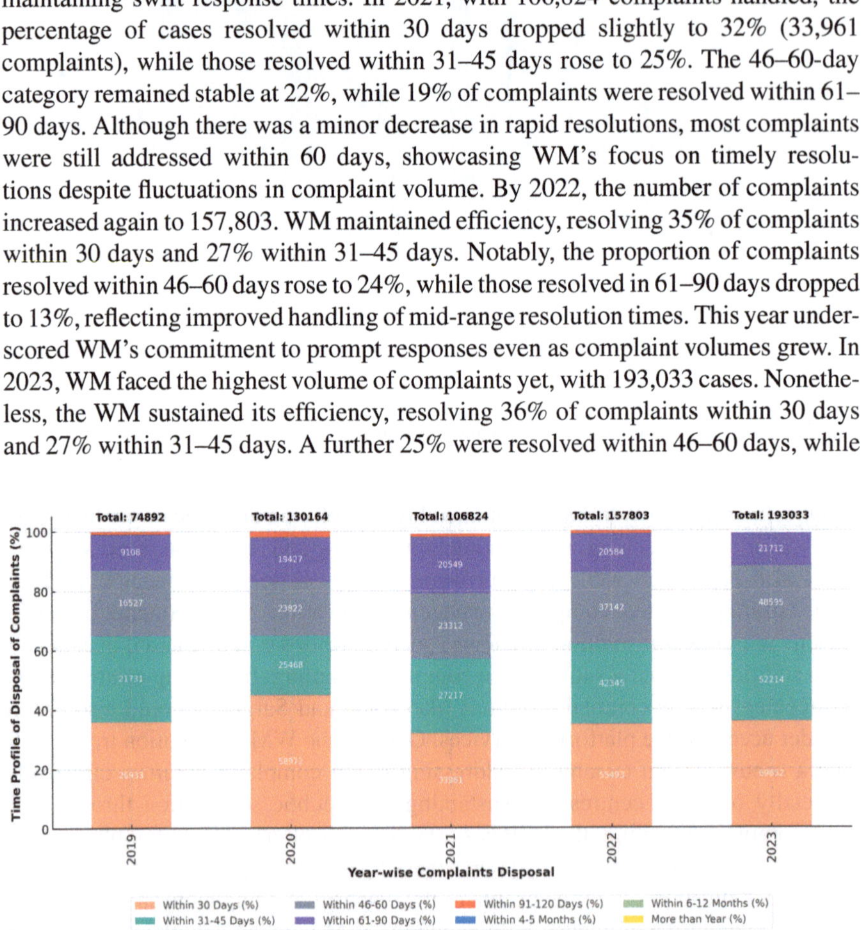


**Fig. 8** Time profile of complaints. Authors' own

only 11% extended to 61–90 days. This year highlighted WM's adaptability, effectively managing a high caseload without compromising response times. Throughout these years, WM demonstrated a consistent ability to resolve complaints quickly. Each year, a significant portion of complaints—ranging from 32 to 45%—were addressed within 30 days, showing a robust commitment to early resolutions. Moreover, mid-range timelines (31–45 and 46–60 days) remained a vital component of WM's response, with 20–29% of complaints resolved annually in each of these timeframes. Extended resolutions beyond 90 days remained rare, constituting only 1–2% each year, reinforcing the efficiency of WM's processes and their emphasis on minimising delays.

As complaint volumes rose from 74,892 in 2019 to 193,033 in 2023, WM's adaptability was evident. The consistently high percentages of resolutions within 30–60 days suggest that WM likely implemented process improvements or allocated additional resources to handle the rising caseload effectively. Even with the peak complaint volume in 2023, WM managed to resolve 36% of cases within 30 days and 63% within 45 days, showcasing improved early resolution rates. Conclusively, WM's complaint resolution trends from 2019 to 2023 illustrate a strong commitment to timely complaint handling, effectively adapting to increasing volumes while maintaining high efficiency. Most complaints were resolved within 60 days each year, and long-term unresolved cases were minimal, underscoring WM's capacity to deliver responsive and reliable service. This efficiency contributes to higher citizen satisfaction and positions WM as a reliable institution for managing public grievances with consistency and adaptability.

Figure 9 presents the analysis of complaints registered with the WM across various government ministries and departments, providing a detailed insight into areas with high citizen grievances, particularly within essential services. The Ministry of Energy & Power Division leads in complaint volumes, with Lahore Electric Supply Company (LESCO) recording the highest number at 3,208 complaints, followed closely by Multan Electric Power Company (MEPCO) with 2,132 complaints. Other power companies like Peshawar Electric Supply Company (PESCO) and K-Electric (KESC) have 803 and 736 complaints, respectively. This trend suggests that public interactions with electric power providers are marked by challenges, possibly around service quality, billing, and reliability issues. Within the Ministry of Petroleum and Natural Resources, Sui Southern Gas Company Limited (SSGCL) is a focal point for grievances, with 1,631 complaints. In comparison, Sui Northern Gas Pipelines Ltd. (SNGPL) has 725 complaints. These figures highlight public dissatisfaction with gas supply services, emphasising the need for more efficient complaint resolution in this critical area.

The Poverty Alleviation and Social Safety Division also sees a substantial number of complaints, particularly related to the Benazir Income Support Programme (BISP), which accounts for 1,649 complaints. This figure underscores the dependency of citizens on social safety nets and the challenges faced in service delivery. Other welfare programs under the Ehsaas Program have comparatively low complaints, at only 8. The Ministry of Interior also experiences a high volume of complaints, predominantly directed at the National Database & Registration Authority (NADRA), which has 639

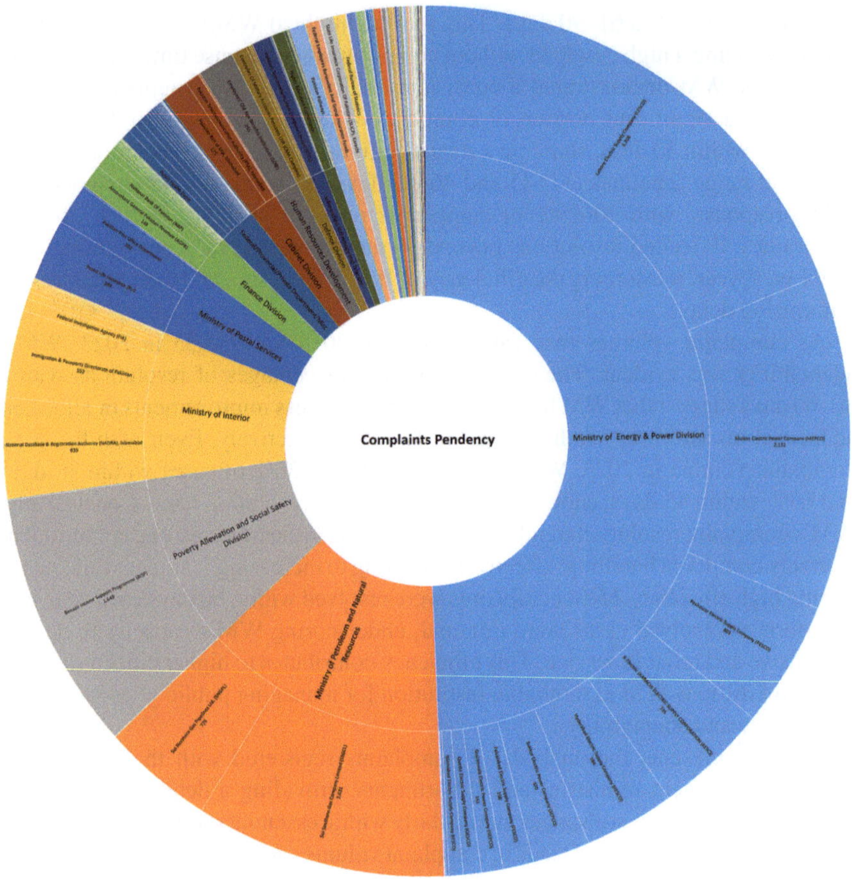


**Fig. 9** Complaints pendency at organization/program level by ministry/division. *Source* Authors' own

complaints. The Directorate of Immigration and Passports records 552 complaints, reflecting difficulties citizens face in accessing essential identification and travel documents. Other areas within the ministry, such as the Federal Investigation Agency (FIA) and Capital Development Authority (CDA), have 80 and 46 complaints, respectively, indicating specific operational challenges in these departments. The Pakistan Post Office Department and Postal Life Insurance received 262 and 389 complaints from the Ministry of Postal Services. These figures suggest that the postal system, while fundamental to many citizens, faces ongoing service and reliability issues. For the Finance Division, the Accountant General Pakistan Revenue (AGPR) leads with 148 complaints, followed by the National Bank of Pakistan (NBP) with 101 complaints. The Central Directorate of National Savings (CDNS) and State Bank of Pakistan (SBP) also have significant complaints, at 42 and 23, respectively. This concentration of complaints reflects public grievances around financial administration, banking services, and savings. The Cabinet Division also appears on the chart with notable complaints against Pakistan Bait-ul-Mal (175 complaints) and Pakistan Telecommunication Authority (PTA) (95 complaints), indicating issues in social welfare distribution and telecom regulation. Human Resources Development sees most complaints directed at the Employees' Old Age Benefits Institution (EOBI), which has 236 complaints, highlighting challenges in pension distribution and benefits administration.

Overall, the data suggest that the primary areas of public grievance relate to essential services like electricity and gas, social welfare, financial services, and identification documentation. The high complaint volumes in these ministries and departments indicate a need for targeted improvements in public service delivery, particularly in complaint resolution processes. This comprehensive breakdown emphasises the critical areas where government responsiveness could significantly enhance citizen satisfaction and service efficiency.

## 5 Conclusions

This chapter underlines the transformative potential of technological innovation in public service delivery through the performance and uptake analysis of Pakistan's Online Public Grievance Redressal Systems (OPGRS) within the broader landscape of digital governance. It examines key OPGRS, such as the PCP and WM, and reveals significant advancements in transparency, accountability, and efficiency in public service delivery. OPGRS has empowered citizens by fostering a more participatory governance approach, allowing the public to hold government bodies accountable and actively influencing the direction of digital governance in Pakistan. Despite resource limitations, OPGRS has also demonstrated its effectiveness in improving administrative responsiveness and public trust, aligning with the UN's Sustainable Development Goal 16 (SDG-16), which promotes peace, justice, and robust institutions.

The study's findings suggest essential recommendations to sustain and expand OPGRS, including ongoing investments in digital infrastructure, robust data security, and capacity-building to ensure high service standards. Integrating citizens' feedback mechanisms is also crucial, enabling these platforms to evolve with citizens' needs and expectations. This adaptability is foundational for a successful digital governance framework, ensuring scalability, cross-departmental integration, and user feedback responsiveness.

Through Pakistan's experience, the chapter offers strategic insights applicable to other EMDEs aiming to implement user-centric, scalable digital solutions to improve public service delivery. By emphasising the importance of a strong technological infrastructure and seamless data sharing across government departments, the study provides a roadmap for EMDEs to enhance service accessibility and inclusivity. Community-based training programs, partnerships with local organisations, particularly in rural areas, and transparent communication on citizen rights and data privacy are highlighted as strategies to reinforce public confidence in such digital governance platforms.

The study's broader implications imply that even resource-constrained developing countries can achieve meaningful improvements in public service delivery by adopting similar initiatives. OPGRS exemplifies how digital governance systems, designed with scalability and inclusivity, can foster more transparent, accountable, and efficient governance. The study encourages policymakers, development agencies, and technology providers to explore digital grievance systems as a viable solution to modernise public administration and support data-driven decision-making. By inspiring digital governance initiatives globally, this study aims to contribute to a more inclusive, responsive, and effective approach to public service delivery that aligns with the 2030 global development agenda.

## References

1. Anthopoulos, L.: Smart governance for smart cities: understanding the role of technology in public service delivery. Cities **91**, 1–10 (2019). https://doi.org/10.1016/j.cities.2018.10.024
2. Asian Development Bank (ADB).: Governance and institutional reforms in South Asia: enhancing public service delivery (2021). https://www.adb.org
3. Bolivar, M.P.R.: Innovation in public sector organizations: the role of digital technologies. Public Adm. Rev. **76**(3), 391–400 (2016). https://doi.org/10.1111/puar.12512
4. Bower, J.L., Christensen, C.M.: Disruptive technologies: catching the wave. Harv. Bus. Rev. **73**(1), 43–53 (1995). https://hbr.org
5. Christensen, T., Lægreid, P.: The importance of trust in public administration reform. Public Manag. Rev. **21**(4), 518–537 (2019). https://doi.org/10.1080/14719037.2018.1544272
6. Christensen, T., Lægreid, P.: The Routledge Handbook to Accountability and Welfare State Reforms in Europe. Routledge (2019). https://doi.org/10.4324/9781315164813
7. Digital Pakistan Initiative.: Transforming public service delivery through digital platforms. Ministry of IT and Telecommunication, Islamabad (2022)
8. Government of Pakistan.: Performance evaluation report: Pakistan Citizen Portal. Prime Minister Delivery Unit, Prime Minister Office, Islamabad (2022)
9. Government of Pakistan: Online public grievance redressal systems (OPGRS) progress report. Government of Pakistan Press, Islamabad (2023)
10. Heeks, R.: Implementing and Managing eGovernment: an International Text. SAGE Publications (2006)
11. Mettler, T., Rohner, P.: Digital transformation in public administration: enablers and barriers. Public Adm. Rev. **80**(5), 857–864 (2020)
12. Misuraca, G., Viscusi, G.: AI-enabled public administration: reconciling efficiency and inclusion. Policy Internet **13**(2), 191–209 (2021). https://doi.org/10.1002/poi3.241
13. Moher, D., Liberati, A., Tetzlaff, J., Altman, D.G., The PRISMA Group.: Preferred reporting items for systematic reviews and meta-analyses: the PRISMA statement. PLoS Med. **6**(7), e1000097 (2009). https://doi.org/10.1371/journal.pmed.1000097
14. OECD.: Government at a Glance 2020. Organisation for Economic Co-operation and Development (2020). https://doi.org/10.1787/4dd50c09-en
15. OECD.: Digital transformation in emerging markets: trends and policy insights. OECD Publishing, Paris (2021). https://www.oecd.org
16. Prime Minister Delivery Unit.: Pakistan Citizen Portal statistics. Government of Pakistan, Islamabad (2023)

17. Shahid, M., Ahmed, R.: The impact of digital public administration on governance in Pakistan. J. Gov. Innov. **15**(3), 45–62 (2021)
18. Transparency International.: Corruption perceptions index 2022: Pakistan. Berlin, Germany: Transparency International (2022). https://www.transparency.org/en/cpi/2022
19. United Nations Department of Economic and Social Affairs (UNDESA).: E-Government Survey 2020: digital government in the decade of action for sustainable development (2020). https://publicadministration.un.org
20. United Nations Department of Economic and Social Affairs.: Member States Questionnaire (MSQ) for the United Nations E-Government Survey 2022. United Nations (2022). https://publicadministration.un.org/en/
21. United Nations Development Programme (UNDP).: Fostering effective governance and institutional integrity. United Nations Development Programme, New York, NY (2021). https://www.undp.org/publications
22. United Nations Development Programme (UNDP).: The impact of corruption on service delivery in Pakistan. United Nations (2021)
23. United Nations Development Programme (UNDP).: Digital governance and citizen engagement in developing countries. UNDP Publications, New York (2022). https://www.undp.org
24. United Nations.: Transforming our world: the 2030 Agenda for Sustainable Development. United Nations General Assembly (2015). https://sdgs.un.org/2030agenda
25. World Bank: World Development Report 2019: the changing nature of work. World Bank, Washington, DC (2019)
26. World Bank: The role of digital governance in public service delivery. World Bank, Washington, DC (2020)
27. World Bank.: World Development Report 2020: trading for Development in the Age of Global Value Chains. The World Bank (2020). https://doi.org/10.1596/978-1-4648-1457-0
28. World Economic Forum.: The Global Risks Report 2020. World Economic Forum (2020). https://www.weforum.org/reports/the-global-risks-report-2020
29. World Economic Forum.: The global risks report 2021. World Economic Forum, Geneva, Switzerland (2021). https://www.weforum.org/reports/global-risks-report-2021

# Industry 4.0 and Sustainable Development Goals: Current Challenges and Solutions

**Saqib Rehman, Nabeela Arshad, and Aman Ullah**

**Abstract** This study aims to explore the potential approaches, factors, and tactics that can be applied to Industry 4.0. Further, it focuses on grouping those approaches that fully allow industries to achieve sustainable development. The United Nations' Sustainable Development Goals (SDGs) is a common, shared, and open agenda that has been applied in industries. With the spice of Industry 4.0, it is easier for industrialists to implement the SDGs in true spirit and achieve the standards of sustainable development. Under the umbrella of scholarly findings, key insights are identified through influential aspects and conceptual analysis. At initial, the study grouped 1972 research findings, after filtration of other disciplines we finalized 1880 research attempts for further proceedings and analysis. We identified the top countries, authors, citation networks, and journals; thereby, we may able to compile the current practices and 4.0 strategies of those regions where they are already being applied. For conceptual analysis, we made an effort to construct the thematic map co-occurrence and keyword analysis. From conceptual analysis, this chapter identified the themes and subthemes, as well as potential, cost-efficient, and eco-friendly activities adopted by various industrial regions and zones. Its analytical findings revealed that the applications of Industry 4.0 have a substantial prospect for current survival, future growth, and competitiveness of businesses. To attain continuous growth, SDGs were built as a universal mechanism that is equally important for the product, process, and operational activities of all industries. Nowadays, Industry 4.0 has revolutionized regular working standards and has technologically controlled working tracks. To best deal with current and future challenges, this chapter is a robust guiding principle for government, current entrepreneurs, business firms, industrialists, and policymakers.

**Keywords** Industry 4.0 · Sustainable development goals · Cost efficient · Internet of things · Challenges of sustainable development · Competitiveness

S. Rehman (✉) · N. Arshad
Department of Management Sciences, Lahore College for Women University, Lahore, Pakistan
e-mail: dr.saqibrehman81@gmail.com; saqib.rehman@lcwu.edu.pk

A. Ullah
Department of Management and Marketing, Faculty of Business and Economics, University of Melbourne, Melbourne, Australia

 
N. D. K. Nguyen et al. (eds.), *Digital Transformation and Sustainability: Methods and Applications*, Information Systems Engineering and Management 47,
https://doi.org/10.1007/978-3-031-91278-8_18

## 1 Introduction

In the early years of research and the Industrial Revolution, sustainability within the organization was considered a relatively less important, generic, and informal phenomenon. Thereby, its external impacts and cost to the environment were almost neglected by the industrialists and even by the environmental regulators [18]. During the entire span of the Industrial Revolution, innovation, automation, and advancement in manufacturing, trading, and services were introduced at private, social, external, and environmental costs [7, 18]. Moreover, industrialists (both large and small-sized functionalists) have little realization or have paid less attention to the future severity of environmental disturbances. Amalgamations and joint negligence by both industrialists and environmental regulators abruptly endured a severe impact on our natural environment. Therefore, following the current era of digital and industrial advancement, past disturbances to the environment become one of the leading challenges for both developed and developing countries.

Under the reign of each industrial revolution, industrialists were opportune through various avenues, methods, transformations, and tactics. Following the various periods and paradigms, such applications have fulfilled their presently demanded needs. Additionally, these opportunities gave a remarkable incline and offered continuous growth while surviving for the future. However, each revolution touched its maturity and was replaced by an upcoming one.

Every industrial revolution opened new avenues, methods, transformations, and tactics under its reign. In the past few decades, the industry at each level has faced intense dynamics, large competitiveness, and transformations, fully allowing the industry to offer massive customizations, transitions, and variants for products and services [16, 29]. Studies like [30] have centered their focus on the current industrial transformations and reported how advancement in operations or working standards are becoming emerging opportunities to take steps forward toward sustainable development. Further, this study addresses the phenomenon of circular economy and fully provides a key mechanism for the optimum usage of resources and efficient means of recycling. By following the exploratory documentary methodology, the study confirmed that the current development of industry holds a shell of limitations and is not sufficient enough to achieve the standards of sustainable development.

Considering this current severity and the possible appearance of upcoming impacts in the near future, the United Nations stood to preserve the environment. They took the responsibility to transform the earth into a peaceful place where future generations can spend their lives in a pure, safe, and natural generic environment. Therefore, an open, common, and shared agenda of 17 Sustainable Development Goals (SDGs) was settled and equally shared with both developed and developing countries. These goals fully acknowledged and listed the environmental aspects, including water usage and consumption, equal and parallel distribution of resources, exemptions of inequalities based on gender, right to get the quality of education, optimum and safe usage of energy, economic growth with sustainable industrial infrastructure, sustainable consumption and production, and abrupt severity of climate borne by land and water life [42]. With realizing the importance of current environmental seriousness, the current study is going to make a little effort to achieve the UN-formulated sustainable development goals and address the industrial zones. This study fully reported how industries are distressed or still distressing the environment and how this current distractedness can be controlled or at least minimized.

Presently, Industry 4.0 and machine learning have made an emerging contribution; simultaneously, they offer a variety of industrial customization, competitiveness, improved precision, and environmental sustainability. The study by Costa et al. [12] worked to enhance green transitions by integrating Industry 4.0 and machine learning (ML). Similarly, Muzammal et al. [40] have illustrated that Industry 4.0 and sustainability are paving the path for a green future by offering acknowledgeable transformative practices in the manufacturing industry. For instance, Industry 4.0 technologies include the Internet of Things, big data analytics, robotics, automation, and other artificial intelligence. Nowadays, the goal of current operating organizations is to achieve the standards of sustainable development and fulfill their social and environmental responsibilities. However, a number of organizations have emerged that have already integrated sustainable practices into their day-to-day operations and declined their ecological stress [56]. According to the study of Oguntola et al. [44], cement manufacturing industries are considered to be a significant consumer of energy, additionally, cement industries adversely create a large ecological and environmental footprint. Such industries that have a direct contribution to economic growth, GDP (Gross Domestic Product), decline of unemployment, and country-level development cannot be restricted or banned from working.

However, industries have taken progressive measures and tried their best to seek high-quality standards. However, with the consideration of large industries, i.e., cement or chemical sector, all we need to develop a precise mechanism and incorporate the greener strategies in their current working operations so these industries may able to decline their presently incurred operational costs and industrial environmental impacts [2]. Following this, there is a dire need to incorporate additive manufacturing approaches, which allow the cement sector to take an immediate initiative toward environmental sustainability.

Therefore, the current evidential attempt aims to explore and list the currently existing factors, potential approaches, and tactics that have applications in Industry 4.0. Further, it focuses on grouping those approaches that fully allow industries to achieve sustainable development. Since 2015, the United Nations has given the Sustainable Development Goals (SDGs), although it was a common, shared, and open agenda whose true implications are demanded by entire sectors of industry, health care units, and clean environmental practices [53]. At the industrial level, these goals are fully helpful in searching for technological, social, and economic hurdles in a net-zero economy.

In each industrial revolution, technology drives survival, growth, competitiveness, and positioning [9, 46]. This constant and frequent transformation of digitalization opportune the operational industries to sustainable development. Industry 4.0 has shortened the past gaps, breaches, and limitations of both physical and digitalization [30, 68].

**RQ1**: How do existing factors, practicing approaches and tactics of works significantly make their contribution to the proposed agenda of the UN?

**RQ2**: In tandem with the current contribution, what are the current challenges that still need to be addressed while achieving the standards of environmental sustainability?

**RQ3**: How do applications of Industry 4.0 pave the path to catching a true spirit of UN-formulated sustainable goals?

**RQ4**: How does Industry 4.0 help organizations in fulfilling the ethical mandate of environmental sustainability?

The remained of the paper is organized as follows. Section 1 introduced the study. Section 2 covers the literature and reports the key findings in the past research. Section 3 proposes the methodology and figures the keyword analysis and tables. Section 4 contains a discussion. Section 5 describes the current limitations and future research avenues. Lastly, Section 6 concludes the study.

## 2 Survey of the Literature

The current industrial transformation is the result of multiple efforts that were made during each industrial revolution [12, 65]. Under the happenings of industrial revolutions and paradigm shifts in production standards, industrial means were lacking in advancement and environmental sustainability. Thereby, multifold attempts addressing various contexts, ecological settings, and zones of the industry have been found. Onile et al. [45] worked on the optimal usage of energy plans and offered a precise set of energy-saving plans. Similarly, another key contribution was delivered by Lara-Guillén et al. [30], which focused on raising the importance of a circular economy and closed-loop supply chain. This attempt at research offered a comprehensive plan to achieve sustainable development goals. Moreover, on a current basis, various aspects and acknowledgeable means have been grouped by the currently existing literature, here, most of them are detailed in this manuscript.

### *2.1 Industry 4.0 and Sustainable Development Goals (SDGs)*

At present, sustainable development goals, i.e., SDG 7 (affordable and clean energy), SDG 8 (decent work and economic growth), and SDG 9 (industry innovation and infrastructure), are coupled with Industry 4.0 and play a key role in current industrial brought-ups. The linkage of Industry 4.0 to industrial transformation has paved the path to achieving the Sustainable Development Goals (SDGs). According to the efforts of Kılıç and Atilla [29], Industry 4.0 seems considered to be the transformative force that allowed the industries to contribute a segment of their efforts to sustainable development. Studies like Delbari and Hof [14] demonstrated that solid waste, i.e., glass waste, is one of the hurdles that ought to be catered for both business and environmental sustainability. This study provided a sustainable solution for the best handling of the wastage of high-quality glass. In the past, there was no solution for its wastage other than to dump it in landfills or recycle it under the open loop chain. However, with the advent of Industry 4.0 and Industry 5.0 and their applications like machine-to-machine learning, the Internet of Things, Artificial Intelligence, and collaborative robotics, this matter was resolved with meaningful outcomes. Advancements in the recycling process have given a cushion to industrialists and allowed them to use recycled high-quality glass in smartphone screens and other high-tech applications.

For instance, the study of Amani and Aghamohammadi [7] found that taking great care of the wastage of resources, especially by addressing the phenomenon of food wastage, the study paid severe attention to the possible food wastage in supply chains. This research has brought a solution for warehouse management, placing or managing perishable foods, e.g., fruits, and the separation of rotten or stale food items from fresh ones. Similarly, Kumar and Agrawal [28] addressed the most sensitive supply chain of perishable agricultural products, where the government and food supply teams set a few classifications, divisions, and quality standards. Therefore, small negligence in handling, timely storage, and care about the supply resulted in various environmental challenges, such as food items rotting due to their higher perishability. To best handle this problem, a sustainable, smooth, and flawless supply of regular food items has been supplied to the demanded zones and regions. Next, Pitakaso et al. [47] inserted their research into the management of disastrous activities and listed the key solutions for better control and management.

In the same vein, to bridge the gap between theory and current practicing technique, the study of Kılıç and Atilla [29] provided a set of comprehensive business models for current business landscapes, equally important for present transformations of industries and having precise solutions and significant applications with current business or environmental challenges. Moreover, this study reported that Industry 4.0 holds valuable insights for business and yielded organizational sustainability.

Unlike large-sized industries, applications of Industry 4.0 have the same level of significance to small and medium-sized enterprises [51, 52]. The study of Görçün et al. [20] has catered to the transformative measures of automotive manufacturing industries and reported how such industries can align their process and proceedings of decision-making to sustainable development goals. Their findings are fully available in applicable form to the current practitioners and displayed an open platform where their routine practices of industries came under the umbrella of sustainability and are part of digital transformations. Similarly, Rehman et al. [49] provided a key map for achieving SDG 8 of decent and economic scope. This study posits the implications of SDG 8 related to the workforce's well-being. The true implications of the application of industrial transformations are that they provide an accessible means of working and get rid of the industrialists from traditional and higher cost-incurred business models.

## 2.2 Industry 4.0 and Artificial Intelligence (AI)

Past social, economic, environmental, technological, and sustainability challenges profoundly reshaped the business functions that are seen today [10]. As a result, Industry 4.0 and artificial intelligence attained an equal level of importance, acknowledgment, and applicability in industrial transformations. Industry 4.0 structured or restricted the strategies of industries, while Artificial Intelligence paved the path for its timely achievement and gave a boost to the process of efficiency [25]. According to Joshi and Sharma [24], to deal with the current environmental and sustainability challenges, industries have to significantly add the phenomenon of low-carbon manufacturing to their current operating styles [54]. Simultaneously, industries have to promote sustainability and zero carbon emissions at their workplaces.

Figure 1 formulates the conceptual framework. It consists of three main segments, i.e., factors, approaches, and tactics. At first, it compiled the ground realities and collected various factors. These factors target the most relevant, current, and fresh form of information, which the industries have to acknowledge on a prior basis; thereby, factors include the present needs of people, product efficiency, and product customization. At present, working against the acknowledgment of such factors in regular working operations requires a smart product service system and digital infrastructure [11]. Secondly, the addition of factors, looking after the present needs, and advanced customization of products need the different approaches that allowed them to be part of the current system. Therefore, this study has grouped different approaches that have a higher important and in-depth functioning at the industrial level. Nowadays, an additive manufacturing approach is gaining great popularity and creating sufficient ease in operations [2]. Similarly, a human-centric approach also appeared with great importance, teaching the industries through its modern means of working and directing them toward Industry 5.0 [21]. Next, another approach to industrial needs is given, which is the green lean approach. More specifically, manufacturing, pharmaceutical, education, and service industries. It delivered modern and smart working techniques by combining green and lean methods [56]. Under the given set of approaches, the energy management approach covers the dire needs of society. It controls industrial wastage, damage of resources, and dumping of waste [34]. It also delivers information about energy management, supply chain management, resource allocation, distribution of resources, and handling of resources.

The third part of the research framework comprised the tactics depicting the possible outcome under the applied approaches. Following the listed approaches of the research framework, industries are offered multiple tactics, i.e., competitiveness, Fin-tech, input-efficient technology, and product variants. The proposed tactics govern the industries and supervise them continuously to achieve sustainable development goals, especially SDGs 8 and 9. Further, tactics have an ample cushion to align industries on the suggested patterns of working shared by the United Nations.

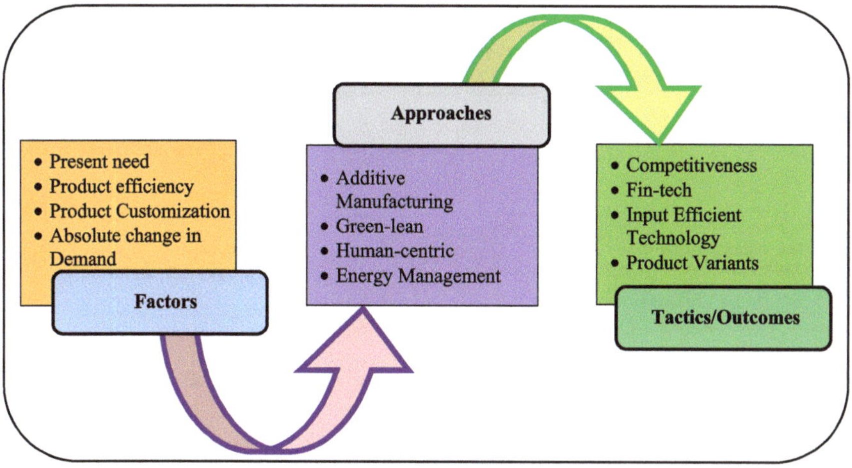


**Fig. 1** Conceptual framework. *Source* Authors' work

## 3 Proposed Methodology

### *3.1 Data Extraction*

In the past few decades, data extraction from Scopus, PubMed, Web of Science, and Google Scholar become popular and has brought several research volumes to academia [4, 17]. These websites offered multiple means of research, e.g., Bibliometric analysis, Systematic Literature Review, and review papers; thus, such research attempts fully allowed a researcher to compile the heterogeneous literature including past findings, research approaches, and key directions for future research [43]. According to Thanuskodi [61], research techniques and methodologies began in the 1960s, however, their frequent use has been seen in the last decade of the current century (Fig. 2).

Similar to the past attempts, the current study also extracted the scholarly findings from the Scopus Website by inserting the search syntax at Scopus. While searching the data, the study found a list of 1972 research papers; later, a manual filtration of other disciplines allowed us to proceed with 1880 research attempts. The search syntax is as:

*"TITLE-ABS-KEY (industry 4.0) OR TITLE-ABS-KEY (sustainable AND development) AND TITLE-ABS-KEY (sustainability) AND TITLE-ABS-KEY (sustainable development AND goals)) AND (LIMIT-TO (SUBJERA "ENVI") OR (LIMIT-TO (SUBJERA "SOCI") OR (LIMIT-TO (SUBJERA "BUSI") OR (LIMIT-TO (SUBJERA "ARTS") AND (LIMIT-TO (DOCTYPE "ar") OR (LIMIT-TO (DOCTYPE "re") OR (LIMIT-TO (DOCTYPE "cp") OR (LIMIT-TO (DOCTYPE "ch") OR (LIMIT-TO (DOCTYPE "bk")) AND (EXCLUDE (LANGUAGE "Spanish") OR (EXCLUDE (LANGUAGE "Chinese") OR (EXCLUDE (LANGUAGE*

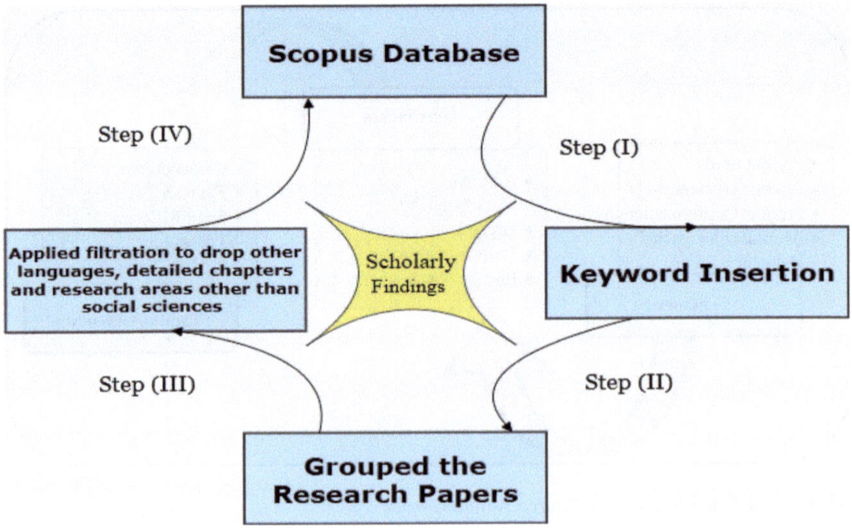


**Fig. 2** Methodology cycle. *Source* Authors' work

*"German") OR (EXCLUDE (LANGUAGE "Portuguese") OR (EXCLUDE (LANGUAGE "Russian") OR (EXCLUDE (LANGUAGE "French") OR (EXCLUDE (LANGUAGE "French") OR (EXCLUDE (LANGUAGE "Persian") OR (EXCLUDE (LANGUAGE "Croatine") OR (EXCLUDE (LANGUAGE "Polish") OR (EXCLUDE (LANGUAGE "Turkish") OR (EXCLUDE (LANGUAGE "Norwegian") OR (EXCLUDE (LANGUAGE "Ukraine") OR (EXCLUDE (LANGUAGE "Korean") OR (EXCLUDE (LANGUAGE "Bosnian") OR (EXCLUDE (LANGUAGE "Slovak") OR (EXCLUDE (LANGUAGE "Lithuanian") OR (EXCLUDE (LANGUAGE "Japanese") OR (EXCLUDE (LANGUAGE "Finnish") OR (EXCLUDE (LANGUAGE "Thai") OR (EXCLUDE (LANGUAGE "Swedish") OR (EXCLUDE (LANGUAGE "Slovenian") OR (EXCLUDE (LANGUAGE "Macedonian") OR (EXCLUDE (LANGUAGE "Dutch") OR (EXCLUDE (LANGUAGE "Arabic") AND (EXCLUDE (SRCTYPE, "k"))*"

As a result, we found that all the research was done under the heads of various subjects and languages. Later, a few filtrations about languages other than English, research areas other than social sciences, and a collection of research papers other than detailed book chapters were applied. Therefore, the current study gathered a list of the relevant research papers by means of language, research areas, or language and imported them to the VOS viewer (Visualization of Similarities) software or keyword clouds and others.

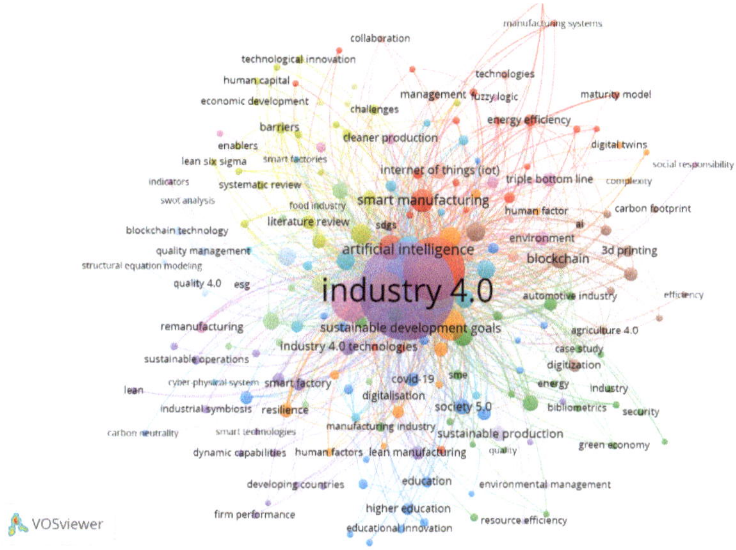


**Fig. 3** A cloud of all keywords. *Source* Authors' work

## 3.2 *Keyword Analysis*

After extracting scholarly findings from the Scopus website, we will find the following themes and their sub-themes.

### 3.2.1 Cloud of All Keywords

See Fig. 3.

### 3.2.2 Indexed Keywords

Figure 3 represents the linkages of all keywords. This massive cloud showed the interconnectivity of multiple aspects of sustainability to the industry 4.0. With the implications of Industry 4.0, industries were getting a competitive advantage and secured a leading position compared with those industrial units operated or still functioning without switching to Industry 4.0 [23]. In this way, the research efforts of Matarneh et al. [38] have raised the importance of establishing a sustainable supply chain, and practices of immediate and fast-track supply chain significantly reduce the possible social, economic, and environmental challenges. Another research done by Khan et al. [26] catered to issues regarding the supply chain of natural resources that provide empirical evidence for fostering sustainable human development, especially in the direction of the mineral industry.

Presently, financial technology (FinTech) and efficient use of energy are also getting a boost from Industry 4.0. Smart applications of Industry 4.0 fully allowed industries to raise their work efficiency to a noticeable extent and give a cushion to use their resources to an optimum level. As a result, industries are getting their entire efficiency with almost no or zero wastage [53]. Following the research work of Joshi and Sharma [24], the algorithms of Industry 4.0 allowed the manufacturing units to work with zero or almost no carbon emissions. Further, this cloud emerged and is interconnected with smart manufacturing, digitalization, society 5.0, sustainable production, and quality 4.0. Therefore, the joint contribution or simultaneous usage of such dimensions of Industry 4.0 direct the industrial units towards the automation, advancement, work with less human efforts, attainment of great efficiency and work with zero wastage.

### 3.2.3 Co-occurrence with Industry 4.0

Nowadays, sustainable development is purely allied with the future survival and exitance of business units (with all types of business units, i.e., manufacturing, trading, and service). Figure 4 displays a cloud of a grouping of indexed keywords where sustainable development appeared with a larger volume about its size and emergence. To compete and sustain with current competitors, both internal and external stakeholders have to pay an additional level of attention to sustainable development. Managers have to upgrade their decision-making process by enhancing the advanced measures of additive manufacturing [6, 9]. Key and supportive evidence has been given by Gift et al. [19], who reported that with the currently enhanced integration and digitalization, manufacturing industries have switched to 3D printing, explored the dynamics of automatic manufacturing, and added the bio-printing of documents.

While developing or making progress in business units, there is a global responsibility that needs to be addressed by each industrialist (all the industrialists either functioning their small or large-sized firms) to address global challenges [23]. For example, the study by Oguntola et al. [44] addressed the cement manufacturing units and gave the automotive means of working in their regular operation. This addition is making an obvious ease in achieving the milestone of sustainable development (Fig. 5).

Next, this research attempt has grouped the co-occurrence of keywords with industry 4.0 and made a keen observation for the emergence relative to its size and sharp brightness among other keywords. Industry 4.0 routed its linkages to the circular economy, industrial economies, and supply chain management [16]. These linkages have a strong bondage with Industry 4.0 and fall under the sharpened part nearer to Industry 4.0. After that, personal, social, global, and economic aspects of industries are getting severe attention. Moving beyond the origin of Industry 4.0, there are multiple aspects and dimensions of sectors that have immensely stronger, average, and smaller connectivity with Industry 4.0. However, industry 4.0 has a broader expansion with almost all aspects of industries.

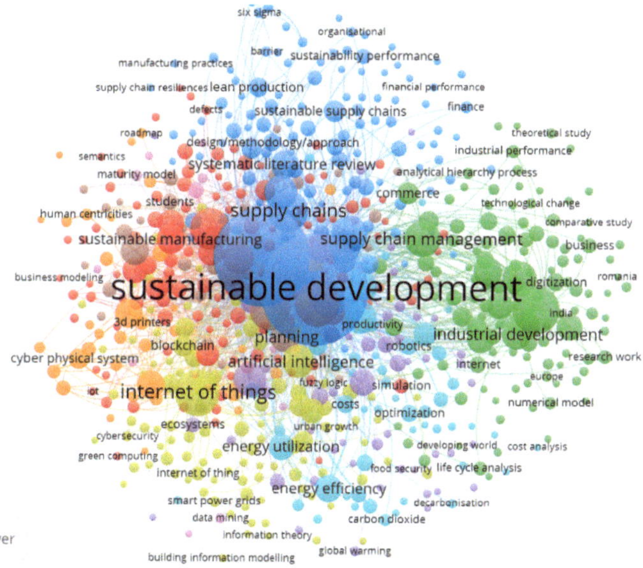


**Fig. 4** A cloud of indexed keywords. *Source* Authors' work

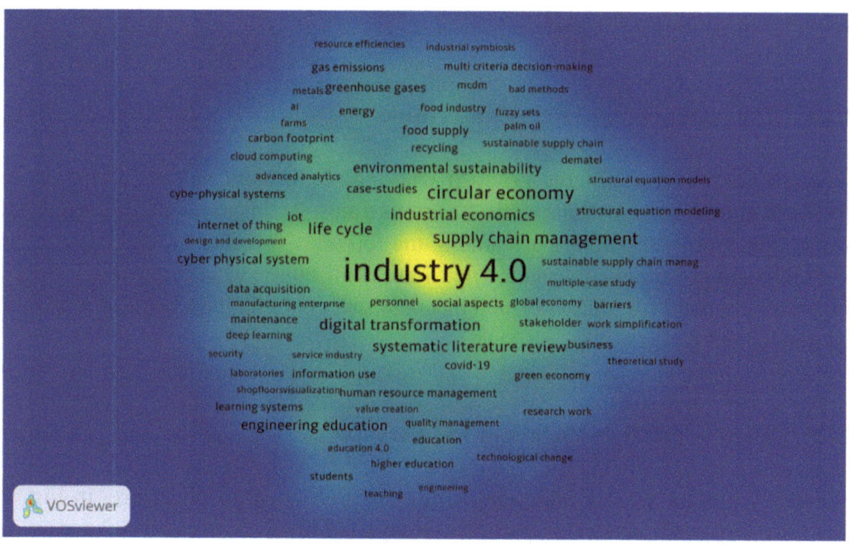


**Fig. 5** A view of the co-occurrence of keywords with Industry 4.0. *Source* Authors' work

The Internet of Things (IoT) is also tying the knot with Industry 4.0; following the research given by [32], we found novel solutions for energy efficiencies and structure the favorable environmental settings for Green Internet of Things (G-IoT). At shopfloors of industries, (G-IoT) is considered a means of easiness at various levels. It includes the responses systematically generated, automated communications, software tools, and hardware equipment, formulated machine-to-machine (M2M) learning environment at less cost and brought the entire state of working under the single roof by networking the low- and higher-level devices.

Prominently, Industry 4.0 is allied with the circular economy and supply chain management; for instance, Lara-Guillén et al. [30] have put their center of the study to the implications of a closed-loop supply chain where perishable products are supplied with the treatment of best handling and secured them from the possible wastage, damage, and loss. Correspondingly, Matarneh et al. [38] suggested and introduced flexibility in the supply chain rather than structure the tightened supply chains. Along with industries, this evidence is getting a similar level of appreciation in the corporate setting of the environment by providing a cushion to achieve sustainable development goals. Considering the work settings, this attempt is making its contribution under SDG 8 and 9 of "decent work and economic growth" and "industry innovation and infrastructure," respectively. Consistent with the circular economy, the research evidence of Varella et al. [64] raised the importance of the circular economy in agricultural yields.

### 3.2.4 Top Authors and Their Consortiums

Figure 6 gives a visual of the grouped working of different authors and gives us details of mutual contributions made at various times. At this moment, it is necessary to mention that these authors are among the top ones with the top publications in reputable and prestigious journals. Their joint contribution to Industry 4.0, Industry 5.0, environmental sustainability, and business sustainability have been admired and acknowledged globally.

### 3.2.5 Themes and Sub-themes of Industry 4.0

Industry 4.0 is the central attention of this research attempt. Therefore, in Fig. 7, the study displayed the strong, partially strong, and relatively weaker linkages with other components that are energized by the emergence and applications of Industry 4.0. At first, in the main display, e.g., Fig. 7, Industry 4.0 is linked with various other keywords that were given by multiple authors in the research work. Following the segment red in color, it shows that Industry 4.0 is an amalgamation and simultaneously important for economic aspects, the manufacturing sector, the best utilization of human capital, and the introduction of new technologies. Next, another grouped networking (blue in color) was found with deep learning, development of smart cities, machine learning, securitizing the networks, maintenance, and energy management.

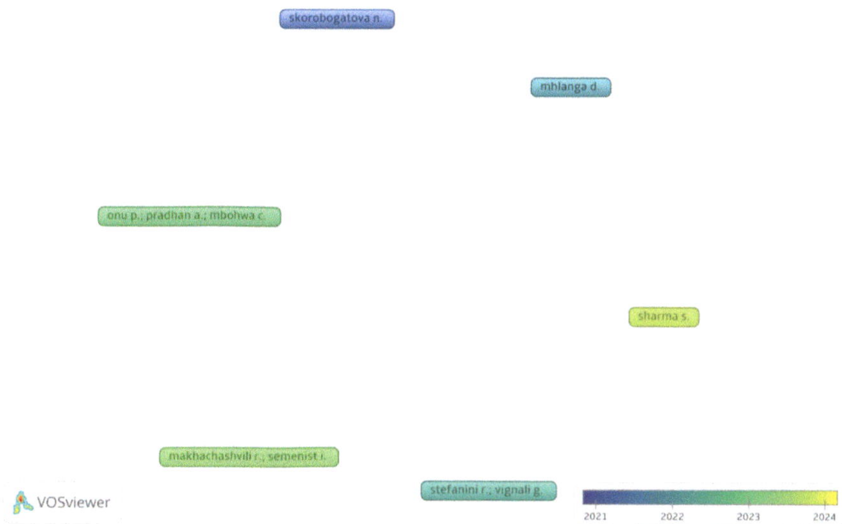


**Fig. 6** Groups of co-authors. *Source* Authors' work

Similarly, Industry 4.0 is aligned with the circular economy, carbon footprints, sustainable industries, carbon emissions, supply chains, and supply chain management. Finally, Industry 4.0 appears to share its connectivity with other aspects. However, these aspects have secondary importance and are available under the partially stronger and relatively weaker linkages. It includes and shows the networking of Industry 4.0 with mass customization, equal sharing of information, precise development, and structuring of information. Further, it covers the phenomenon related to environmental performances, engineering education, 3D printing, and the cyber security of physical systems.

The first sub-theme (Fig. 7a) illustrates that Industry 4.0 gives a boost to sustainable practices and holds an ample cushion to adjust the sustainable practices in their regular operations [3]. Correspondingly, Industry 4.0 formulates the mechanism for sustainability in terms of manufacturing, learning, engineering, continuous innovation, and the introduction of smart means of working. The study of [13] has proven the development of smart infrastructure through Industry 4.0.

Next to Industry 4.0, the sub-theme of sustainable development appeared to have more dominance and ranked second (Fig. 7b). Under the context of industrial development and its current exerting environmental impacts, the prior research attempts have found a steady and sustainable development for industries by paving the path for sustainability. In this way, the Research efforts of Marconnet et al. [37] gave a shred of empirical evidence for industries and directed them to follow sustainable practices (described by the United Nations) for current survival, future growth, business expansion, and sustainability.

The sub-theme of Industry 4.0 is available in Fig. 7c, which shows an important and deep connectivity with engineering education, personal training, computer-aided

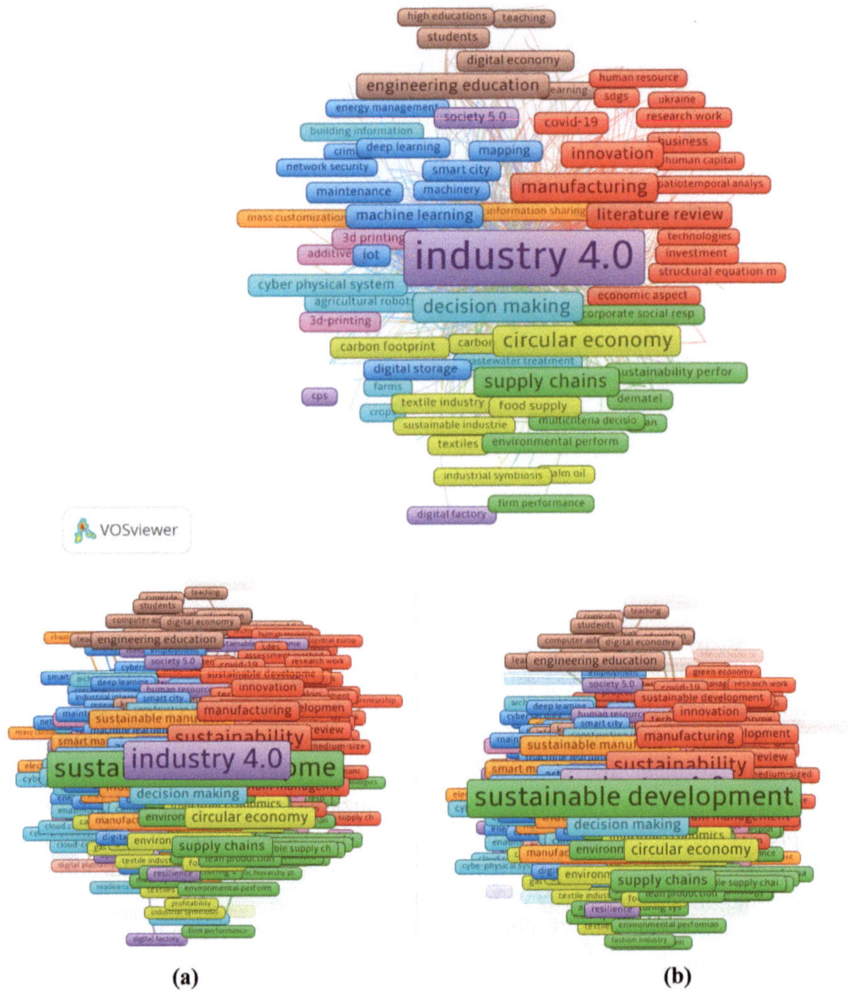


**Fig. 7** A combined amalgamation of current scholarly themes. **a** Sub theme of industry 4.0. **b** Sub theme of sustainable development. **c** Sub-theme of engineering education. **d** Sub-theme of sustainability. **e** Sub-theme of Internet of things. *Source* Authors' work

instructions, and curricula of sustainability. In addition, Fig. 7d shows the dependence of sustainability on the circular economy, sustainable manufacturing, innovations, and smart decision-making that is proven or given significance from prior literature. Likewise, the research efforts of Narkhede et al. [41] provided comprehensive modeling for small and medium enterprises, depicting that following the sustainability standards, enterprises may introduce smartness in their major operating functions like new product development, supply chain management, internal logistic management, quality control, production and execution planning, and other

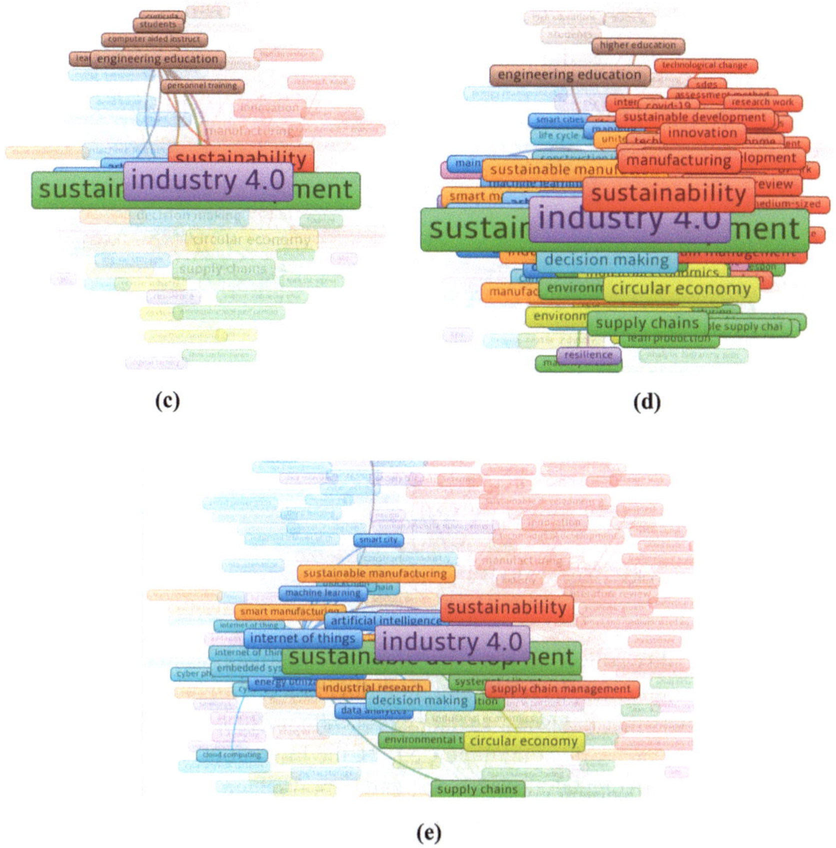


**Fig. 7** (continued)

maintenance management. Combining and linking this study's research farmwork ensures the development of sustainability in both large and small-sized enterprises.

Figure 8 displays the keywords that were introduced by prestigious authors and extends the literature through a timely addition of a valid pool of words. Particularly, this type of grouping allows the researchers to get the different theoretical perspectives and schools of thought of subject experts and environmental experts. At the view of this cloudy amalgamation, it shown that Industry 4.0 has routed its directions to additive manufacturing [57], digitalization [8], remanufacturing [67], circular manufacturing [39], industrial revolution [60, 53, 65], blockchain technology [59], cleaner production [34], society 5.0 [31, 35] and Internet of Things (IoT) [1, 50]. This grouping gives validity and positive significance to the current formulated research farmwork and assured its stream of propagation.

Lastly, industrial connectivity has been found with Industry 4.0 and displayed its representation in Fig. 9. Further, following Fig. 9, its sub-themes are available. At

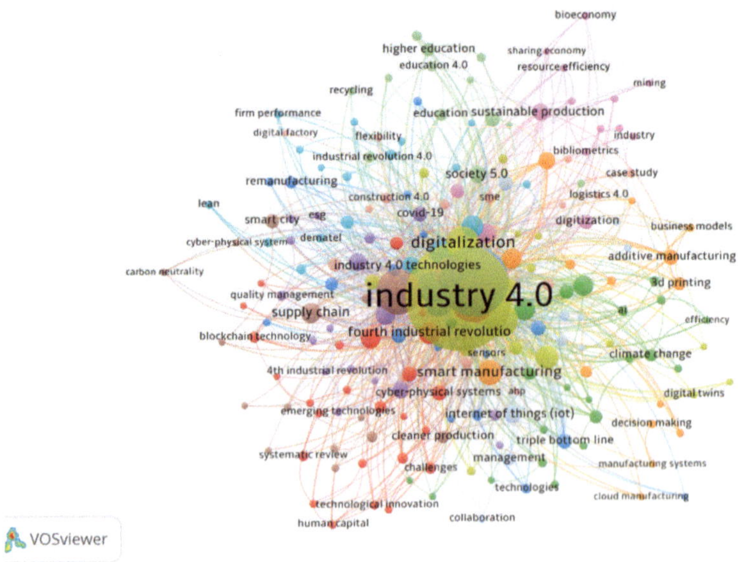


**Fig. 8** Authors keywords. *Source* Authors' work

initial, sustainability collectively grouped the entire set of keywords. However, from following the sub-themes, its roots, and joint direction got higher clearance. Firstly, the Fig. 9a depicts that industrial sustainable development is an amalgamation and has a dependency on artificial intelligence [47], the Internet of Things (IoT) [1, 50], industry 5.0 [48], data analytics [33, 55] and supply chain management [22, 63]. These themes are fully covered under the research frameworks that were structured under the lightening paths of past studies and literature.

Moreover, the sub-theme of sustainability (Fig. 9b) shows its route to efficient management of supply through the applications of artificial intelligence [31, 36], machine learning [66] and energy efficiency [27, 33]. Subsequently, sustainability is showing its dependency on the Internet of Things (IoT) (Fig. 9c), resulting in the emergence and efficiency of energy efficiency [58], engineering education [27], artificial intelligence [5] and industry 5.0 [21]. Comparatively to sub-themes of Fig. 9d, sustainability has a large association and capturing the segment of the supply chain [28], industrial economies [15], and industrial research [62]. Resultantly, the core theme of sustainability in Fig. 9 and its sub-themes listed a maximum of keywords that proved the process of sustainability at the industrial level.

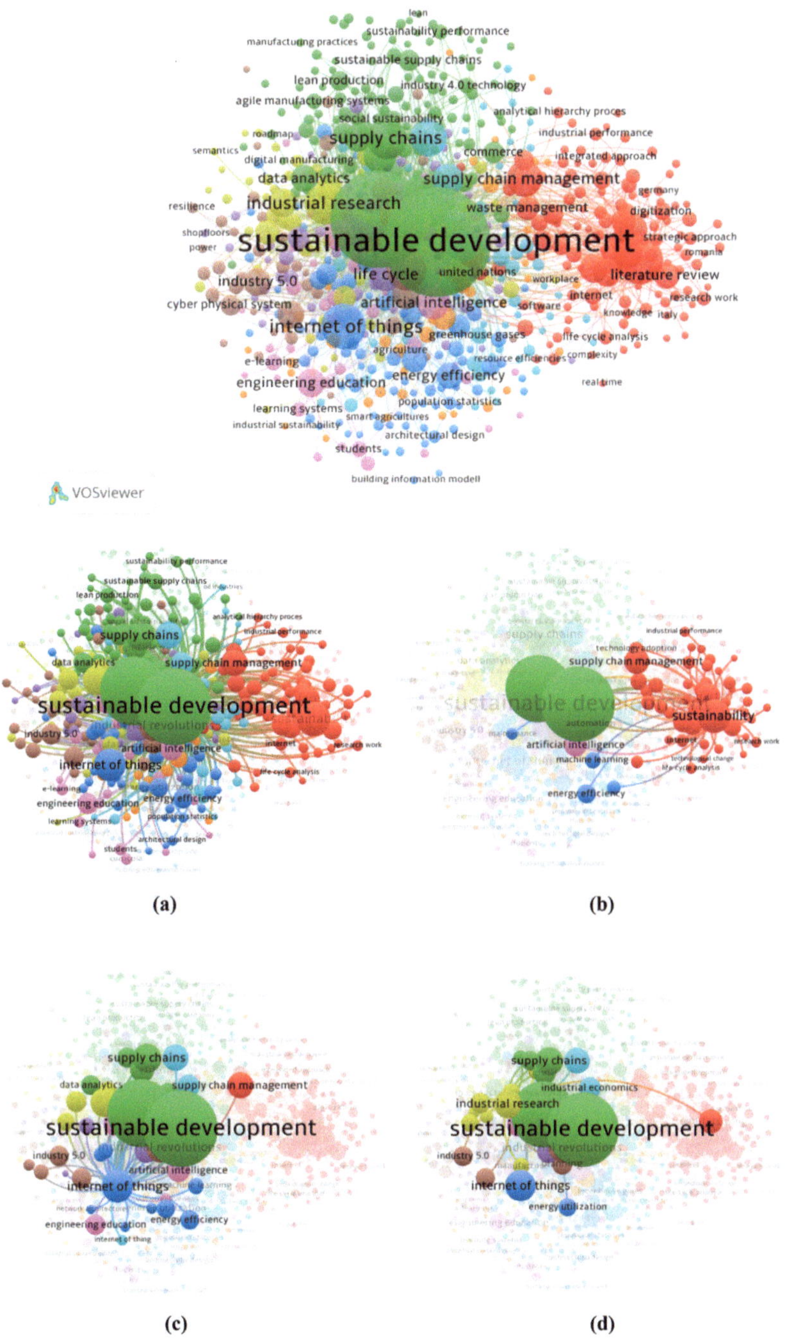


**Fig. 9** A core theme of sustainable development at the industrial level. *Source* Authors' work. **a** Sub-theme of sustainable development. **b** Sub-theme of sustainability. **c** Sub-theme of internet of things. **d** Sub-theme of industrial research

## 4 Discussion

At the initial stages of industrial research and revolution, industrialists intentionally ignored sustainability and environmental costs. During that duration, attention was set to building industrial zones that were centered around the progressive measures of the industries [7]. However, the abrupt effects of these progressive measures were seen later and are currently the leading challenges for both developed and developing countries.

While meeting environmental challenges, industries seek automation, come across multiple transformations, have opportunities with different avenues, group different approaches, and learn different tactics. Following these efforts, the study grouped the factors, approaches, and possible tactics and formulated a research framework (referring to Fig. 1). Firstly, the model scrutinizes the higher impacting factors that mold and remold the current operations of industries. Next, various approaches are listed, which allow the industrialists to deal best and work for the grouped factors. Lastly, the model structured the possible outcomes and advancement through industries that can gain competitiveness and remain in a win–win state. Moreover, the study provided a theoretical mechanism for entire industries to make major or minor contributions to sustainable development goals (SDGs). It is an open, common, and equally shared agenda in which industries and other sectors have to give their part from their regular routines and operational work.

To achieve the desirous aims, the study formulated a set of research questions, which were answered in various sections of the attempt. Through the discussion of possible outcomes of Industry 4.0, the study mentioned the approaches and tactics that permit the industries to cater to a maximum of current prevailing factors. Resultantly, the first stated research question got its answer. Next, the study mentioned the current challenges and environmental abruptness, which demanded the current and severe attention of researchers, academicians, policymakers, and environmental regulators. As a result of this, the study answered its second stated research question.

Moving next to research question three, the study worked for the current and future applications of Industry 4.0 through keyword analysis. Here, the study listed the findings of past research through the keyword clouds. As a result, the study answered research question three. Lastly, the present research studies detailed applications and benefits in different industries and regions. In addition, the studied stuff was taken from peer-reviewed papers; the level of authenticity was confirmed by the strong, partially strong, and weak connectives within the keyword clouds and their proposed clusters. Thus, research question four got a satisfactory answer.

To compile the data set of the most relevant research studies, we came across different search engines, i.e., Web of Sciences, Scopus, PubMed, and Google Scholar. For the current attempt, we used Scopus and obtained the datasheet by inserting various keywords, such as Industry 4.0, sustainability, sustainable development, and industrial sustainability. Furthermore, a filtration about the exemption of other languages, research areas, and detailed book chapters was applied; thereby, the study obtained the required data by dropping the filtered material. Later on, the downloaded data was imported to VOS viewer software for the visual representations of keyword placement and strength of connectivity.

The study provides key directions and mechanisms to make a valid and considerable contribution to achieving sustainable development goals. Meanwhile, the research attempt of Kılıç and Atilla [29] stated that Industry 4.0 is a transformative force to achieve sustainable development. According to the research efforts of Delbari and Hof [14], the sustainability of industries is purely dependent on handling and best dealing with waste. Kumar and Agrawal [28] put our attention to structuring a sustainable supply chain even for perishable items.

Likewise, Madhavan et al. [35] and Santos et al. [51] dealt with the sustainable practices of small and medium enterprises and also incorporated their contribution to the sustainable development goals. Khalid [25] studied AI applications for business sustainability. Khan et al. [26] mentioned the FinTech approaches to sustainability. Pitakaso et al. [47] emphasized the control management of resources suggested by AI. As a whole, several past researchers worked to bring sustainability to industries via different routes of digitalization and transformations. Upon grouping and collecting the relevant material from the Scopus website and its visuality in keyword clouds, i.e., all keywords, author keywords, indexed keywords, and co-occurrences, the study provided robust empirical evidence to researchers, industrialists, policymakers, practitioners, environmental regulators, and governing bodies.

## 5 Current Limitations and Future Research Avenues

Despite having a considerable contribution, the current study holds some limitations, which may be the guiding principles for upcoming or young scholars. At first, the study dealt with the entire set of industries without segregating them according to their sector or the nature of their work. Therefore, the study calls for further research by applying a sectorial filtration of industries. Thereby, the contribution of each sector to SDG can be studied. Subsequently, at collective, we may get brief views, statistical figures and contributions of industrial sectors to the entire set of 17 SDGs. Second, the current study has grouped the research studies of both developed and developing countries. In the future, a comparative study is suggested. So, a differentiated contribution can be studied and weightage can be assigned to the most collaborative countries. Finally, the study makes an appealing call to acknowledge and study the industrial contribution to SDGs other than 7, 8, and 9. In the future, a mechanism can be structured for industries where they opt for the practices of SDGs other than 7, 8, and 9, and make their contribution to this open, shared and common agenda of the UN.

23. Kabra, G., Ramesh, A., Akhtar, P., Dash, M.K.: Understanding behavioural intention to use information technology: insights from humanitarian practitioners. Telematics Inform. 34(7), 1250–1261 (2017). https://doi.org/10.1016/j.tele.2017.05.010
24. Alanazi, M.A., Tully, M.P., Lewis, P.J.: Prescribing errors by junior doctors- a comparison of errors with high risk medicines and non-high risk medicines. PLoS ONE 14(1), e0211270 (2019). https://doi.org/10.1371/journal.pone.0211270

9. Smith, B., Kistruck, G., Cannatelli, B.: The impact of moral intensity [illegible] stakeholder [illegible] social enterprises. J. Bus. Ethics 133, 677[illegible] https://doi.org/10.1007/s10551-0[illegible]
10. Rechtman, [illegible]
[illegible]

## Websites


1. https://en.wikipedia.org/wiki/History_of_money
2. [illegible]
3. [illegible]
4. [illegible]
5. [illegible]
6. [illegible]
7. [illegible]
8. [illegible]
9. [illegible]
10. [illegible]

and calculated priorities for constraints falling under ‘Individual effort,’ ‘Infrastructure,’ ‘Resources,’ and ‘Management’, and the priorities of constraints pertaining to ‘Technology Provision’ and ‘Economic Provision’.

The calculations of AHP outcomes for Expert-1 are shown in detail below:

Adequate IoT Resources = $(0.691 \times 0.667 \times 0.421) + (0.691 \times 0.333 \times 0.579) + (0.218 \times 0.8 \times 0.535) + (0.218 \times 0.2 \times 0.58) + (0.091 \times 0.5 \times 0.404) + (0.091 \times 0.5 \times 0.55) = 0.4893$.

Professional Association = $(0.691 \times 0.667 \times 0.267) + (0.691 \times 0.333 \times 0.270) + (0.218 \times 0.8 \times 0.270) + (0.218 \times 0.2 \times 0.308) + (0.091 \times 0.5 \times 0.443) + (0.091 \times 0.5 \times 0.281) = 0.1963$.

Operational Environment = $(0.691 \times 0.667 \times 0.181) + (0.691 \times 0.333 \times 0.072) + (0.218 \times 0.8 \times 0.120) + (0.218 \times 0.2 \times 0.053) + (0.091 \times 0.5 \times 0.099) + (0.091 \times 0.5 \times 0.108) = 0.1281$.

Strategic Planning = $(0.691 \times 0.667 \times 0.121) + (0.691 \times 0.333 \times 0.078) + (0.218 \times 0.8 \times 0.075) + (0.218 \times 0.2 \times 0.059) + (0.091 \times 0.5 \times 0.054) + (0.091 \times 0.5 \times 0.062) = 0.095$.

The AHP decision hierarchy for the first phase analysis, as evaluated by Expert-1, has been visually represented in Fig. 1. The AHP priorities from all twenty-two experts involved in the initial phase of analysis have been determined. The average priority percentages for the factors "adequate IoT resources," "professional association," "operational environment," and "strategic planning" are reported as 55%, 24%, 14%, and 9% respectively. In addition, the standard deviation (SD) figures for these percentages are recorded as 4%, 6%, 1%, and 3% respectively.

### 3.3 *Potential Solutions for the Integration of IoT in Rural Healthcare Management*

Interviewees participating in the first phase were requested to recommend possible solutions to overcome the constraints based on their opinions and real-world experiences. The outcomes of the first phase analysis were shown to the key informants participating in the second phase. With the help of the second phase questionnaire, ten individual key informants were asked to recommend possible solutions for mitigating the major constraints to facilitate the integration of IoT in the rural healthcare sector of Bangladesh. The responses of the interviewees were studied critically. Finally, six possible solutions for the integration of IoT in the rural healthcare sector of Bangladesh were found in the second phase which were addressed among the three responsible categories (Organization X, Organization Y, and Organization Z). The six possible solutions are: Need effective associations and resource empowerment among the organizations (A), Need proper professional training about new technology and industry attachment with academic institutes (B), Set up IoT incubator

66. Ammar, Alhanwar, … Mohammed, K.: Design development of robotic devices for rehabilitation. Lecturia Univ. Prc. 45(…) (2024) [illegible]
67. Shepherd, K., Reddy, A.: Improving preventive care and health outcomes for patients with chronic diseases [illegible]
[illegible]
[illegible]
[illegible]
[illegible]
[illegible]
[illegible]
[illegible]
[illegible]

15. Harry, I.: Stress management and employee performance. Eur. J. Hum. Resour. Manag. Stud. 37–41 (2020)
16. Lian, S.Y., Tam, C.L.: Work stress, coping strategies and resilience: a study among working females. Asian Soc. Sci. [illegible] (2014)
17. [illegible]: Role of work stress and coping strategies [illegible]
18. [illegible]
19. [illegible]
20. [illegible]
21. [illegible]: stress management and employee performance. [illegible]
22. [illegible]
23. [illegible]
24. [illegible]
25. [illegible]
26. [illegible]
27. [illegible]

17. Shahid, M., Ahmed, R.: The impact of digital public administration on ... in Pakistan. J. Gov. Innov. 15(3), 45–62 (2021)
18. Transparency International: Corruption perceptions index 2022: Pakistan. Berlin, Germany: Transparency International (2022) [illegible]
[illegible]

# 6 Conclusion

The study searched for the basic factors, approaches, and possible tactics for the current applications of Industry 4.0. Further, the study aligned the approaches and tactics to the current existence, sustainable development, and competitiveness of businesses. From the past scholarly findings, the study visualized the keywords cloud and their sub-themes. Further, we studied the current existing research contribution in circular economy [14], sustainable resource management [26], smart manufacturing [54], supply chains [22], Fin-Tech [26, 62], additive manufacturing [2] green-lean approach [56] and Internet of Things (IoT) [32, 50]. As a result, we viewed the most relevant grouping of Industry 4.0, sustainability, and business transformations in industrial or, especially, manufacturing units. By giving a comprehensive model and views of prior researches, the study gave empirical evidence to the current and future industrials; it is equally important for the researchers, practitioners, policymakers and environmental regulatory bodies.

# References


1. Abaee, M., Saeedi, M., Taherdoost, H.: IoT and cloud computing for sustainable development goals in industry 4.0: a bibliometric analysis. In: Reshaping Environmental Science Through Machine Learning and IoT, pp. 98–118 (2024)
2. Abdulrahman, K.O., Baba, R.A.: Additive manufacturing technologies for on-demand production of customised Goods in the Era of the 4th industrial revolution. In: Advances in Industrial Engineering in the Industry 4.0 Era, pp. 127–146. CRC Press (2024)
3. Alazab, M., Alhyari, S.: Industry 4.0 innovation: a systematic literature review on the role of blockchain technology in creating smart and sustainable manufacturing facilities. Information **15**(2) (2024). https://doi.org/10.3390/info15020078
4. Ali, I., Nguyen, N.D.K., Gupta, S.: A multi-disciplinary review of enablers and barriers to cloud ERP implementation and innovation outcomes. J. Enterp. Inf. Manag. **36**(5), 1209–1239 (2023). https://doi.org/10.1108/jeim-08-2022-0273
5. Aliyev, A.G., Shahverdiyeva, R.O., Salimkhanova, S.A.: Management problems of information support of technical and economic systems based on artificial intelligence technologies. In: Artificial Intelligence, Medical Engineering and Education (2024). https://doi.org/10.3233/atde231329
6. Alves, A.S.F., Oliveira, J.P., Godina, R.: Advancing sustainable decision making in additive manufacturing: a comprehensive review of multi-criteria decision making approaches. Clean Technol. **6**(2), 646–661 (2024)
7. Amani, M.A., Aghamohammadi, N.: A novel technology to monitor effects of ethylene on the food products' supply chain: a deep learning approach. Int. J. Environ. Sci. Technol. **21**(5), 5007–5018 (2023). https://doi.org/10.1007/s13762-023-05328-3
8. Asif, M., Naeem, G., Khalid, M.: Digitalization for sustainable buildings: technologies, applications, potential, and challenges. J. Clean. Prod. **450** (2024). https://doi.org/10.1016/j.jclepro.2024.141814
9. Avilés-Sacoto, S.V., Velasco-Tapia, K.C., Avilés-Sacoto, E.C., Argüello-Herrera, J.I.: Innovating for impact: proposing quality 4.0 integration to achieve sustainable development goals in microenterprises. Bus. Strategy Dev. **7**(2), e371 (2024)
10. Aziz, N.A., Al Mamun, A., Reza, M.N.H., Naznen, F.: The impact of big data analytics on innovation capability and sustainability performance of hotels: evidence from an emerging

economy. J. Enterp. Inf. Manag. **37**(3), 1044–1068 (2024). https://doi.org/10.1108/jeim-07-2023-0354
11. Bag, S., Gupta, S., Chan, H.-L., Kumar, A.: Building smart product-service systems capabilities for circular supply chains in the Industry 4.0 era. Transp. Res. Part E: Logist. Transp. Rev. **188** (2024). https://doi.org/10.1016/j.tre.2024.103625
12. Costa, F., Alemsan, N., Bilancia, A., Tortorella, G.L., Portioli Staudacher, A.: Integrating industry 4.0 and lean manufacturing for a sustainable green transition: a comprehensive model. J. Clean. Prod. **465** (2024). https://doi.org/10.1016/j.jclepro.2024.142728
13. Daios, A., Kladovasilakis, N., Kostavelis, I.: Mixed palletizing for smart warehouse environments: sustainability review of existing methods. Sustainability **16**(3) (2024). https://doi.org/10.3390/su16031278
14. Delbari, S.A., Hof, L.A.: Glass waste circular economy-advancing to high-value glass sheets recovery using industry 4.0 and 5.0 technologies. J. Clean. Prod. 142629 (2024)
15. Detwal, P.K., Agrawal, R., Samadhiya, A., Kumar, A., Garza-Reyes, J.A.: Research developments in sustainable supply chain management considering optimization and industry 4.0 techniques: a systematic review. Benchmark.: Int. J. **31**(4), 1249–1269 (2023). https://doi.org/10.1108/bij-01-2023-0055
16. Duong Thi Binh, A., Akbari, M., Le Thi Cam, H., Nguyen Canh, L., Truong Quang, H.: Forging pathways to circular economy excellence: integrating industry 4.0 with quality management. Sustainability **16**(7), 3053 (2024)
17. Ellegaard, O., Wallin, J.A.: The bibliometric analysis of scholarly production: how great is the impact? Scientometrics **105**(3), 1809–1831 (2015). https://doi.org/10.1007/s11192-015-1645-z
18. Gajdzik, B., Siwiec, D., Wolniak, R., Pacana, A.: Approaching open innovation in customization frameworks for product prototypes with emphasis on quality and life cycle assessment (QLCA). J. Open Innov.: Technol. Market Complex. **10**(2) (2024). https://doi.org/10.1016/j.joitmc.2024.100268
19. Gift, M.D.M., Senthil, T.S., Hasan, D.S., Alagarraja, K., Jayaseelan, P., Boopathi, S.: Additive manufacturing and 3D printing innovations: revolutionizing industry 5.0. In: Technological Advancements in Data Processing for Next Generation Intelligent Systems, pp. 255–287. IGI Global (2024)
20. Görçün, Ö.F., Mishra, A.R., Aytekin, A., Simic, V., Korucuk, S.: Evaluation of Industry 4.0 strategies for digital transformation in the automotive manufacturing industry using an integrated fuzzy decision-making model. J. Manuf. Syst. **74**, 922–948 (2024). https://doi.org/10.1016/j.jmsy.2024.05.005
21. Holzinger, A., Fister, I., Fister, I., Kaul, H.-P., Asseng, S.: Human-centered AI in smart farming: toward agriculture 5.0. IEEE Access **12**, 62199–62214 (2024). https://doi.org/10.1109/access.2024.3395532
22. Jiang, M., Jia, F., Chen, L., Xing, X.: Technology adoption in socially sustainable supply chain management: towards an integrated conceptual framework. Technol. Forecast. Soc. Chang. **206** (2024). https://doi.org/10.1016/j.techfore.2024.123537
23. Jonathan, H., Magd, H., Khan, S.A.: Artificial intelligence and augmented reality: a business fortune to sustainability in the digital age. In: Navigating the Digital Landscape: Understanding Customer Behaviour in the Online World, pp. 85–105. Emerald Publishing Limited (2024)
24. Joshi, S., Sharma, M.: Intelligent algorithms and methodologies for low-carbon smart manufacturing: review on past research, recent developments and future research directions. IET Collab. Intell. Manuf. **6**(1) (2024). https://doi.org/10.1049/cim2.12094
25. Khalid, M.: Energy 4.0: AI-enabled digital transformation for sustainable power networks. Comput. Ind. Eng. **193** (2024). https://doi.org/10.1016/j.cie.2024.110253
26. Khan, I., Khan, I., Ikram, R.M.A., Ali, M.A.: Integrating FinTech with Industry 4.0 for sustainable mineral resource management: exploring the interplay of technology, regulation, and economic dynamics. Resour. Policy **92**, 104971 (2024)
27. Kheirkhah Ravandi, Z., Bozorgmehry Boozarjomehry, R., Babaei, F., Pishvaie, M.R.: Consensus-based dynamic optimization of the integrated energy-to-product networks through

an ontologically-aware multi-agent system. Eng. Appl. Artif. Intell. **133** (2024). https://doi.org/10.1016/j.engappai.2024.108626
28. Kumar, A., Agrawal, S.: A quality-based sustainable supply chain architecture for perishable products using image processing in the era of industry 4.0. J. Clean. Prod. **450** (2024). https://doi.org/10.1016/j.jclepro.2024.141910
29. Kılıç, C., Atilla, G.: Industry 4.0 and sustainable business models: an intercontinental sample. Bus. Strategy Environ. **33**(4), 3142–3166 (2023). https://doi.org/10.1002/bse.3634
30. Lara-Guillén J, Méndez-Aparicio MD, Jiménez-Zarco AI (2024) Circular economy and closed-loop supply chains in Industry 4.0: importance to achieve sustainable development. In: Digital Transformation for Improved Industry and Supply Chain Performance, pp. 299–333. IGI Global (2024)
31. Leng, J., Zhu, X., Huang, Z., Li, X., Zheng, P., Zhou, X., Mourtzis, D., Wang, B., Qi, Q., Shao, H., Wan, J., Chen, X., Wang, L., Liu, Q.: Unlocking the power of industrial artificial intelligence towards Industry 5.0: insights, pathways, and challenges. J. Manuf. Syst. **73**, 349–363 (2024). https://doi.org/10.1016/j.jmsy.2024.02.010
32. Liu, L., Mishra, A.R.: Enabling technologies challenges of green internet of things (IoT) towards sustainable development in the era of Industry 4.0. Technol. Econ. Dev. Econ. **30**(2), 344–375 (2024)
33. Ma, S., Huang, Y., Cai, W., Leng, J., Xu, J.: Integrated sustainable benchmark based on edge-cloud cooperation and big data analytics for energy-intensive manufacturing industries. J. Manuf. Syst. **74**, 1037–1056 (2024). https://doi.org/10.1016/j.jmsy.2024.05.010
34. Ma, S., Ding, W., Liu, Y., Zhang, Y., Ren, S., Kong, X., Leng, J.: Industry 4.0 and cleaner production: a comprehensive review of sustainable and intelligent manufacturing for energy-intensive manufacturing industries. J. Clean. Prod. **467** (2024). https://doi.org/10.1016/j.jclepro.2024.142879
35. Madhavan, M., Sharafuddin, M.A., Wangtueai, S.: Measuring the Industry 5.0-readiness level of SMEs using industry 1.0–5.0 practices: the case of the seafood processing industry. Sustainability **16**(5) (2024). https://doi.org/10.3390/su16052205
36. Manta, A.G., Bădîrcea, R.M., Doran, N.M., Badareu, G., Gherţescu, C., Popescu, J.: Industry 4.0 transformation: analysing the impact of artificial intelligence on the banking sector through bibliometric trends. Electronics **13**(9) (2024). https://doi.org/10.3390/electronics13091693
37. Marconnet, B., Gaha, R., Assuad, C., Martinsen, K., Eynard, B.: Selecting sustainability indicators for smart product design based on industry 4.0/5.0 technologies: analysis and proposal of a methodological framework. Proc. Des. Soc. **4**, 1369–1378 (2024). https://doi.org/10.1017/pds.2024.139
38. Matarneh, S., Piprani, A.Z., Ellahi, R.M., Nguyen, D.N., Le, T.M., Nazir, S.: Industry 4.0 technologies and circular economy synergies: Enhancing corporate sustainability through sustainable supply chain integration and flexibility. Environ. Technol. Innov. **35**, 103723 (2024)
39. Mesjasz-Lech, A., Kemendi, Á., Michelberger, P.: Circular manufacturing and Industry 5.0 assessing material flows in the manufacturing process in relation to e-waste streams. Eng. Manag. Prod. Serv. **16**(1), 114–133 (2024). https://doi.org/10.2478/emj-2024-0009
40. Muzammal, S.M., Tayyab, M., Zahra, F., Jhanjhi, N.Z., Ashraf, H.: Smart factories greener future: a synergy of Industry 4.0 and sustainability. In: Digital Transformation for Improved Industry and Supply Chain Performance, pp. 70–92. IGI Global (2024)
41. Narkhede, G., Mahajan, S., Narkhede, R., Chaudhari, T.: Significance of Industry 4.0 technologies in major work functions of manufacturing for sustainable development of small and medium-sized enterprises. Bus. Strategy Dev. **7**(1) (2023). https://doi.org/10.1002/bsd2.325
42. Nations, U.: Transforming Our World: the 2030 Agenda for sustainable development (2015). sustainabledevelopment.un.org. Accessed 02 Feb. 2023
43. Nguyen, N.D.K., Ali, I., Gupta, S., Chen, R., Naresho, B.S.: Bridging the nexus between cloud ERP and enterprise resilience. J. Glob. Inf. Manag. **32**(1), 1–22 (2024). https://doi.org/10.4018/jgim.336556
44. Oguntola, O., Boakye, K., Simske, S.: Towards leveraging artificial intelligence for sustainable cement manufacturing: a systematic review of AI applications in electrical energy consumption optimization. Sustainability **16**(11), 4798 (2024)

45. Onile, A.E., Petlenkov, E., Levron, Y., & Belikov, J.: Smartgrid-based hybrid digital twins framework for demand side recommendation service provision in distributed power systems. Future Gener. Comput. Syst. **156**, 142–156 (2024).
46. Peldon, D., Banihashemi, S., LeNguyen, K., Derrible, S.: Navigating urban complexity: the transformative role of digital twins in smart city development. Sustain. Cities Soc. **111** (2024). https://doi.org/10.1016/j.scs.2024.105583
47. Pitakaso, R., Srichok, T., Khonjun, S., Golinska-Dawson, P., Sethanan, K., Nanthasamroeng, N., Gonwirat, S., Luesak, P., Boonmee, C.: Optimization-driven artificial intelligence-enhanced municipal waste classification system for disaster waste management. Eng. Appl. Artif. Intell. **133** (2024). https://doi.org/10.1016/j.engappai.2024.108614
48. Ramírez-Márquez, C., Posadas-Paredes, T., Raya-Tapia, A.Y., Ponce-Ortega, J.M.: Natural resource optimization and sustainability in society 5.0: a comprehensive review. Resources **13**(2) (2024). https://doi.org/10.3390/resources13020019
49. Rehman, S., Hamza, M.A., Nasir, A., Ullah, A., Arshad, N.: Impact of COVID-19 and consortium factors on mental health: role of emotional labor strategies in achieving sustainable development goals. Front. Psychol. **13**, 795677 (2022). https://doi.org/10.3389/fpsyg.2022.795677
50. Rup, C., Bajic, E.: Green and sustainable industrial internet of things systems leveraging wake-up radio to enable on-demand IoT communication. Sustainability **16**(3) (2024). https://doi.org/10.3390/su16031160
51. Santos, A.D.M., Becker, A.M., Ayala, N.F., Sant'Anna, Â.M.O.: Industry 4.0 as an enabler of sustainability for small and medium enterprises. Academia Revista Latinoamericana de Administracíón **37**(2), 204–226 (2024). https://doi.org/10.1108/arla-07-2023-0118
52. Santos, A.D.M., Sant'Anna, Â.M.O.: Industry 4.0 technologies for sustainability within small and medium enterprises: a systematic literature review and future directions. J. Clean. Prod. **467** (2024). https://doi.org/10.1016/j.jclepro.2024.143023
53. Shafik, W.: Industry revolution 4.0 and beyond: abilities and perils for sustainable zero net economy development. In: Powering Industry 5.0 and Sustainable Development Through Innovation, pp. 292–316. IGI Global (2024)
54. Sharma, M., Vadalkar, S., Antony, R., Chavan, G., Tsagarakis, K.P.: Can Industry 4.0-enabled smart manufacturing help firms in emerging economies move toward carbon-neutrality? Comput. Ind. Eng. **192**, 110238 (2024)
55. Sheela, S.C., Srivastava, S., Kumar, S.V., Aravinda, K., Singh, N., Yusupov, A.: Strategic implementation of big data analytics for Industry 5.0's sustainable expansion. In: Powering Industry 5.0 and Sustainable Development Through Innovation, pp. 276–291. IGI Global (2024)
56. Siegel, R., Antony, J., Garza-Reyes, J.A., Cherrafi, A., Lameijer, B.: Integrated green lean approach and sustainability for SMEs: from literature review to a conceptual framework. J. Clean. Prod. **240**, 118205 (2019). https://doi.org/10.1016/j.jclepro.2019.118205
57. Silva, C.M.A., Gomes, V.C., Ferreira, A.B.S., Nascimento, L.D., Leite, A.B., Santos, V.S., Acruz, C.S., Marques, F.R.V., Leão, A.P.D.S., Tourem, R.V.: Industry 4.0 and sustainability: a study on the applicability of additive manufacturing as a tool for minimizing environmental impacts on production processes. Revista de Gestão Social e Ambiental **18**(7) (2024). https://doi.org/10.24857/rgsa.v18n7-014
58. Simbolon, S., Simbolon, D.M.: Sustainable workforce in a green era: Indonesia's energy sector transition. Int. J. Energy Econ. Policy **14**(3), 702–710 (2024)
59. Spigarelli, F., Compagnucci, L., Lepore, D.: Blockchain unlocking collaborative opportunities for environmental sustainability through innovation intermediaries. J. Technol. Transf. (2024). https://doi.org/10.1007/s10961-024-10106-5
60. Sánchez-García, E., Martínez-Falcó, J., Marco-Lajara, B., Manresa-Marhuenda, E.: Revolutionizing the circular economy through new technologies: a new era of sustainable progress. Environ. Technol. Innov. **33** (2024). https://doi.org/10.1016/j.eti.2023.103509
61. Thanuskodi, S.: Journal of social sciences: a bibliometric study. J. Soc. Sci. **24**(2), 77–80 (2017). https://doi.org/10.1080/09718923.2010.11892847

62. Trotta, A., Rania, F., Strano, E.: Exploring the linkages between FinTech and ESG: a bibliometric perspective. Res. Int. Bus. Financ. **69**, 102200 (2024)
63. Ülkü, M.A., Bookbinder, J.H., Yun, N.Y.: Leveraging Industry 4.0 technologies for sustainable humanitarian supply chains: evidence from the extant literature. Sustainability **16**(3) (2024). https://doi.org/10.3390/su16031321
64. Varella, W.A., Oliveira Neto, G.C.D., Stefani, E., Costa, I., Monteiro, R.C., Conde, W., da Silva Junior, W., Baptestone, R.C., Goes, R.D.S., Riccotta, R.: Integrated service architecture to promote the circular economy in agriculture 4.0. Sustainability **16**(6), 2535 (2024)
65. Wanyama, J., Bwambale, E., Kiraga, S., Katimbo, A., Nakawuka, P., Kabenge, I., Oluk, I.: A systematic review of fourth industrial revolution technologies in smart irrigation: constraints, opportunities, and future prospects for sub-Saharan Africa. Smart Agric. Technol. 100412 (2024)
66. Xu, F., Wang, J.: Harnessing hybridized machine learning algorithms for sustainable smart production: a case study of Solar PV energy in China. J. Knowl. Econ. (2024). https://doi.org/10.1007/s13132-024-02006-8
67. Yu, H., Sun, X.: Uncertain remanufacturing reverse logistics network design in industry 5.0: opportunities and challenges of digitalization. Eng. Appl. Artif. Intell. **133**, 108578 (2024)
68. Zhao, X., Nakonieczny, J., Jabeen, F., Shahzad, U., Jia, W.: Does green innovation induce green total factor productivity? Novel findings from Chinese city level data. Technol. Forecast. Soc. Chang. **185** (2022). https://doi.org/10.1016/j.techfore.2022.122021

MIX
Papier aus verantwortungsvollen Quellen
Paper from responsible sources
FSC® C105338

If you have any concerns about our products,
you can contact us on
**ProductSafety@springernature.com**

In case Publisher is established outside the EU,
the EU authorized representative is:
**Springer Nature Customer Service Center GmbH**
**Europaplatz 3, 69115 Heidelberg, Germany**

Printed by Libri Plureos GmbH
in Hamburg, Germany